Composites Science and Technology

Series Editor

Mohammad Jawaid, Laboratory of Biocomposite Technology, Universiti Putra Malaysia, INTROP, Serdang, Malaysia

This book series publishes cutting edge research monographs comprehensively covering topics in the field of composite science and technology. The books in this series are edited or authored by top researchers and professional across the globe. The series aims at publishing state-of-the-art research and development in areas including, but not limited to:

- Conventional Composites from natural and synthetic fibers
- Advanced Composites from natural and synthetic fibers
- Chemistry and biology of Composites and Biocomposites
- Fatigue damage modelling of Composites and Biocomposites
- Failure Analysis of Composites and Biocomposites
- Structural Health Monitoring of Composites and Biocomposites
- Durability of Composites and Biocomposites
- Biodegradability of Composites and Biocomposites
- Thermal properties of Composites and Biocomposites
- Flammability of Composites and Biocomposites
- Tribology of Composites and Biocomposites
- Applications of Composites and Biocomposites

Review Process

The proposal for each volume is reviewed by the main editor and/or the advisory board. The chapters in each volume are individually reviewed single blind by expert reviewers (at least two reviews per chapter) and the main editor.

Ethics Statement for this series can be found in the Springer standard guidelines here - https://www.springer.com/us/authors-editors/journal-author/journal-author-helpdesk/before-you-start/before-you-start/1330#c14214

Sarat K. Swain
Editor

Nanohybrid Materials for Water Purification

Springer

Editor
Sarat K. Swain
Department of Chemistry
Veer Surendra Sai University of Technology
Burla, Sambalpur, Odisha, India

ISSN 2662-1819 ISSN 2662-1827 (electronic)
Composites Science and Technology
ISBN 978-981-19-2331-9 ISBN 978-981-19-2332-6 (eBook)
https://doi.org/10.1007/978-981-19-2332-6

This Springer imprint is published by the registered company Springer Nature Singapore Pte Ltd.
The registered company address is: 152 Beach Road, #21-01/04 Gateway East, Singapore 189721, Singapore

The editor dedicates this book to his beloved mother…

Mrs. Malati Manjari Swain

Preface

Nanohybrid materials or nanocomposites, especially polymer-based nanocomposites, have attracted a lot of attention as they play a vital role in the field of material science. Starting from the production of packaging materials to biomedical devices, from environmental remediation to industrial applications, advanced nanocomposite materials have created their own niche, owing to their outstanding features and performances. Various nanostructured materials have been successfully established as suitable reinforcement materials for nanocomposite formation, keeping sustainability in mind. Materials with multiple functionalities are of utmost importance in addressing various research problems, especially for solving various environmental issues that have become a concern worldwide. Water decontamination or wastewater management has posed a great problem across the globe, and tremendous efforts are being made to design materials and devices to serve the purpose. When various nanomaterials are reinforced in matrix materials, several properties of the matrix are accentuated significantly, namely permeability, mechanical properties, antifouling, selective barrier properties, etc. Nanocomposites make up efficient membranes and other devices for water purification via adsorption, filtration or catalysis activities.

The present book aims at presenting a broad overview of the proposed topic by combining the past, present and future of wastewater treatment and water purification. The book also provides detailed information regarding the versatile applications of nanocomposites and other smart nanostructured hybrid materials for water purification purposes, focusing on the new generation separation techniques, use of various nanomaterials for adsorption or degradation of the pollutants present in water. The book further explores the structural characterization as well as assessment of properties of the advanced materials useful for water purification. I hope that the selected 13 chapters included in the book will surge interest in readers regarding new age materials and technologies used for water treatment applications.

I express my gratitude to all the contributors of the chapters who have spared time to share their ideas regarding this rising concern via this edited book. I deeply thank all the authors from around the globe for sharing their expertise and knowledge in the concerned field. I convey thanks to two of my Ph.D. scholars, namely Anuradha

Biswal and Swapnita Patra for constant help during the editing process of this book. I am very much thankful to the Springer team for their unparalleled cooperation and support at each stage of production of this book.

Sambalpur, India Prof. Sarat K. Swain

Contents

Editor and Contributors

About the Editor

Sarat K. Swain is currently working as a Professor of Chemistry at Veer Surendra Sai University of Technology (VSSUT), Burla, Sambalpur, India since 2011. Prof. Swain was working as a Post-Doctoral Fellow in Department of Polymer Engineering, University of Akron, Akron, OH, USA after receiving his Doctoral degree from Utkal University, Bhubaneswar, India. He was also working as a visiting researcher at Indian Association for the Cultivation of Science (IACS), Kolkata and Jawaharlal Nehru Centre for Advanced and Scientific Research (JNCASR), Bangalore. Added to his credits, Prof. Swain was the former Dean (Academic Affairs), Dean (PG studies & Research) and Head of the Department of Chemistry of the VSSUT. He has more than twenty five years of experience in teaching and research in Organic Chemistry, Materials Science, Nanotechnology and Polymer Chemistry at UG, PG and Ph.D. levels. His area of research interest includes hybrid nanomaterials, reinforced polymer nanocomposites, advance materials like graphene, nanoclay, CNT, CNF, CQD for improving of gas barrier, fire retardant, thermal, mechanical and chemical resistance properties of the composites. He has designed various kinds of nanostructured materials for waste water treatment, packaging, anti-corrosion performance, superconductor properties, sensing behaviors and biomedical applications. Sixteen scholars have been successfully awarded Ph.D. degrees along with completion of more than fifty master degree thesis under the supervision of Prof. Swain. He has published more than 100 research papers in different SCI indexed international journals of repute. So far he has authored/edited 04 books, 35 book chapters, along with invention of two patents (One USA and One Indian). Prof. Swain has received BOYSCAST Fellowship, DAE-Young Scientist Research Award and Samanta Chandra Sekhar Award for various academic achievements.

Contributors

Mohamed A. Abdelwahab Chemistry Department, Faculty of Science, Tanta University, Tanta, Egypt

Md. Ahmaruzzaman Department of Chemistry, National Institute of Technology, Silchar, Assam, India

W. L. Ang Department of Chemical and Process Engineering, Universiti Kebangsaan Malaysia UKM, Bangi, Selangor, Malaysia;
Centre for Sustainable Process Technology, Universiti Kebangsaan Malaysia UKM, Bangi, Selangor, Malaysia

Hirakendu Basu Analytical Chemistry Division, Bhabha Atomic Research Centre, Mumbai, India

Anindita Bhuyan Department of Chemistry, National Institute of Technology Silchar - 788010, Assam, India

Anuradha Biswal Department of Chemistry, Veer Surendra Sai University of Technology, Sambalpur, Odisha, India

Eliazer Bobby Naidoo Institute of Chemical & Biotechnology, Southern Gauteng Science and Technology Park, Vaal University of Technology, Vanderbijlpark, South Africa;
Department of Chemistry, Vaal University of Technology, Vanderbijlpark, South Africa

Rachid Bouhfid Moroccan Foundation for Advanced Science, Innovation and Research (MAScIR), Composites and Nanocomposites Center, Rabat Design Center, Rabat, Morocco

P. V. Chai Department of Chemical and Petroleum Engineering, UCSI University, Cheras, Kuala Lumpur, Malaysia

Soulaima Chkirida Moroccan Foundation for Advanced Science, Innovation and Research (MAScIR), Composites and Nanocomposites Center, Rabat Design Center, Rabat, Morocco;
Faculty of Science, Laboratoire de Chimie organique et Hétérocyclique, Mohammed V University Mohammed in Rabat, Rabat, Morocco

S. F. Chua Department of Chemical and Process Engineering, Universiti Kebangsaan Malaysia UKM, Bangi, Selangor, Malaysia

De la Cruz-Marquez Galilea I. Unidad Académica de Ciencias Químicas, Universidad Autónoma de Zacatecas, Ejido la Escondida, Zacatecas, México

Durón-Torres Sergio M. Unidad Académica de Ciencias Químicas, Universidad Autónoma de Zacatecas, Ejido la Escondida, Zacatecas, México

Vishal Gadore Department of Chemistry, National Institute of Technology, Silchar, Assam, India

Mridula Guin Department of Chemistry and Biochemistry, Sharda University, Greater Noida, India

Preeti Jain Department of Chemistry and Biochemistry, Sharda University, Greater Noida, India

C. Y. Jong Department of Chemical and Petroleum Engineering, UCSI University, Cheras, Kuala Lumpur, Malaysia

Michael Klink Department of Chemistry, Vaal University of Technology, Vanderbijlpark, South Africa;
Department of Biotechnology, Vaal University of Technology, Vanderbijlpark, South Africa

Suresh Kumar Kailasa Department of Applied Chemistry, S. V. National Institute of Technology, Surat, Gujarat, India

Rakesh Kumar Singhal Analytical Chemistry Division, Bhabha Atomic Research Centre, Mumbai, India

Anny Leudjo Taka Institute of Chemical & Biotechnology, Southern Gauteng Science and Technology Park, Vaal University of Technology, Vanderbijlpark, South Africa;
Department of Chemistry, Vaal University of Technology, Vanderbijlpark, South Africa

Blanca L. Martínez-Vargas Centro de Estudios Científicos y Tecnológicos, Instituto Politécnico Nacional, Zac., Zacatecas, México

Maie Mousa Chemistry Department, Faculty of Science, Tanta University, Tanta, Egypt

Fanyana Mtunzi Institute of Chemical & Biotechnology, Southern Gauteng Science and Technology Park, Vaal University of Technology, Vanderbijlpark, South Africa;
Department of Chemistry, Vaal University of Technology, Vanderbijlpark, South Africa

Swapnita Patra Department of Chemistry, Veer Surendra Sai University of Technology, Burla, Sambalpur, Odisha, India

Peralta-Hernández Juan M. Departamento de Química, Región de Ciencias Naturales y Exactas, Universidad de Guanajuato, Pueblito de Rocha, Guanajuato, México

Picos-Benítez Alain R. Centro de Estudios Científicos y Tecnológicos, Instituto Politécnico Nacional, Zac., Zacatecas, México

Abou el kacem Qaiss Moroccan Foundation for Advanced Science, Innovation and Research (MAScIR), Composites and Nanocomposites Center, Rabat Design Center, Rabat, Morocco

Soumya Ranjan Mishra Department of Chemistry, National Institute of Technology, Silchar, Assam, India

Gyanaranjan Sahoo Department of Basic Science and Humanities, Majhighariani Institute of Technology and Science, Rayagada, Odisha, India

Krishna Manjari Sahu Department of Chemistry, Veer Surendra Sai University of Technology, Burla, Odisha, India

Nehal Salahuddin Chemistry Department, Faculty of Science, Tanta University, Tanta, Egypt

Niladri Sarkar Department of Chemistry, Centurion University of Technology and Management, Odisha, Jatni, India;
School of Chemistry, University of Hyderabad, Gachibowli, Telangana, India

N. B. Singh Department of Chemistry and Biochemistry, Sharda University, Greater Noida, India;
Research and Development Cell, Sharda University, Greater Noida, India

Shweta Singh Analytical Chemistry Division, Bhabha Atomic Research Centre, Mumbai, India

Sarat K. Swain Department of Chemistry, Veer Surendra Sai University of Technology, Burla, Sambalpur, Odisha, India

Xavier Yangkou Mbianda Department of Chemical Sciences, University of Johannesburg, Doornfontein, Johannesburg, South Africa

Nadia Zari Moroccan Foundation for Advanced Science, Innovation and Research (MAScIR), Composites and Nanocomposites Center, Rabat Design Center, Rabat, Morocco

Water Pollution Issues and Monitoring the Problems

Swapnita Patra and Sarat K. Swain

Abstract Water is indispensable for entire existing life forms of the Universe. Being the backbone for climate change and pivotal for economic, social and energy development in every respect, the management of water resource should be the premier responsibility for all human beings. Reduction of water quality due to natural and man-made pollution and failures of conventional water treatments for purification of water can frequently lead to many contamination incidents resulting serious disease outbreaks. Recently, nanohybrid materials attracted attentions of researchers to design and develop many purification techniques by nanohybrid materials adopting different approaches such as removal and sensing of dyes, heavy metals and oily wastes, desalination, antifouling agent, designing of nano-membranes, magnetic, catalytic and photocatalytic nanohybrid materials. This chapter delineated in words and pictures about many important facts and terms related to water along with the sources, effects, and prevention of water pollution by conventional and the present fashions by nanohybrid materials. Ultra-modern outfits for wastewater treatment by nanohybrid materials are environment friendly, low cost and effective for the capitalization purpose.

Keywords Water pollution · Nanohybrid materials · Dye removal · Sensing · Nano-membranes · Antifouling · Oily waste

1 Introduction

The principal constituent in Earth's hydrosphere and the fluids of entire life forms water (H_2O) is the "solvent of life" acts as a "universal solvent". Till now it is the most analysed polar inorganic chemical compound which is a transparent, odourless, tasteless and colourless liquid with an instinctive hint of blue at room temperature [1]. It is the third superabundant (after molecular hydrogen and carbon monoxide) and exclusive universal substance on Earth's surface to exist in three natural states

S. Patra · S. K. Swain (✉)
Department of Chemistry, Veer Surendra Sai University of Technology, Burla, Sambalpur, Odisha 768018, India
e-mail: skswain_chem@vssut.ac.in

S. K. Swain (ed.), *Nanohybrid Materials for Water Purification*, Composites Science and Technology, https://doi.org/10.1007/978-981-19-2332-6_1

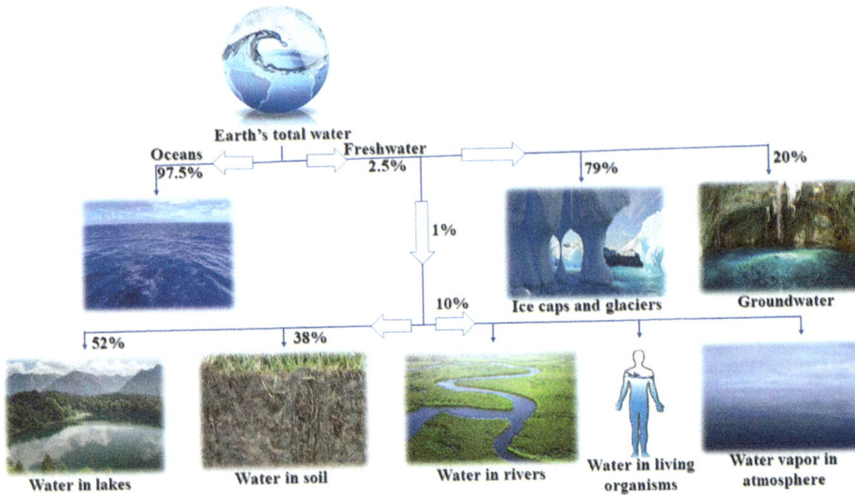

Fig. 1 Distribution of Earth's total water

as liquid, solid, and gas. The exceptionally uneven distribution of total Earth's water is schematically portrayed in Fig. 1. Approximately 70.9% water of Earth's surface spreads over oceans and seas whereas 1.7% present in the ice caps and glaciers of Greenland and Antarctica, another 1.7% present as groundwater and 0.001% exists as precipitation, vapor and clouds (contains suspended liquid water and ice in air). A famous British poet W. H. Auden once noted that, "Thousands have lived without love, not one without water", clearly mentioning the importance of water for the world at large.

A remarkable character in world economy is played by water. As a major part of economy, agriculture consumes about 70% of freshwater used by humans. One of the extensive sources of food is provided by fishing in fresh and salt water bodies in different areas of the world. Transportation of numerous numbers of manufactured products, oils, and natural gas usually carried out through rivers, seas, lakes and canals by boat. Huge amounts of ice, water and steam are utilized for heating and cooling processes in industries. Along with industries use, water mainly use as a solvent for large varieties of mineral and organic substances. Again, being the central to several forms of sports and entertainment like boating, swimming, diving and surfing, water contributes frequently to economic growth. Supremely, cooking and washing are the unavoidable use of water.

The availability and scarcity of water raise many political conflicts around the globe called water politics or hydropolitics defined as "the systematic study of conflict and cooperation between states over water resources that transcend international borders" by Arun P. Elhance's. Keeping an eye upon the availability and demand, predictions show that clean water may become the "next oil" alerting many possible wars over water. Water resource policies are dealing with water management that control the use and disposal of water. Whereas, hydropsychology act towards

Fig. 2 Hydropolitics, hydropsychology and dihydrogen monoxide parody

using water in smaller scale at individual level. Hydropolitics is a top-down approach whereas, hydropsychology is mentioned as a bottom-up approach. Hydropolitics and hydropsychology operate like the two sides of the same coin. The dihydrogen monoxide parody is one type of chemophobia which includes calling water in an unconventional chemical name "dihydrogen monoxide" (DHMO) and describes water in a disturbing manner, such as assisting corrosion and suffocation due to drowning. The parody demands to ban and label water as dangerous, showing how scientific illiteracy and overemphasized analysis can create illogical terror. In 1983, April Fools' Day edition of a weekly newspaper in Durand, Michigan published that the water pipes of city contain "dihydrogen oxide" which could release blistering vapours and shouldn't be inhaled [2]. A general idea about hydropolitics, hydropsychology and dihydrogen monoxide parody is depicted in Fig. 2 for better understanding.

Pollution is the release of unfavourable substances called pollutants into the environment. Water pollution is a principal issue which is defined as reduction of water quality and introducing toxicity causing harm to both humans and environment owing to contaminated river, stream, lake, ocean and other water bodies. Even after knowing water is a prime source of life, 80% of entire world's polluted water is thrown out into environment contaminating water bodies without thinking about the finiteness of accessible drinkable freshwater i.e., lower than one percent. Between the two types of water pollution, first one is organic pollution which causes by bacteria, viruses and microorganisms developed from vegetable and animal waste. And second one is chemical pollution caused by animal and human drugs, phosphates and nitrates of pesticides, acids, hydrocarbons and heavy metals released from industries. Although natural phenomenon such as floods, landslides etc. partly contribute to water pollution, human activities are predominantly responsible for degradation of water characteristic. The chemical and biological agents responsible for water pollution are referred as water pollutants. The natural purification and recovery are faster in larger water bodies which are attributable to accelerated natural degradation, quick diffusion and dissipation of pollutants.

1.1 *Sources of Water Pollution*

Grounded on origin open water adulteration classified as two orders as point sources and non- point sources, displayed in Fig. 3.

1.1.1 Point Sources

Point source water contamination termed as pollutants those enter a raceway from single, identified source, similar to a gutter. Exemplifications of these sources are discharges from a “sewage treatment factory”, a factory or a megacity storm drain.

1.1.2 Non-point Sources

Non-point source (NPS) profanity refers to circulate pollutants of air or water which does not appear from a single source. This type contamination is frequently accretive the effects of small quantities of contaminants accumulated in a large surface. This is in discrepancy to point source pollution which are resulted from single source. NPS contamination generally a rush, land runoff, atmosphere deposit, drain system, seep-aging, or hydrologic refashioning (cloudburst) where it is tricksy for taking back the

Fig. 3 Point and non-point sources of water pollution

entire pollutants. Whilst point source originates all the above adulterants, the great-distance transport capability with multiple sources to the contaminant comprises NPS of contamination.

1.2 *Effects of Water Pollution*

Water pollution has veritable negative goods on public health. Every year, about 1 billion people get sick due to unsafe water. Waterborne pathogens are complaint-raising bacteria and contagions from bio waste, which are present in the defiled drink water causing diseases. A bunch of serious conditions attributable by drinking or being in contact with polluted water, mainly diarrhea, typhoid, or skin infections. In regions where there is scarcity of drinking water, the obvious main threat is dehumidification. Arsenic and mercury containing chemical adulterants used to produce fungicides and nitrate fertilizers are frequently introduced into the water inventories. Ingestion of this contaminated water to human body can create fatal, cancer causing diseases and hormonal disturbances which can even change redid brain function. This is also more harmful for pregnant women and kids.

Water pollution admittedly harms biodiversity and submarine ecosystems. The poisonous chemicals can alter the colour of water and accelerate the quantum of minerals, further causes enhancement of eutrophication which has a bitter influence on life span in water. Detriment to any of these organisms can produce series of effects, venturing all marine surroundings.

1.3 *Prevention of Water Pollution*

Pollution is not an ungovernable disaster; it is controllable and frequently avoidable, but yet vastly ignored. Answers by government to the questions on pollution failed to address the big case scenario. Meanwhile, better consciousness, indispensable consumption and product models, as well as ingenious technological results, indicate that numerous nations, metropolises, and businesses are formerly successfully diving a series of dangerous pollution issues. A pollution-free earth is by far and down is essential for the survival and well-being of ongoing and unborn generations of humans and ecosystems. Responsibility is to help dismissing adulterants and detoxify the polluted regions all around the earth. Problems should to be considered, ballooned, and gauged up in order to remove risking another exposure of humans and ecosystems to current and unborn pollutions as well as accelerating the prices of clean-up. Bettered threat estimate of new pollution heads is again an urgent demand. Perpetration gaps are in particular due to absence of coffers, shy executives, fiscal, financial, institutional unavailability. Lack of inter-ministerial collaboration and political will are crucial causes of fruitlessness.

1.4 Water Pollution Monitoring by Nanohybrid Materials

Water deficit all over the earth hangs the welfare of human society and arising smart technology targeted to open up this challenge are encouraging. In this context, nanomaterials are playing a pivotal part. Nanomaterials are those accoutrements which have dimensions of 1–100 nm size range, exposing outstanding futuristic technologies for water sanctification. This includes nanosized adsorbents, nano-membranes, photocatalysts, etc. The water sanctification applications are mainly owing to rare structure characteristics, properties along with size chart of nanomaterials, the membranes of nanostructured material and adsorbents with high effectiveness, by lower energy consumption with price. There are variety of ways in nanotechnology which uses nano patches for furnishing reliable drinking water with exorbitant productiveness. Some ways have come capitalized. The main features that make nanoparticles strong for water treatment are more superficial area and small volume. The advanced the surface area and volume, the particles come stronger, more stable and durable. Accoutrements may change electrical, optical, physical, chemical, or natural parcels at the nano position, making chemical and natural responses of the whole nanohybrid material more compelling. This chapter describes colourful operations of nanomaterials in omitting different types of contaminations from polluted water.

2 Properties of Nano Hybrid Materials for Water Purification

Due to the unique properties like size, surface layout, composition and function, nanoparticle-based adsorbents are able to adsorb pollutants of very low concentration level. Adsorbents are classified into two categories: one is conventional and another is nonconventional. Conventional type of adsorbents includes silica gel, sawdust, wood whereas the example of nonconventional adsorbents is hydrogel, clay, biosorbents. Dye elimination through adsorption phenomenon can easily be understood by considering the interaction between the adsorbent and adsorbate. The mechanisms responsible for adsorption are physisorption, ion-exchange, precipitation and chemisorption. Reusability, selectivity, economical and biodegradability are the necessary requirements for the utility and preparation of excellent nano-adsorbent.

Nano-adsorbent has gained its attention as a potential adsorbent system in pollution management application. Being one of the prominent functional units of nanoscience, nano adsorbent has the competence to adsorb organic and inorganic materials. It has also manifested its capacity to eliminate pollutants from both air and water by adsorption phenomenon. The effective adsorbents that are used to eradicate organic and inorganic effluent from wastewater must satisfy the following conditions.

a. The toxicity level of adsorbent should be negligible.
b. The adsorbents must have high ability to adsorb contaminants at low concentration.
c. The removal of contaminants from the surface of adsorbent should follow easy pathway.
d. The reusability and regeneration of adsorbent should be high [3].

Among various nanocomposites, polymer nanocomposite (PNC) has been regarded as one of the significant materials in present-day research, as it has promising selectivity for decontamination of water pollutants. In last few years, PNCs have gained recognition by the researchers for involvement in wastewater treatment process due to its high surface area, ease to use, cost-efficient characteristics. The dimension variable, film forming ability and presence of active functional groups have made PNC more advantageous than other materials. Along with above properties, the catalytic property is also considered during the designing of nano-catalysis for the purification of water.

3 Approaches Towards Water Treatment by Nanohybrid Materials

This chapter describes the outstanding contributions of nanohybrid materials towards water purification by different remarkable approaches (Fig. 4).

3.1 Magnetic Nanohybrid Materials

Magnetic nanohybrid materials exhibit all unique types of property of nano ranged materials including magnetic property which makes the above advanced materials an outstanding adsorbent and catalyst in water decontamination and purification

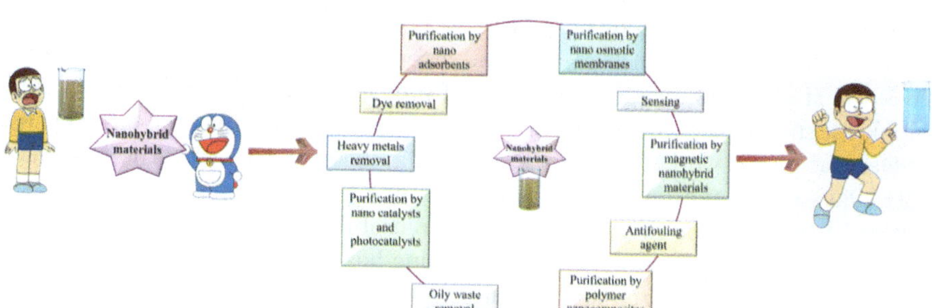

Fig. 4 Different approaches towards water purification by nanohybrid materials

process. Various nanoscopic magnetic materials like FeO, Fe_3O_4, γ-Fe_2O_3, $CuFe_2O_4$ have been predominantly investigated in different applications related to catalysis and adsorption. Although magnetic nano ranged materials have advantages as they are economical, less toxic and can be easily separated from water but using this material without any stabilizer lowers its efficiency as they start to aggregate. Another drawback of nano ranged materials is, they easily lose their integrity in aquatic environment due their high reactive and oxidation potential. Consequently, fabrication of nanohybrid materials based on magnetic nanosized particles with greater stability is one of the foremost steps to enhance performance and viability of magnetic nanosized particles [4]. The utility of magnetic nanosized particles has been escalated in last few decades because they have the ability to interact with several functional groups and easily get attached to the target materials in wastewater [5]. Maa et al. has investigated the adsorption of both heavy metals and dyes by fabricating a magnetic hybrid nanomaterial based on cyclic oligosaccharides, graphene oxide (GO) and magnetic iron nanosized particles using one-step reverse phase co-precipitation method. To fabricate (β-CD/MGO) nanohybrid materials, they have used GO and β-cyclodextrin (β-CD) as carrier and modifier, respectively. The most accustomed iron sources such as ferrous sulfate heptahydrate ($FeSO_4 \cdot 7H_2O$) and ferric chloride hexahydrate ($FeCl_3 \cdot 6H_2O$) are used for the preparation of magnetic iron oxide nanoparticles. They concluded that the prepared hybrid material has the ability to adsorb 279.21, 51.29 and 93.97 $mg \cdot g^{-1}$ of Pb (II), Cu (II) and methylene blue (MB), respectively [6]. Reddy et al. has reported the adsorption potential of various spinel ferrites derivatives for both organic and inorganic effluents including pollutants released from pharmaceutical industry [7]. Heavy metal ions not only responsible for contaminating surface water but also one of the major components to pollute groundwater. Groundwater contaminated by arsenic metal ions has jolted the entire health system of South and South East Asia regions. World Health Organization has reported that arsenic becomes more toxic and highly carcinogenic once it exceeds 0.01 mg/L concentration limit in consumable product. The noxious arsenic metal ions exist with two oxidation states: one is in +5 oxidation state called arsenate and another one is arsenite with +3 oxidation state. Different magnetic nanosized particles such as iron oxide materials and ferrite materials have been used to extirpate highly toxic arsenic metal ions from wastewater [8, 9]. Kumar et al. has engineered a ferrite based nano hybrid material by using $MnFe_2O_4$ with single layered GO for the eradication of heavy metals such as lead and arsenic from contaminated water. They have also studied the magnetic behavior of prepared materials by using vibrating sample magnetometer [10]. Another ferrite based nano hybrid material has prepared by Dai and his co-workers for the selective elimination of Pb (II). The fabricated PEHAPhos-GO/$MnFe_2O_4$ material has shown its remarkable properties for adsorption and feasibility for easy separation and reusability [11]. Like inorganic effluents, organic effluents become a life-threatening hazardous contaminant for human health as they retain its original structure without undergoing any further degradation in environment. One of notable organic pollutant that known for its stability in aqueous phase is Perfluorooctanoic acid (PFOA). Elanchezhiyan et al. has synthesized a bimetallic magnetic hybrid material by utilizing lanthanum and GO to remove perfluorooctanoic acid form contaminated water using adsorption phenomenon [12]. Siddiqui and his co-researchers have investigated antibacterial activity and adsorption capacity by developing manganese-based ferrite hybrid material. Co-precipitation preparation method has been used for the synthesis of $MnFe_2O_4$/BC composite by allowing nucleation of $MnFe_2O_4$ on the surface of Nigella Sativa plant seeds (Black Cumin). This material has the capability to inhibit the of growth both Gram-positive and Gram-negative bacteria (Escherichia coli and Staphylococcus aureus) and also has the potential to adsorb MB from aqueous solution [13]. The adsorption ability of adsorbent for the removal dye not only depends upon the availability of adsorption sites but also the functional groups present on the surface of adsorbent. Based on charge types of reactive functional groups present on dyes, it is divided into three categories such as negatively charged, positively charged and neutral dyes. So, the choice of materials that are used for the preparation of adsorbent becomes a key factor in determining the effectiveness of removal process. Rong et al. has fabricated a three-dimensional magnetic sulfur/nitrogen co-doped reduced graphene oxide (3D-MSNG) with a greater number of pores. 3D-MSNG has the potential to adsorb all three kind of dyes such as positively charged MB, negatively charged Congo red (CR) and neutral red (NR) and the equilibrium is reached within 15 m. The experimental value shows the perfect correlation between adsorption isotherm and Langmuir adsorption model. The highest adsorption value of MB, CR and NR by the composite are 171.53 mg/g, 909.09 mg/g and 877.19 mg/g, respectively [14].

3.2 *Heavy Metals Removal*

Materials at nanoscale level show exceptionally high enhanced properties, which gives freedom to researchers for the exploration and investigation in various fields with the help of nanotechnology. Nanosized materials possess a wide range of unique properties such as high surface effect, small size effect, quantum effect, and macro quantum tunnelling effect that makes them advantageous over other material for adsorption applications. Over the past few decades, numerous nanosized materials have been developed and investigated as one of the promising substitutes for decontamination of wastewater by elimination heavy metal ions through adsorption because of their distinct surface properties. Among inorganic effluents, heavy metal ions in both contaminated surface and groundwater are responsible for causing immense level of environmental pollution and ended up affecting both aquatic and human life because of their non-biodegradable nature and high toxicity level. The standard density value that considers metal ion as heavy metal ion is 6 g cm^{-3} and their consumption is greatly infect human health system because of their high degree of toxicity [15]. With the growth of industrialization, the quantity of release of deleterious heavy metal ions to the surroundings has been increasing rapidly. The innocuous character of natural habitat sustains even after addition of some number of contaminants but after certain value it becomes a main factor for human illness. The most perilous heavy metal ions like mercury, cadmium and lead are the chief sources

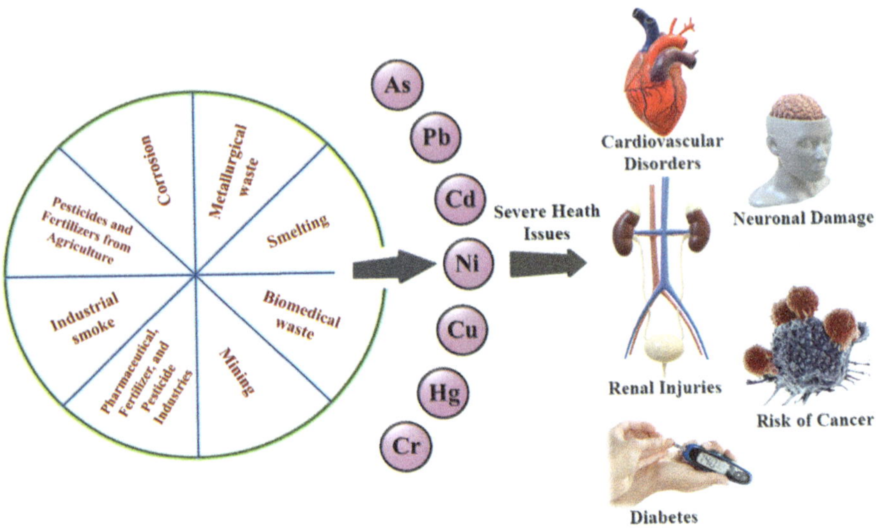

Fig. 5 Sources of heavy metal ions in water causing serious health issues

behind the fatal human diseases. Different sources of heavy metals ions in water cause severe health issues given in Fig. 5. There are also other heavy metal ions that cause illness and are vicious to ecosystem. For instance, cadmium with +2 oxidation state induces critical kidney and renal complications and bones lesions whereas Cu (II) ions are poisonous to aquatic life. Nickel metal ions are responsible for several skin irritation problems arises on human body. To avoid such high risked diseases, Zeng and his co-researchers have developed a hybrid nano composite using biodegradable materials by cross-linking chitosan/rectorite for the adsorption of Cd (II), Cu (II) and Ni (II) ions from polluted water [16]. Park et al. has investigated adsorption rate of different metal ions by fabricating reduced GO (rGO) based hybrid materials, namely rGO/magnetite, rGO/silver, and rGO/magnetite/silver. They have taken several ions of copper, cadmium, nickel, zinc, cobalt and palladium metals for experimental evaluation purpose. The end result of experiments follows the same order as single metal adsorption study i.e. (Pb (II), Cd (II)), <Co (II) < Ni (II) < (Cu (II), Zn (II)) [17]. The USEPAs recommended standard value of Cu (II) must be less than 1.3 mg L^{-1} in water that is used for drinking. Taimur et al. has developed an amidoxime eco-friendly material to eradicate copper ions from polluted water through chelation technique [18]. Chromium is one of those heavy metals that has potential to participate in blood circulatory system once it enters the body. The root sources behind the disclosure of poisonous chromium ions into natural habitat are leather tanning, electroplating, textile, metallurgy, steel fabrication, nuclear power plants, and photography. In nature, chromium metal exists in two oxidation states i.e., +6 and +4. Cr (VI) metal ion shows pronounce risk to ecosystem as it has high soluble affinity and ability to sustain in its original state [19, 20]. Due to the above facts, these metal ions create difficulties in human health by inducing several allergies and cancer and also it may lead to neurological disorder. Kanani-Jazi et al. has engineered a hybrid material using a clay mineral i.e., halloysite and polymer. The prepared halloysite/poly (amidoamine) dendritic material has been evaluated for the effective elimination of hexavalent chromium ions form water phase [21]. The industries related to manufacturing of battery, smelting and printing are the principal sites for the disclosure of lead ions. The lead metal ions have been regarded as most harmful heavy metal contaminant because of its carcinogenic nature. So, the researchers have started taking interest to deal with depollution of wastewater by preparing nano hybrid material as a novel candidate. Advanced nano hybrid materials have shown their capability to reduced pollution level to world health organization recommended pollution level for drinking water by adsorbing pollutants. Wang et al. has synthesized an adsorbent based on magnetic ferrite and carbon dot for the effective eradication of Pb (II) metal ions [22]. The development and designing of adsorbents with enhanced porous and functionalized structure has gained attention of many scientists for effective removal of metal ions. The most potent adsorbent among other porous material is metal organic framework (MOF) composite because it has capacity to adsorb heavy metal ions though host guest interactions mechanism develops between positively charged metal ions and functional groups present on the surface of MOF. Esrafili et al. has successfully engineered a dual functionalized TMU-81 MOF with sulfonyl and amide functional groups for the removal of metal ions like cadmium, copper and chromium. The highest adsorption value i.e., 526 mg/g of cadmium (II) ions is shown by a MOF related material such as TMU-81 [23]. Nanosized adsorbent produced from metal oxide and hydroxide has been dominantly employed for elimination of metal ions because they are cost effective, excellent adsorbent and eco-friendly. Mohan et al. has fabricated a nano adsorbent using GO and magnesium oxide by precipitation method. This hybrid material fulfils the requirement for the removal of Pb (II) [24].

3.3 Dye Removal

Dyes are coloured materials that are designed to give different shade of colours to various dyeable materials such as fabrics, paper etc. Most significant classification of dyes is based on origin and depending upon this type of categorization, dyes are divided into two types: natural and synthetic. Dyes that derived from natural resources like animal, vegetable and mineral are called natural dyes whereas synthetic dyes are generally organic compound that prepare in laboratory. Natural dyes are not suitable for industrial uses so the synthetic dyes are fabricated to use mainly in textile industries for the coloration of fabrics and in other industry for dyeing different materials. With the increase in demand, sundry kinds of dyes are being used extensively in industries based on textile, leather, paper, plastic, cosmetics, pharmaceutical, printing, and food processing. One of the hazardous and fatal water pollutants that cause negative impact not only on human but also on entire ecosystem because of its toxic

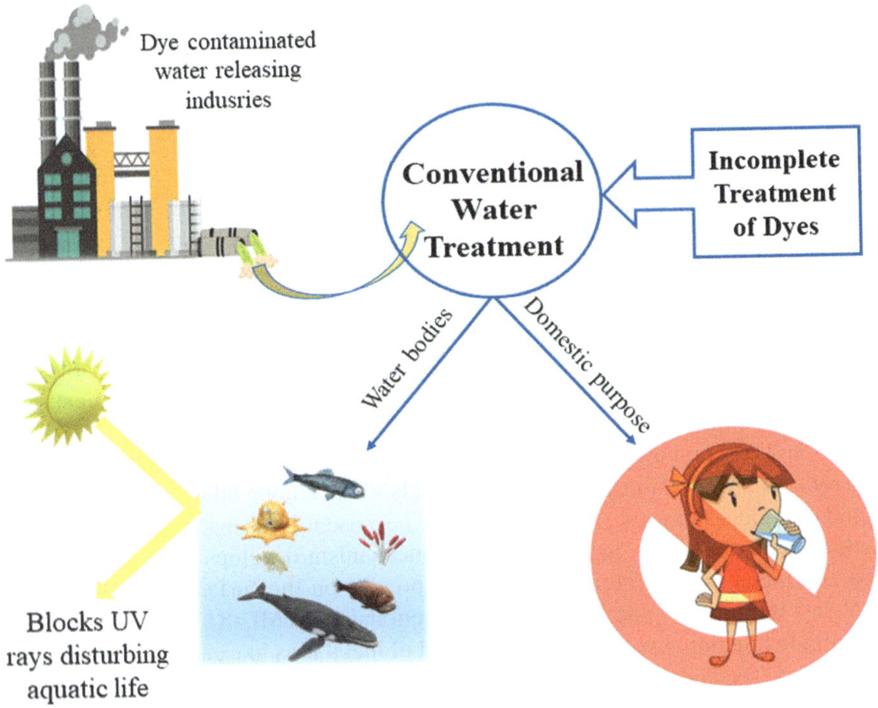

Fig. 6 Disturbing consequences of dye contaminated water

and non-biodegradable nature is dye. These effluents induce negative influence on photosynthesis process of aquatic plants and enhance the risk of carcinogenic disease in living beings. Disturbing consequences of dye contaminated water releasing from industries which are not completely treated by conventional methods are picturized in Fig. 6. Due to the above reasons, eradication of organic effluents like dye from wastewater became a crucial step to provide high quality drinkable water for living beings and also to meet the freshwater demand for various industrial uses [25, 26]. It has shown that, about 1×10^4 types of dyes and pigments are generated every year [27]. The release of dyes from different industries to environment is also increasing day by day and according to recent data the amount of dye pollutants that discharge from industries in each year is about 7×10^5 tons [28]. About 15% of discharged dyes are responsible for direct contamination of drinking water which leads to assimilation of dye into living beings through skin, lungs and digestive system and caused acute complication in their bodies. Azo dye is the most released dye from fabric industry. This type of dye effluents is easily converted into aromatic amine with genotoxic and carcinogenic properties. Sulfur anthraquinone, indigoid, and phthalocyanine derivatives are the most used dyes that have been used in industry after azo dyes. The eradication of this type of organic effluents is a strenuous process because of their high soluble efficiency in water. Owing to the above fact, researchers have suggested various kind of techniques to eradicate dyes from wastewater through various mechanisms. The techniques that are used to degrade dyes are photocatalytic, aerobic and anaerobic microbial degradation and other methods such as adsorption, coagulation and membrane technology is used to remove dyes through physical method. Adsorption is predominant over all other technologies because it is a cost effective and most efficacious method to discard dyes form polluted water. The most potent and productive adsorbent that has been widely use in removal both organic and inorganic effluents is PNC. Nano ranged materials such as metal oxide nanoparticles, carbon nanotubes (CNTs), graphene-based materials and magnetic nanosized particles are used as adsorbents in wastewater treatment to eradicate synthetic dyes. In recent years, various hybrid nanomaterials are investigated to reduce environmental pollution and to make water available for drinking and other purposes by eliminating toxic dyes and metal ions from wastewater.

3.4 Sensing

In the broadest definition, sensor is a device that response to the external stimuli and converts the input single to quantifiable output signal. For the sensing purposes, nanosized material is reported as one of the most active and highly effective transducer materials. The efficacy of nanosized material has substantially been increased than bulk material due to the unique characteristics of nanomaterials like electrical, optical, catalytic properties and thermal and mechanical stability. These materials are used in construction of sensor to detect and monitor the contamination level in air, water and soil. For the quantity measurement of poisonous metal ions, pesticides, chemicals and carbon, gold and silver nano ranged materials of different shape have been tremendously explored and are used for this purpose. Graphene has been emerged as a dominant sensing material for the detection of numerous materials in different applications. For instance, in biomedical field it is employed to detect glucose, cholesterol, H_2O_2, dopamine, ascorbic acid and uric acid where as in case of food industry it has the capability to sense the presence of some microbes, namely erythromycin and Staphylococcus aureus. Graphene has also manifested its crucial role in detection and elimination of heavy metals and pesticides from polluted water to make water reusable. The credibility of a sensor is determined by various aspects such as degree of sensing and detection capability, reusability and reproductive nature. For the treatment of wastewater, eradication of heavy metals is one of the essential parts since heavy metal caused acute diseases. To achieve this goal, researchers have already designed highly selective sensors and technologies [29–31]. The detection of deleterious heavy metal has become an important part in sensor application to reduce or avoid its harmful effect on environment. There are some technologies that have the ability to detect heavy metal even beyond its standard concentration level. For this purpose, large variety of detection methods like spectroscopic, chromatographic and electrochemical methods have been introduced. The employed methods are ion chromatography, chemiluminescence, atomic absorption spectroscopy, X-ray absorption and fluorescence spectroscopy. Graphene and its derivatives have successfully shown their potential in the sensing application as a chemical sensor due to the presence of large number of double bonds that helps in easy mobility of electron at high speed. Up until the present time, other than metal nanosized material which are reported for the sensing of heavy metals are DNAzymes, fluorescent aptamers, quantum dots. It is revealed that nanoscopic materials are extensively used not only as chemical sensor but also as biosensor. Some of the high sensitive metal oxide nanomaterials are zinc oxide (ZnO), titanium dioxide (TiO_2), iron oxide (Fe_2O_3), cuprous oxide (CuO), nickel oxide (NiO), tin dioxide (SnO_2), cerium oxide (CeO) [32].

Among manifold perilous pollutants, 4-Nitrophenol (4-NP) is regarded as one of the adverse water pollutants that impose negative influence on ecosystem due to its toxic and preserving nature. It has harmful effect on human since it is a cancer-causing agent and teratogen and mutagen material. In consideration to above facts, the need of explicability to detect and eradicate of 4-NP from environment should be given the top most priority. It is also necessary to engineered a cost effective and high selective natured system for the removal of tiny traces of 4-NP. Khan and co-workers have fabricated an excellent nanocomposite with high selectivity magnitude towards 4-NP. The prepared CeO_2–Cu_2O and CeO_2–Cu_2O/CH nanocomposites show detection of 4-NP up to 2.03 μM [33]. Wang et al. has prepared FeO_x/TiO_2@mC composite material to detect 4-NP electrochemically. The nanosized materials i.e., FeO_x and TiO_2 are evenly dispersed throughout the mesoporous carbon matrix to avoid agglomeration. The composite is prepared from Fe/Ti-based metal–organic framework by utilizing pyrolysis technique. They have taken three different temperature variations for pyrolysis process i.e., 500, 700, and 900 °C [34].

Another hazardous phenol derivative that adversely affect environment is 4-aminophenol. The main reason for the disclosure of 4-aminophenol to the ecosystem is the food industry as it utilizes 4-aminophenol as flavouring and fragrance agent in food, soap, cosmetics and pharmaceutical. It is also used in preparation of lauryl aldehyde, lauric acid, light-green dye. Besides these, it is used as solvent, plasticizer and low temperature lubricate. The presence of phenol derivative in food above the recommended level leads to cause several deadly diseases in human and also disturb the natural processes of environment [35]. Rahman et al. has fabricated a nanocomposite by utilizing polyelectrolyte and carbon nanosized materials for the detection of hazardous 4-aminophenol. They concluded that the synthesized Polyaniline/graphene/carbon nanotubes nanocomposites have the sensing and detection limit for 4-AP are ~2.1873 $\mu A\ cm^{-2}\ \mu M^{-1}$ and 63.4 pM, respectively [36].

4-Nitrochlorobenzene (4-NCB) possesses low biodegradability and high toxic property and is responsible to injure nervous system and cause hepatopathy, hemolytic anemia in human body. Since 4-NCB is used in numerous production applications such as dye, pesticide, herbicide etc. its disclosure amount is very high. Zhu et al. has synthesized GO based nanohybrid sensing system with carbon nanosized material for the detection of 4-NCB [37]. Due to the excessive use of pesticides to improve productivity, the agricultural fields are converted into less fertile production field and this also cause the contamination of ground water. Tian and his co-workers have engineered a nano enzyme-based sensor for the multiple sensing material by utilizing $CuO–TiO_2$. The sensing mechanism they have used for this detection process is electrochemical which mainly detect methyl parathion. The prepared sensing device has also potential to detect organophosphorus in ground water [38].

3.4.1 Nano-Membranes

The membranes are designed to use as a selective barrier in separation process to eliminate particles, ions and molecules from a solution mixture. Depending upon the diameter and size of pores, membranes are employed in different filtration techniques and based on this, filtration techniques are classified as conventional filtration, micro-filtration (MF), ultrafiltration (UF), nanofiltration (NF) and reverse osmosis (RO). The above-mentioned filtration methods have their own significant role in water purification application since each membrane is responsible to separate different level of impurities from wastewater. To eliminate large sized sand particles, conventional method is used. The pollutants having diameter of 0.1–10 μm like microorganism and macromolecules of 10–1000 Å diameter size are separated by using MF and UF membrane, respectively. The best outcome can be achieved in separation process of ions and salts by using NF and RO technologies [39]. Membrane separation method is advantageous than other water purification method since it does not require additional chemicals, generation of spent media and extra thermal energy to carry out separation process [40]. The membrane filtration methods have availed the high-level proficiency to produce purified water by eliminating pollutants at ionic levels. The above-mentioned membrane filtration methods are based on pressure-driven mechanism and are more conveniently used in wastewater treatment plants than other membrane-based method such as distillation, dialysis and electrodialysis [41]. Membrane filtration techniques have shown greater accountability to deal with water insufficiency. The reliance over polymer made membrane has increased as a potent candidate for the utilization in membrane technology because of its selective property. Since 1980, the use of membrane with nanosized pores as competent filtration apparatus has become an essential process in numerous industries such as food, pharmaceutical, petroleum, electronic and paper industry. To enhance the structural efficiency, performance and to minimize the drawbacks shown by nanocomposite polymeric membrane during water purification process, recently, researchers have incorporated inorganic fillers as nanosized material, namely silica, silver, TiO_2, aquaporins and CNTs within the polymeric nanocomposite membrane system. The other nanosized materials that are used as filler in developmental process of nanocomposite membranes for wastewater treatment are ZnO, CuO, alumina (Al_2O_3), magnetite (Fe_3O_4), cobalt (Co), zirconia (ZrO_2), GO, and zeolites.

The change in climatic and environmental regularity has already started to make its impact on the salinity level of sea water and other water reservoir which ultimately leads to the decrease in availability of fresh water for household, agricultural and industry usage. Due to the constant surge in global warming, ice present on polar region is melting and cause the mixing of fresh water with saline water. The

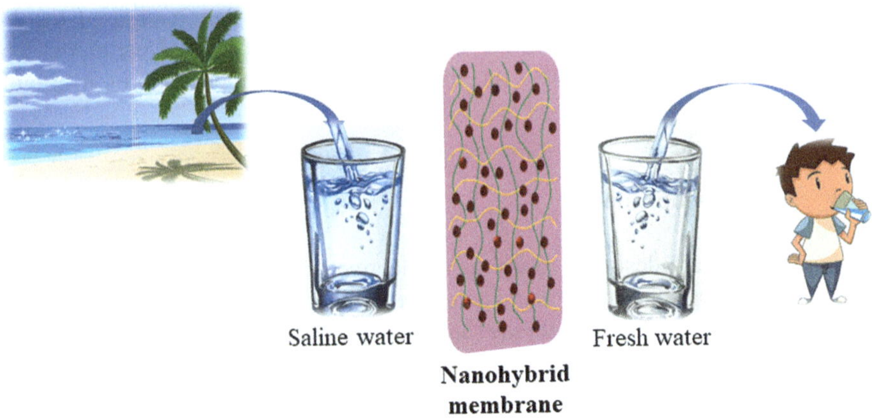

Fig. 7 Nanohybrid membrane for desalination

adsorption phenomena greatly help in wastewater purification but cannot be used in desalination process. In consideration to above fact, the development of techniques that not only use in separation of pollutants but also help in desalination process is required. In this regards, nano-membrane technology has gained popularity due to its ability of pollutants filtration and salinity removal nature (Fig. 7). There are some membrane technologies such as forward osmosis, membrane distillation and capacitive deionization have been introduced to decrease the salinity level of water but the utilization of these techniques is a practically challenging process [42]. Among the reported membrane techniques, RO process has showed greater efficacy in desalination process. It is also revealed that, globally, about 19% of desalination plants have successfully accomplished the installation and use of RO membrane to desalinate brackish water [43].

3.5 Antifouling Behaviour

Pollutants release form various industries have different level of negative impact on freshwater which ultimately leads to use of distinctive methods and materials to purify water having dissimilar effluents. Unlike textile industry, polluted water release from petroleum and natural gas refinery, printing and food industry has a greater number of oxidisable chemical and hence, the wastewater contains pollutants from these industries have high chemical oxygen demand (COD). The polluted water with high COD value initiates fouling and obstructs the flow of water. To avoid or minimizing fouling, membrane having nanoscopic materials with augmented mechanical, physical, chemical properties and high functional characteristic has been designed and investigated. A wide range of nanosized particles, namely TiO_2, ZnO, SiO_2, Al_2O_3 and silver have shown favourable features for the improvement of permeability, selectivity and antifouling characteristics of membrane. Li and his co-workers have designed silica-based silicone polyacrylate/SiO_2/HDPE nano hybrid composite material to minimize fouling inside the industrial polluted water pipe [44]. Nano clays have also shown its beneficial effect in the production of membrane with enhanced anti-clogging property. Dehghankar et al. has developed a polymer (polyvinylidene Fluoride)-nanoclay (Cloisite 30B and Palygorskite) composite matrix membrane with anti-fouling property [45]. Fouling can be categorized into various types; out of these bio-fouling creates heavy economical loss in maritime industry and destroy the harmony of sea habitat. For fouling prevention, typically, anti-fouling paints have been used but its toxicity property affect the whole sea life. Selim et al. has prepared a nanostructured material with greater surface property using silicone and graphene for the avoidance of marine fouling [46].

3.6 Oily Waste

Meanwhile, wastewater produced from petroleum plants an admixture of grease, lube, heavy substance with other deadly adulterants. Discharge from these unctuous wastewaters (OWW) to surrounding previous from any sanctification or pre-treatment is probable to endure disastrous outcomes on humans and biosphere. So, oil junking from unctuous wastewater will be a prime concern for alleviating dangerous belongings found in OWW. Numerous decontamination ways, includes "booms using mooring system", "air in dissolved floatation", "filtration of membrane", along with other adsorption and bioremediation are extensively developed. Still, these strategies have drawbacks like lower effectiveness on separation, erosion, lofty handling expenditure, high energy-requirement and recontamination. Sharma et al. narrate the synthesis of an efficient, biodegradable, magnetic nanosorbent polydimethyl-siloxane@zincoxide tetrapod@iron oxide nanorod nanohybrid which is operated by a hydrothermal procedure. The synthesized nanohybrid is operated as a magnetic nanosorbent in oil junking process in polluted water [47]. Barroso-solares et al. has applied fiber mat of electrospun nanocomposite with "hydrophobic" and "oleophilic" to get assembled characteristics and these can be used like oil absorbents in emulsions from steady water. Primarily two polymers are mixed for the preparation of mats, and the common solvent contain "poly(methyl methacrylate)" and polycaprolactone" [48]. During the manufacturing of crude oil and gas the by -product produced can be expressed commonly as prepared water. Manufacturing of a barrel oil is calculated to produce approximately three number of barrels for a classic reservoir of prepared water and for a declining oil field bout ten barrels [49]. Formed water, a complex blend of number of organic and inorganic composites, the compounds will be different between oil fields which depends on geographical position of oil well. Chemicals with In-contact hydrocarbon mixed while production and drilling operation. Treatment of formed water involving in de-oiling and de-mineralizing procedures before recycling for satisfying strict environment rules. Membranes of hydrophobic polymeric similar to 'polysulfone', 'polyvinylidene fluoride (PVDF)', polyacrylonitrile (PAN) have been extensively studied for the unctuous wastewater treatment [50].

4 Green Approach Towards Water Purification by Nano Hybrid Structures

Nanosized materials have been employed in several applications such as drug administration, water purification and food science. To reduce the release of hazardous by-products and increase the efficacy of reaction, green approach has been introduced in various chemical approaches including nanotechnology. Most of the water treatment techniques are innovative and green processes. By combining MF and UF membrane techniques a green membrane bioreactor can be formed [51]. The preparation of nanoparticles (NPs) through green process is economical, well-structured and most importantly environmental friendly compared to other traditional methods. Globally, the quality of water can be improved by the green synthesized NPs as they show high beneficial outcome in eradication of heavy metal from polluted water. Although the NPs produced from green approach have numerous advantages, the reusability of the material is the major drawback to use in commercial level. The most liable nanostructure materials like silver NPs incorporated in cyclodextrin or in aloe vera plant extract, metal oxide coated with biopolymer and photocatalytic active titania have been used in water pollution control process. The use of various biosynthesized systems offers favourable, substantial and frugal water decontamination process.

5 Conclusion

The percentage of dinking water available in the earth is very negligible as compare to the total water content. It is a big challenge to keep preserve the water from the pollution. Present problem can be overcome by design new materials for removal of unwanted pollutants from the water. Various techniques are already developed for water treatment with preparation of novel materials. Nanostructured hybrid materials are potential candidates to meet the present crisis in removal of unwanted pollutants. Present chapter gives an overall idea regarding issues related to water pollution and monitoring of problems.

References

1. Bruinsma J (2009) By how much do land, water and crop yields need to increase by 2050? The resource outlook to 2050. Presented at the Food and Agriculture Organization of the

United Nations Expert Meeting entitled "How to Feed the World in 2050", Rome, Italy, June 24–26:2009
2. World Water Development Report 4 (2014) Managing water under uncertainty and risks. Accessed July 24, 2014
3. Awasthi A, Jadhao P, Kumari K (2019) Clay nano-adsorbent: structures, applications and mechanism for water treatment. SN Appl Sci 1:1076
4. Pang Y, Yu J, Tang L, Zeng G, Zhu C, Wei X (2019) Chapter 1-Magnetic nanohybrid materialsfor water-pollutant removal, Nanohybrid and nanoporous materials for aquatic pollution control 1–30
5. Lu AH, Salabas EL, Schüth F (2007) Magnetic Nanoparticles: Synthesis, Protection, Functionalization, and Application. Angew Chem 46:1222−1244
6. Maa YX, Shaoa WJ, Suna W, Koua YL, Lia X, Yang HP (2018) One-step fabrication of β-cyclodextrin modified magnetic graphene oxide nanohybrids for adsorption of Pb(II), Cu(II) and methylene blue in aqueoussolutions. Appl Surface Sci 459:544–553
7. Reddy DHK, Yun YS (2016) Spinel ferrite magnetic adsorbents: Alternative future materials for water purification?. Coordination Chemistry Reviews 315:90–111
8. Sahu UK, Mahapatra S, Patel RK (2018) Application of Box–Behnken design in response surface methodology for adsorptive removal of arsenic from aqueous solution using CeO_2/Fe_2O_3/graphene nanocomposite. Mater Chem Phy 207:233–242
9. Silva GC, Almeida FS, Ferreira AM, Ciminelli VST (2012) Preparation and application of a magnetic composite (Mn_3O_4/Fe_3O_4) for removal of As (III) from aqueous solutions. Mater Res 15:403–408
10. Kumar S, Nair R, Pillai PB, Gupta SN, Iyengar MAR, Sood AK (2014) Graphene Oxide−$MnFe_2O_4$ Magnetic Nanohybrids for Efficient Removal of Lead and Arsenic from Water. ACS Appl Mater Interfaces 6:17426−17436
11. Dai K, Liu G, Xu W, Deng Z, Wu Y, Zhao C, Zhang Z (2020) Judicious fabrication of bifunctionalized graphene oxide/MnFe2O4 magnetic nanohybrids for enhanced removal of Pb(II) from water. J Colloid Interface Sci 579:815–822
12. Elanchezhiyan SSD, Prabhu SM, Kim Y, Park CM (2020) Lanthanum-substituted bimetallic magnetic materials assembled carboxylate-rich graphene oxide nanohybrids as highly efficient adsorbent for perfluorooctanoic acid adsorption from aqueous solutions. Appl Surface Sci 509:144716
13. Siddiqui SI, Chaudhry SA (2018) Nigella sativa plant based nanocomposite-$MnFe_2O_4$/BC: An antibacterial material for water purification. J Cleaner Prod 200:996–1008
14. Rong Y, Huang Y, Jin P, Yang C, Zhong Z, Dong C, Liang D (2020) Highly efficient removal of cationic, anionic and neutral dyes by hierarchically porous structured three-dimensional magnetic sulfur/ nitrogen co-doped reduced graphene oxide nanohybrid. J Water Proc Eng 37:101345
15. Chowdhury P, Elkamel A, Ray AK (2015) Chapter 2-Photocatalytic processes for the removal of toxic metal ions, Heavy metals in water: presence, removal and safety. Royal Soc Chem 25–43
16. Zeng L, Chen Y, Zhang Q, Guo X, Peng Y, Xiao H, Chen X, Luo J (2015) Adsorption of Cd(II), Cu(II) and Ni(II) ions by cross-linkingchitosan/rectorite nano-hybrid composite microspheres. Carbohydrate Polymers 130:333–343.
17. Park CM, Wang D, Han J, Heo J, Su C (2019) Evaluation of the colloidal stability and adsorption performance of reduced graphene oxide–elemental silver/magnetite nanohybrids for selected toxic heavy metals in aqueous solutions. Appl Surface Sci 471:8–17
18. Taimur S, Hassan MI, Yasin T (2017) Removal of copper using novel amidoxime based chelating nanohybrid adsorbent. European Polymer J 95:93–104
19. Zhang X, Tong J, Hu B X, Wei W (2018) Adsorption and desorption for dynamics transport of hexavalent chromium (Cr(VI)) in soil column. Environ Sci Pollut Res 25:459–468
20. Li Z, Pan Z, Wang Y (2019) Enhanced adsorption of cationic Pb(II) and anionic Cr(VI) ions in aqueous solution by amino-modified nano-sized illite-smectite clay. Environ Sci Pollut Res 26:11126–11139

21. Kanani-Jazi MH, Akbari S, Kish MH (2020) Efficient removal of Cr (VI) from aqueous solution by halloysite/poly(amidoamine) dendritic nano-hybrid materials: kinetic, isotherm and thermodynamic studies. Advan Powder Technol 31:4018–4030
22. Wang J, Zhang W, Yue X, Yang Q, Liu Y, Wang Y, Zhang D, Li Z, Wang J (2016) One-pot synthesis of multifunctional magnetic ferrite–MoS2–carbon dot nanohybrid adsorbent for efficient Pb(II) removal. J Mater Chem A 4:3893
23. Esrafili L, Firuzabadi FD, Morsali A, Hu ML (2021) Reuse of Predesigned Dual-Functional Metal Organic Frameworks (DF-MOFs) after Heavy Metal Removal. J Hazard Mater 403:123696
24. Mohan S, Kumar V, Singh DK, Hasan SH (2017) Effective removal of lead ions using graphene oxide-MgO nanohybrid from aqueous solution: isotherm, kinetic and thermodynamic modeling of adsorption. J Environ Chem Eng 5:2259–2273
25. Milani D, Bartlett AJ, de Solla SR, Parrott JL, Intini KD, Legault D, Unsworth J, Balakrishnan VK (2018) Comparative toxicity of azo dyes to two infaunal organisms (Hexagenia spp. and Tubifex tubifex) in spiked-sediment exposures. Environ Sci Pollu Res 25(7):6937–6950
26. Marie S, Heinz HS, Eva F, Petr H, Vaclav M (2014) Enzymes oxidizing the azo dye 1-phenylazo-2-naphthol (Sudan I) and their contribution to its genotoxicty and carcinogenicity. Curr Drug Metab 15:829–840
27. Shah MP (2018) Bioremediation-waste water treatment. Journal of Bioremediation & Biodegradation 9(1):1–10
28. Zubair M, Daud M, McKay G, Shehzad F, Al-Harthi MA (2017) Recent progress in layered double hydroxides (LDH)-containing hybrids as adsorbents for water remediation. Appl Clay Sci 143:279–29
29. Vuković J, Avidad MA, Capitan-Vallvey LF (2012) Characterization of disposable optical sensors for heavy metal determination. Talanta 94:123–132
30. Ajith MP, Priyadarshini E, Rajamani P (2020) Effective and selective removal of heavy metals from industrial effluents using sustainable Si–CD conjugate based column chromatography. BioresTechnol 314:123786
31. Li K, Yang H, Yuan X, Zhang M (2021) Recent developments of heavy metals detection in traditional Chinese medicine by atomic spectrometry. Microchem J 160:105726
32. Tawfik AS, Fadillah G (2019) Recent trends in the design of chemical sensors based on grapheneemetal oxide nanocomposites for the analysis of toxic species and biomolecules. Trends in Analytical Chem 120:115660
33. Khan SB, Akhtar K, Bakhsh EM, Asiri AM (2019) Electrochemical detection and catalyticremoval of 4-nitrophenol using CeO_2-Cu_2O and CeO_2-Cu_2O/CH nanocomposites. Appl Surface Sci 492:726–735
34. Wang M, Liu Y, Yang L, Tian K, He L, Zhang Z, Jia Q, Song Y, Fang S (2019) Bimetallic metal–organic framework derived FeOx/TiO_2 embedded in mesoporous carbon nanocomposite for the sensitive electrochemical detection of 4-nitrophenol. Sens Actuat B: Chem 281:1063–1072
35. Shi B, Zhong Y, Guo L, Li G (2015) Two dimethylphenyl imidazole dicarboxylatebased lanthanide metal–organic frameworks for luminescence sensing of benzaldehyde. Dalton Trans 44:4362
36. Rahman M, Hussein MA, Alamry KA, Al-Shehry FM, Asiri AM (2018) Polyaniline/graphene/carbon nanotubes nanocomposites for sensing environmentally hazardous 4-aminophenol. Nano-Structures Nano-Objects 15:63–74
37. Zhu G, Sun H, Zou B, Liu Z, Sun N, Yi Y, Wong K (2018) Electrochemical sensing of 4-nitrochlorobenzene based on carbon nanohorns/graphene oxide nanohybrids. Biosens Bioelectro 106:136–141
38. Tian X, Liu L, Li Y, Yang C, Zhou Z, Nie Y, Wang Y (2018) Nonenzymatic electrochemical sensor based on CuO-TiO_2 for sensitiveand selective detection of methyl parathion pesticide in ground water. Sen Actuat B 256:135–142
39. Thakur VK, Voicu SI (2016) Recent advances in cellulose and chitosan based membranes for water purification: A concise review. Carbohydrate Polymers 146:148–165

40. Pendergast M, Hoek EM (2011) A review of water treatment membrane nanotechnologies. Energy Environ Sci 4:1946–1971
41. Khin M, Nair AS, Babu VJ, Murugan R, Ramakrishna S (2012) A review on nanomaterials for environmental remediation. Energy Environ Sci 5:8075–8109
42. Goh PS, Ismail AF, Ng BC (2013) Carbon nanotubes for desalination: Performance evaluation and current hurdles. Desalination 308:2–14. https://doi.org/10.1016/j.desal.2012.07.040
43. International Desalination Association (IDA) (2008) Desalination in 2008 global marker snapshot, 21st GWI/International Desalination Association Worldwide Desalting Plant Inventory
44. Li M, Zhao R, Ma S (2021) Preparation of an anti-fouling silicone polyacrylate/SiO2/HDPE composite for industrial wastewater pipe. Materials Letters 304:130661
45. Dehghankar M, Mohammadi T, Tavakolmoghadam M, Tofighy MA (2021) Polyvinylidene Fluoride/Nanoclays (Cloisite 30B and Palygorskite) Mixed Matrix Membranes with Improved Performance and Antifouling Properties. Ind Eng Chem Res 60:12078−12091
46. Selim Mohamed S., Fatthallah Nesreen A., Higazy Shimaa A., Hao Zhifeng, Jing Mo Ping (2022) A comparative study between two novel silicone/graphene-based nanostructured surfaces for maritime antifouling. J Colloid Interface Sci 606:367–383. https://doi.org/10.1016/j.jcis.2021.08.026
47. Sharma M, Joshi M, Nigam S, Avasthi DK, Adelung R, Srivastava SK, Mishra YK (2019) Efficient oil removal from wastewater based on polymer coated superhydrophobic tetrapodal magnetic nanocomposite adsorbent. Appl Mater Today 17:130–141
48. Barroso-Solares S, Pinto J, Nanni G, Fragouli D, Athanassiou A (2018) Enhanced oil removal from water in oil stable emulsions using electrospun nanocomposite fiber mats. RSC Adv 8:7641
49. Ebrahimi M, Kerker S, Schmitz O, Schmidt A, Czermak P (2018) Evaluation of the fouling potential of ceramic membrane configurations designed for the treatment of oilfield produced water. Separ Sci Technol 53:349–363
50. Alammar A, Park SH, Williams CJ, Derby B, Szekely G (2020) Oil-in-water separation with graphene-based nanocomposite membranes for produced water treatment. J Membr Sci 603:118007
51. Neoh CH, Noor Z, Mutamim NSA, Lim CK (2016) Green technology in wastewater treatment technologies: Integration of membrane bioreactor with various wastewater treatment systems. Chem Eng J 283:582–594

Nanohybrid Materials

Anuradha Biswal and Sarat K. Swain

Abstract Nanohybrid materials or nanocomposites are a special class of new age advanced materials that have gained wide scale popularity because of their multifunctionalities. The nanocomposites offer a set of unique and outstanding properties like biodegradability, flexibility, processibility, good mechanical properties, and optical properties. The history of materials is established starting from stone age to present stage. Present chapter reveals an overall idea regarding the introduction to nanohybrid materials.

Keywords Nanohybrid materials · Classification · History · Applications

1 Introduction

Material science or material science and engineering (MSE) is an interdisciplinary branch dealing with exploration, designing of novel materials and further study of their characteristic properties. The origin of material science roots from the "Age of Enlightenment" when scientists and research personals initiated the combining of knowledge from physics and chemistry with engineering to analyse and explain different observations in mineralogy and metallurgy [1]. Materials are used to define the vital turning points of evolution of civilizations. Hence, material science is considered as an important branch of science as it is based on understanding the principles behind the formation of different materials and their functionalities. Investigation of the primitive structure of matter, beginning from atomic to macro scale, material scientists design new materials by combining elemental knowledge to provide novel properties. The field of material science saw a new turning point in 1959, when renowned American physicist, Richard Feynman delivered his most revolutionary keynote lecture, "There's plenty of room at the bottom" in the annual conference of American Physical Society, emphasizing on the efficiency of the unprecedented potential of materials at molecular level [2]. This hypothesis opened a new window in

A. Biswal · S. K. Swain (✉)
Department of Chemistry, Veer Surendra Sai University of Technology, Burla, Sambalpur, Odisha 768018, India
e-mail: skswain_chem@vssut.ac.in

S. K. Swain (ed.), *Nanohybrid Materials for Water Purification*, Composites Science and Technology, https://doi.org/10.1007/978-981-19-2332-6_2

material science to construct nanoscale devices with enhanced properties and unparalleled applications. Following this, the world entered the era of nanotechnology, which evolved as a new research field attracting the attention of material scientists to explore its vast expanse that envelops almost every application known today.

Hybrid nanostructured materials or nanohybrid materials or nanocomposites are defined as multiphase systems with minimum one nano-dimensional phase. The resultant material exhibits enhanced characteristics, as compared with bulk materials, as the inherent properties of materials change drastically when the size of the constituent component drops below critical level [3]. Reducing to nano-level allows interfaces for active interactions between different phases, which is essential for the exhibition of enhanced properties. The vast possibility of combining different phases with unique characteristics to form a single material have provided an opportunity for researchers to design materials with desired set of properties for targeted applications. This results in formulation of materials possessing tuneable functionalities and structure. The nanocomposite materials possess unique properties like mechanical strength, catalytic, optical, electric and magnetic properties, biodegradability, flexibility and processibilities [4]. These unique features offer nanohybrid materials applicability to a vast number of fields including catalysis, biomedical, sensing, energy storage, optical and electronic materials [5]. The aim of the present chapter is to provide a brief discussion regarding the past, present research and applications of nanocomposite materials along with highlighting the properties and importance of nanocomposites over traditional materials.

1.1 History of Nanocomposite Materials

Materials have played a vital role in defining the evolutionary path of mankind since the dawn of civilization. The phases of civilization are marked by the use of certain material tools and weapons at that period. The evolution of material science shows how the use of various materials has shaped the culture of different civilizations and lifestyle of the people. In short, the history of material science can be briefed as the journey and development of different materials throughout the history of earth. The total history can be broadly divided into six ages, primarily named according to the predominantly used material during the particular time (Fig. 1), namely,

- Stone Age
- Bronze Age
- Iron Age
- Porcelain Age
- Steel Age
- Silicon Age

The "**Stone age**" refers to the wide prehistoric era that lasted for about 3 million years, where the only materials used for making of tools and weapons were wood and stones with an edge. It lasted from 9600 to 3000 B.C. Gold and copper were the

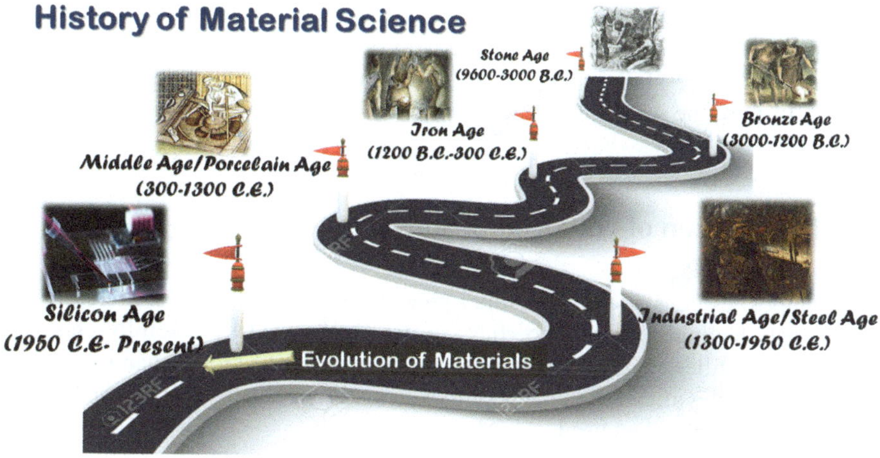

Fig. 1 Evolution of material science from prehistoric era till present

only metals used during this period for the purpose of making ornaments. Further, the process of smelting marked the end of the Stone age [6]. The "**Bronze age**" was stretched from 3000 to 1200 B.C. marked by the use of bronze as the primary material. It is considered as the second principal period and the period is characterized by the production of bronze by the process of smelting copper with other metals like tin and arsenic. The "**Iron age**" started with the advent of replacement of bronze tools and weapons with those made of iron [7]. It led to mass production of iron and marked the beginning of modernization. This period lasted from 1200 B.C. to 300 C.E. **Porcelain age** was marked from 300 to 1300 C.E most predominantly in China. The wares and tools are approximated to be heated from 1200 to 1300 °C [8]. Porcelain was first invented in Tang Dynasty which led to the strategical development of kilns to enhance the quality of produced porcelain. The "**Steel age**" ranged from 1300 to 1950 C.E., where the production of steel from iron was predominant. The tools and weaponry were shifted to steel due to its strength, durability and resistance to corrosion. This era witnessed the start of industrialization and urbanization.

The "**Silicon age**" (1950-present) refers to the age of information. Materials science became a more sought out discipline with the onset of silicon age. The production of new age transistors, semiconductor devices became more pronounced. This era witnessed the birth of nanotechnology and nanoscience which became an integral part of modern material science. The uses of various nanoparticles and nanocomposite materials have dominated the market with its versatile functionalities. As Richard Feynman's revolutionary lecture marked the beginning of the age of nanotechnology, the first polymer nanocomposite was used in 1990 when Toyota used nylon-6/clay nanocomposites for covers for timing belts, followed by the same GDI cover nanocomposite engines used by Mitsubishi. After that, nanocomposites were used in different applications for their versatile and tuneable properties. This era has marked the dawn of modern advanced materials and will shape the future with its ever-increasing demand. Now, the terms nanotechnology and nanocomposites have become coin words in advanced research field of material science and engineering and are being successfully used in addressing almost all problems and challenges posed in front of the scientific community.

2 Classification of Nanocomposites

Nanocomposites consist of two distinct phases, matrix and fillers. Different types of materials having desired properties can be synthesized by carefully choosing the matrix and fillers. Figure 2 illustrates the broad classification of nanocomposite materials. The classification of nanocomposite materials can be done on the basis of types of matrix and fillers used.

2.1 *On the Basis of Type of Matrix*

Based on the types of matrix materials opted, hybrid materials can be broadly classified into three broad categories, namely polymer matrix nanocomposites, metal matrix nanocomposites and ceramic matrix nanocomposites.

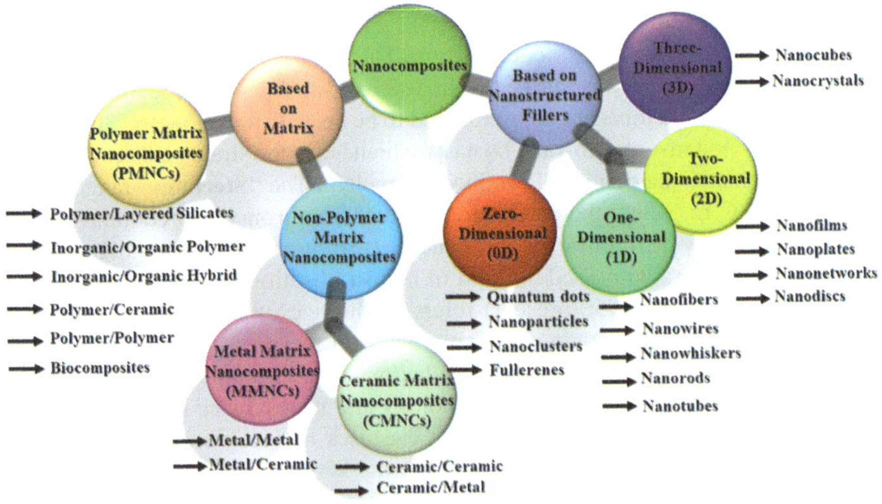

Fig. 2 Brief classifications of nanocomposites

2.1.1 Polymer Matrix Nanocomposites

Polymeric matrix nanocomposites (PMNCs) are defined as materials in which nano-dimensional inorganic and organic materials are distributed in polymer matrix. The resulting material exhibits improved physical and mechanical attributes in comparison with virgin polymers or traditional polymer composites [9]. The use of nanofillers along with polymeric matrices enhances the optical and electrical properties and aids the production of durable transparent coatings and membranes [10]. The reinforcement leads to enhancement of chemical, thermal properties, flame retardancy, and solvent resistance [11, 12]. The PMNCs are further classified into polymer/layered silicates, inorganic/organic polymer, inorganic/organic hybrid, polymer/ceramic, polymer/polymer and biopolymeric nanocomposites.

2.1.2 Metal Matrix Nanocomposites

The metal matrix nanocomposites (MMNCs) consist of nano sized reinforcements dispersed in metal matrix. This variant of nanocomposites can be of two types, having continuous or non-continuous fillers. Many metals as well as alloys have been easily utilized as matrix and require stable reinforcing materials that can withstand a wide range of temperature. Ti, Mg and Al are extensively used as matrix for synthesis of nanocomposites having wide applications in the field of aerospace. Higher mechanical strength of the MMNCs are resulted from the choice of reinforcing materials having high Young's modulus. This increases the strength to weight ratio of the overall nanocomposite greater than alloys [13]. Many metal matrices, like CNT reinforced MMNCs, BN reinforced MMNCs, have gained popularity for many applications [14]. The metal matrix can be further divided into metal/metal nanocomposites and metal/ceramic nanocomposites.

2.1.3 Ceramic Matrix Nanocomposites

In ceramic matrix nanocomposites (CMNCs), the matrix is composed of a ceramic material whereas the nanophase filler may be nanotubes, nanoparticles (Si_3N_4, SiC) nanoplatelets (BN, graphene) or combination of any of these [15]. The idea of ceramic matrix nanocomposite was introduced in 1991 by Niihara et al. [16]. They established the enhancement of mechanical strength of the ceramics by introduction of different nanostructured fillers into the ceramic matrix. The nanofillers inhibited the growth of cracks by playing the role of bridging element, thereby improving the mechanical properties. Furthermore, the incorporation of nanostructured materials as fillers enhances certain attributes of the nanocomposites like shock resistance, thermal and electrical conductivity, bioactivity, and wear resistance [17]. The ceramic nanocomposites are further classified into ceramic/ceramic and ceramic/metal nanocomposites.

2.2 On the Basis of Types of Fillers

The nano-dimensional fillers used in synthesis of various nanostructured composite materials play a major role of imparting different properties to the resulting hybrid material. The shape, size as well as dimensional attributes of the fillers affect the overall properties to a great extent. In 2007, Pokropivny and Skorokhod established a new scheme for classification on the basis of dimension [18]. They classified the nanofillers into: zero dimensional (0D), one- dimensional (1D), two-dimensional (2D), and three-dimensional (3D) nanoparticles [19]. The 0D nanoparticles can be single/polycrystalline, made from single or multiple chemical compositions and can exist in different forms and shapes. 1D nano-dimensional materials refer to the nanostructured materials having longer length and width in nanoscale. The electrons are confined in the two dimensions and hence their mobility is restricted. Nanotubes, nanowires, nanofibers, nanorods, nanowhiskers fall into this category [20, 21]. The 1D nanomaterials can be ceramic-, polymer- or metal-based and are extensively used in eyeglass coatings and computer circuits and chips. Two-dimensional nanomaterials have large aspect ratio and the electronic mobility is confined to one dimension only and are actively used in electronics, sensors, optoelectronics and biomedical applications. 2D structures like nanoplates, nanonetworks, nanodiscs, nanosheets, nanoribbons, nanofilms are placed under this category [22, 23]. The three-dimensional nanomaterials include equiaxed nanoscale grain like materials in which all the regimes do not fall in nano-dimension and are termed as bulk nanomaterials. They possess all the three dimensions above 100 nm. The electrons are delocalized in all directions providing free mobility throughout the materials [24, 25].

3 Synthetic Techniques

Various techniques have been adopted for synthesis of nanocomposite materials (Fig. 3). Different types of nanocomposites are synthesized and processed using different synthetic methods. Some of the important and commonly used synthetic methods of nanocomposites have been discussed in this section.

3.1 Solution Blending

It is a very convenient and extensively used technique for synthesis of polymeric nanocomposites. Its basic principle relies on mixing of the polymeric and colloidal solutions of the fillers dispersed uniformly in appropriate solvent. Commonly used solvents for this technique includes water, dimethyl formamide, toluene, and chloroform [26]. The process of blending of solutions and subsequent elimination of the solvent results the formation of nanocomposites. The final product can be acquired

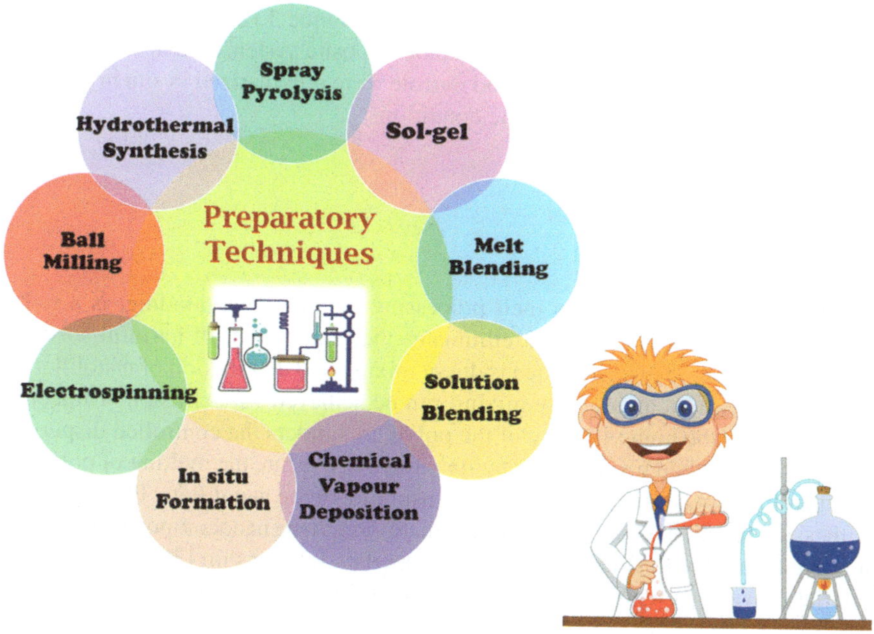

Fig. 3 Various synthetic methods for nanocomposite fabrication

by simple method of precipitation followed by filtration. However, selection of an appropriate solvent, which is to be removed at final stage, as well as toxicity associated with solvents pose some limitations to this method for fabrication of nanocomposite. Nanostructured hybrid materials composed of layered nanofillers like graphene/graphene oxide nanoplatelets embedded in polymeric matrix of polymethyl methacrylate, polystyrene, polyacrylamides, and polyimides are known to be fabricated using this technique [27]. Manafi et al. designed polyacrylamide/graphene-based nanocomposite systems for water purification applications using solution blending method [28].

3.2 *In Situ* Polymerization

This method is used to obtain polymeric nanocomposites with uniform distribution of fillers throughout the polymeric matrix. In this technique, a mixture consisting of monomer and concerned fillers is prepared, which is subsequently polymerized by basic polymerization routes like stirring, ultrasonication, photoinitiation, free radical mechanism etc. This particular synthetic method allows the formulation of nanocomposites with non-covalent/covalent linkages among constituents. The polymers which are difficult to process via melt blending and solution mixing techniques can be easily handled by in situ polymerization technique. Li and co-workers synthesized Fe_3O_4/PPy and Fe_3O_4/PPy/PANI nanocomposite systems via in situ polymerization technique [29]. Both aniline and pyrrole were polymerized in situ in presence of Fe_3O_4 nanoparticles under specific conditions.

3.3 Melt Blending

Melt Blending, also termed as melt processing or melt compounding, is a rather easy and economical method that eliminates the requirement of harmful solvents. The principle behind this technique is direct mixing and distribution of nanofillers in the melted polymeric phase. The mixing can be achieved via extrusion or injection moulding. But, the high viscosity of the polymers hinders the controlled dispersion of fillers. The method needs higher temperatures to execute the melting of polymers as well as high shear forces for uniform distribution of the nanofillers. Platnieks et al. synthesized functional poly(butylene succinate)/graphene nanocomposite using melt blending technique to study its structure-properties relationship [30]. Several other materials have been reported to be synthesized using this technique [31, 32].

3.4 Spray Pyrolysis

Spray pyrolysis refers to the technique which involves the spraying of a solution, leading to deposition of a film on the surface of a heated substrate on which the constituent materials react with each other to yield a chemical compound. The precursor materials are chosen carefully so that the by-products formed become volatile at the selected deposition temperature. The synthetic steps involve: (1) dissolving the selected precursors in desired solvent to obtain the spraying liquid; (2) formation of mist from the liquid source by employing an ultrasonic atomizer; (3) utilization of a carrier gas for transportation of the mist to the preheated chamber; (4) vaporization of the mist droplets in the preheated chamber along with triggering their decomposition to yield corresponding oxides; (5) reduction of metal oxides to give corresponding metallic materials [33]. Gharbi and co-workers designed ZnO-NiO nanocomposite thin films via spray pyrolysis on a glass surface at 480 °C for waveguiding applications [34]. Further, Janakiraman et al. deposited Ta_2O_5/SnO_2 nanocomposite films on the surface of glass substrate by spray pyrolysis method [35].

3.5 Chemical Vapour Deposition

Chemical Vapour Deposition (CVD) is a chemical method of nanocomposite synthesis which is used to generate quality, highly efficient solid materials applications in semiconductor industry. The process involves chemical interactions between halide or organometallic compounds and other gases to generate non-volatile thin films on surface of substrates. CVD leads to a multidirectional kind of deposition of substrate [36]. CVD is widely employed for the preparation of nanohybrid materials. The benefits of CVD as compared to other techniques involves: (1) flexibility of selection of a wide range of nanoparticle solutions and chemically suitable precursors; (2) wide applicability in industrial sector like glazing, microelectronics etc.; (3) ease of execution that doesn't need critical handling steps and ageing time [37].

3.6 Sol–Gel Method

Sol–gel method of nanocomposite synthesis is extensively used for fabrication of polymer nanocomposites. The method involves the dissolution of nanomaterial precursor in appropriate solvent and polymeric matrix. The solution phase (sol) is activated by the means of radiation or heating to form a solid–liquid phase (gel). Following this the gel is cured and the nanocomposite is obtained. In this, the polymer plays the role of initiating the nucleation process and thereby the formation of required nanomaterials. This synthetic technique does not involve steady energy requirement for distribution of the nanomaterials in the polymeric matrix. Recent work of Behnam et al. involved the fabrication of carbon nanotubes embedded polyurethane nanohybrid materials via sol–gel technique. They claimed the use of above said pathway allowed the formation of monodispersed nanocomposites at very low cost [38]. Lee et al. also stated the same results while preparing titanium dioxide nanoparticles incorporated polystyrene/poly(vinylpyrrolidone) hybrid material [39].

3.7 Powder Process

This method is used to synthesize various ceramic nanohybrid materials which involves compression, rolling and extrusion for processing the hybrid materials. In the compression step, the powder is allowed to flow into a two walled tube and is exposed to repeated vertical compression. The obtained compact mass is sintered in a furnace. The powder is also subjected to rolling to give sheets. The metallic powder is transferred into a rolling mill which is subsequently transformed into compact strips and are sintered in furnace. Then the strips are subjected to further rolling and sintering to obtain the desired nanohybrid materials. Powder extrusion is broadly divided into two types, one in which the powder is blended with plasticizer or binder in room temperature and another involving extrusion at higher temperatures without fortification. This method has no demerits related to the vast variety of alloys and metals to be extruded, provided the pressure and temperature conditions are well within the capacity of the die materials [40]. Gholipour et al. synthesized nanoclay reinforced aluminium composites using powder extrusion method [41].

3.8 *Electrospinning Technique*

This technique is employed to fabricate polymer nanocomposite fibers using high voltage application. The nanomaterial and polymer are solubilized in an appropriate solvent and sonicated or stirred to get a homogeneous mixture. This is fed into a syringe and under high voltage conditions, nanocomposite fibers are extruded. This is highly opted as the resulting nanofibrous structures show uniform morphology and size. Burke et al. formulated nano iron oxide trapped poly(ethylene oxide)/poly(vinyl pyrrolidone) fibers via electrospinning method [42].

4 Properties of Nanocomposites

Advanced nanocomposite materials have shown unprecedented characteristics and functionalities in many applications. They have offered better performances and results in almost all fields due to their unique properties, as shown in Fig. 4. Apart from features like flexibility, processibility, biodegradability, durability, chemical

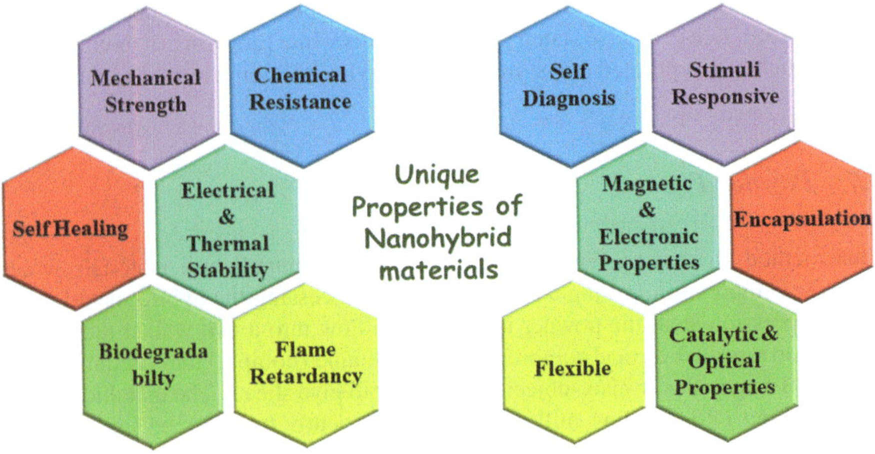

Fig. 4 Unique properties of nanocomposite materials

resistance, etc., some of the important characteristic properties of nanocomposite materials are discussed below.

4.1 Thermal and Fire Retardancy

Many consumable products require enhanced thermal properties to meet the market standards. This is of utmost importance for designing of advanced materials for aerospace applications in which improved functionality and reliability are essential. Hence, incorporation of nanomaterials into classical composite matrix has opened a new window for producing robust materials to ensure durability and stability at elevated temperatures. Especially, layered nanomaterials like nanoclay and other ceramic materials have been efficiently used in these fields to get good performance materials [43]. Addition of clay platelets improves the heat barrier properties as well as supports in the process of char formation during thermal decomposition. Recently, Bel et al. designed graphene nanoplatelets exfoliated polymethylmethacrylate nanocomposite via in situ polymerization technique which showed improved thermal stability at elevated temperatures due to the reinforcement of 2 wt% graphene [44]. Further, Baniasadi et al. formulated polyamide 614 nanocomposites exfoliated with nanoclay to study its thermal, structural and mechanical attributes. The loading of nanoclay improved the thermal properties significantly with single step degradation at above 350 °C [45].

4.2 Electronic Properties

The advancement of nanotechnology research is very pronounced in fields like data storage, electronics, communications, and computing. In this context, carbonaceous nanomaterials have gained a great deal of attention, especially CNTs. Carbon nanotubes are considered as electrically conductive fillers possessing high aspect ratio. Their morphology allows them to create a networking structure to facilitate conduction within the polymeric matrix, even at lower loading percentages. Nanotubes provide even conductivities throughout the material along with minimal warp [46]. The use of multi-walled CNTs results in formation of robust materials as they are highly resistant to wear and tear. They form powerful mechanical bearings as the gaps between the consecutive nanotubes are in the order of nanometers. However, their properties depend on many other factors such as rate of dispersion, aspect ratio, synthesis technique, nature of network formed by fillers etc. Charitos et al. designed GO/CNF and GO/CNT dispersed low density polyethylene (LLDPE) matrix to investigate its thermomechanical and electrical properties for concerned applications. They reported that the loading of hybrid nanofillers resulted in the enhancement of electrical properties of the resulting material as compared to the virgin matrix [47].

4.3 Mechanical Properties

The mechanical attributes, such as stiffness and strength, of a composite material are greatly influenced by addition of reinforcing material into its matrix. Polymer matrix nanocomposites have been widely investigated for their significant physical and mechanical properties. This is caused by the large interfacial area for interaction between the polymer and nanofillers. The effect of huge interfacial interaction due to the dispersion of nano-dimensional reinforcements leads to the exhibition of outstanding physical and mechanical properties as compared to traditional composites [48]. Incorporation of graphene in polymeric matrices showed outstanding enhancement in mode-II fracture toughness as well as shear strength of the resulting nanocomposites [49]. Furthermore, the use of nanofibers has also been widely investigated for improving the mechanical strength of conventional composite materials. The consequence of embedment of cellulose nanofibers (CNF) on the overall thermal expansion and mechanical properties of biocomposites has been reported earlier. Oishi et al. prepared CNFs reinforced PVA biocomposites to establish CNF as suitable reinforcing agents [50]. Detailed investigation of structural and mechanical properties of the bionanocomposites revealed a 48% elevation in Young's modulus as a result of incorporation of the fibrous nanofillers.

4.4 Magnetic Properties

Two categories of composites exhibit magnetic characteristics, namely composites reinforced with metal nanoparticles and other reinforced with ferrite nanoparticles. Generally, nanoparticles are devoid of hysteresis which yields superparamagnetic materials. The previous report gives information about the magnetic behaviour of polymer nanocomposites containing several metal oxide nanoparticles for dye removal purpose. They reported the material to be optically transparent and hysteresis free at room temperature [51]. Ramesan et al. formulated NiO nanoparticles incorporated polyvinyl cinnamate nanocomposite and revealed its ferromagnetic nature. Additionally, the embedment of NiO nanoparticles increased the magnetic response, remanence and hysteresis values [52].

4.5 Self-Healing

The need for smart self-healing nanocomposites have been met by different nanocomposites especially with polymeric matrix ranging from thermosetting/thermoplastic polymers to rubbers [53]. Combination of special attributes of the nanocomposites like efficient nanofillers and reversible polymeric networks in matrix has led to the formulation of many self-healing nanocomposites [54]. The main feature of these intelligent nanocomposite systems is the presence of stimuli-responsive interactions that initiates self-healing of the material [55]. Hence, self-healing materials provide an effective and efficient alternative to address environmental issues generated by the use of traditional polymers. They are extensively utilized in various fields like electronics [56], automobile industry [57], textile industry [58], etc. The self-healing mechanism of these nanocomposites is aided by two pathways, namely intrinsic and extrinsic mechanisms. Nanocomposites with intrinsic self-healing properties usually consist of polymeric backbones having active groups that exhibit reversible stimuli-responsive non-covalent and covalent bond interactions [59]. The intrinsic nature of the materials allows the materials to self-repair many times. The extrinsic healing mechanism is a rather temporary phenomenon that is rendered to the material by micro-vascular and capsular systems containing self-healing agents [60]. These repairing agents undergo polymerization to heal the damage [61]. The exhaustion of the repairing agents devoid the material of healing ability.

4.6 Optical Properties

Nanocomposites reinforced with metallic and semiconductor-based nanomaterials exhibit unique photoluminescence effect, linear absorption and non-linear optical attributes. nanofillers possessing smaller particle size distribution show remarkable optical properties because of the effect of quantum confinement. Investigation of optical properties of nanocomposites with reinforcement of nanomaterials with anisotropic shapes has gained a lot of attention. Extensive research efforts have been given for formulation of advanced materials having applications in the field of photonics and optoelectronics. Sharma et al. fabricated lignocellulose-based nanofibers incorporated cellulose acetate composite films and studied the effect of nanofiller loading on the structural and optical properties of the nanocoposite [62]. Abdelrazek et al. reported the effect of Au nanoparticles on the optical properties of PEO/PVP composites [63].

5 Applications of Nanocomposite Materials

The main attraction of nanocomposite materials is its multifunctionality. The features like easy processibility into different forms and tuneable properties have allowed nanocomposites to be used in a wide range of applications (Fig. 5). Some of the most pronounced applications of nanocomposites have been discussed in this section.

Fig. 5 Different applications of nanocomposites

5.1 Environmental Remediation

Some of the commonly used nanomaterials for the application in environmental remediation are enzyme nanoparticles, polymeric nanoparticles and metal oxide and metallic nanoparticles. Metallic nanoparticles like Fe, Cu, Ni, Co etc. are being actively used in many environmental remediation applications. This is attributed to their smaller size and high specific surface area [64]. Hence, these nanoparticles and their corresponding nanocomposites with ceramic, polymeric, carbon, silica-based matrix are considered suitable for the purpose of environmental remediation [65]. Nanoparticles are successfully loaded into polymeric matrices, namely cellulose, chitosan and their carboxymethyl derivatives, alginates, resin etc., [66] which results in exhibition of excellent thermal and mechanical properties. Selecting material for organic applications, chemical resistance, hydrophilic-hydrophobic balance, biodegradability and biocompatibility, chemical performance, and optoelectronic properties should be taken into account [67].

Polymeric nanocomposites have a wide set of applications in environmental protection like removal and degradation of pollutants from water like organic dyes, compounds, heavy metals etc., wastewater treatment, air purification, sensing of soil pollutants, oil spill restoration etc. [68]. Trujillo-Reyes et al. published a review in 2014 stating the effect of nanomaterials and new age nanocomposites in addressing the environmental issues [69].

Photocatalytic degradation using nanocomposites has become a widely used technique for removal of toxins such as organic dyes like malachite green, methylene blue, etc. [70]. Gupta et al. combined Zr(IV) selenotungstophosphate with pectin to formulate polymeric nanocomposite for adsorption and photocatalytic degradation of the organic dyes [71]. Different combinations of polymeric matrices and nanofillers have been reported for the purpose of removal of pollutants from air, soil and water bodies [72]. After the removal of effluents, pathogens like bacterial, protozoans, viruses are still present and can be removed by a series of processes like chlorination, catalytic disinfection or UV treatment. Metallic Ag, Ti and Zn have been used for inhibition of pathogen activity in the water disinfection process [73]. The nanocomposites provide an eco-friendly alternative associated with minimal waste production, substantial pollutant removal ability, and negligible energy needs [72].

5.1.1 Wastewater Treatment

Wastewater treatment occupies a major part of environmental remediation. As our lifestyle is highly dependent on consumption of water, availability of clean drinking water is the need of the hour. But the alarming increase in water pollution has made the available water (31% of the total volume of fresh water is accessible to us) unfit for consumption. Various materials and devices have been designed for purification of water but conventional techniques fail to deliver the desired results. Introduction of nanotechnology provided a new scope for formulation of new age technologies to serve the purpose. Advanced hybrid materials and nanocomposites offer a relatively safer, eco-friendly, cost-effective way to restore the integrity of drinking water. Several nanocomposite-based hydrogels, films, membranes have been designed in the past few decades. The use of polymer nanocomposite membranes along with some chemical and physical alterations in the matrix interface have been highly encouraged for their enhanced selectivity, permeability, foul resistance, mechanical strength and many other features [74]. Specifically, polymeric or inorganic nanoparticles are incorporated in polymeric matrix phase in order to achieve superior efficiency of removing pollutants from water. In this direction, great focus has been laid on polymer/carbon, polymer/polymer, polymer/metal, and polymer/clay nanocomposites for their effective filtration capabilities. The main role of polymer-based nanocomposite membranes are separating the pollutants from water samples by the process of reverse osmosis, ultrafiltration, microfiltration, desalination and other relevant methods; adsorption of organic and inorganic pollutants via physisorption, and chemisorption method; photo-/electro-catalytic degradation of the pollutants; and antifouling or prevention of pore clogging of the membrane structure by the use of inorganic nanomaterials like CNTs and Ag nanoparticles [75].

5.2 *Energy Storage*

The production of renewable form of energy and its subsequent storage is of importance to mankind in present day. Polymer nanocomposites have been efficiently used for this purpose. Polymeric nanocomposites have been successfully used for storage of electrochemical energy in the form of batteries, supercapacitors, fuel cells and other modes [76]. The interface provided by the conducting polymers, due to its larger surface area, gives a better platform for electrochemical reactions in the supercapacitors. In lithium-ion batteries, the polymeric nanocomposites show high selectivity towards the mobility of Li^+ ions across the separators and hence, improve the columbic performance of the batteries [77].

5.2.1 Nanocomposites in Supercapacitors

Polymers like polythiophene, polypyrrole, poly(3,4-ethylene dioxythiophene), polyacrylonitrile play a major role in designing of supercapacitors. The conjugated polymeric nanomaterials are combined with allotropes of carbon or metal oxides to make electrodes that improve the electrochemical action. In a nutshell, the generation of polymer-nanomaterial interface minimizes distortion of polymeric molecules accentuates the rate capability and cyclic stability of the supercapacitor [78].

5.2.2 Nanocomposites in Lithium-Ion Batteries

These are the most commonly used devices to store electrochemical energy efficiently. Various nanomaterials have been used for the designing of Li-ion batteries due to their effect on flexibility of design as well as overall efficiency. Li-ion batteries based on polymeric nanocomposites offer minimal toxicity, huge capacity, good cycling life, and large working voltage and hence are used in rechargeable energy storage devices [79].

5.3 *Construction*

Nanomaterials like nano alumina, nanoclay, nano iron and nano zirconites have been successfully reinforced into concrete matrix to enhance the formation of C-S–H gels, durability, porous structure and other physico-mechanical properties of the concrete that is one of the most important components of civil construction. Incorporation of fly ash improves the tensile properties of concrete but elongates the strengthening time of the concrete. This demerit is addressed by reinforcing nanosilica in the concrete mixture which further accentuates the density and structure of the concrete. Fiber/cement composites reinforced with natural fibers are economical, eco-friendly alternatives to traditional concrete composites [80]. Attachment of nanoscale bacterial cellulose on natural fiber's surface is a favourable method of surface modification which in turn regulates the matrix-filler interaction. This technique uses the nature of the fibers to absorb water molecules in the bacterial nanocellulose dispersion and nanocellulose is attached to the surface of fibers along with the water molecules [81]. Sakr et al. synthesized concrete mixture based on nonclay and nano silica reinforced silane/methyl methacrylate composites. The resultant nanocomposites exhibited good transport/barrier attributes and durability on exposure to salt and cyclic environmental conditions [82].

5.4 Biomedical Applications

Nanostructured hybrid materials have been extensively used for various biomedical applications like bioimaging and biosensing, drug delivery, tissue engineering, wound healing, gene therapy and many more. The main reason behind choosing these materials is the enhanced accuracy and efficiency in diagnosis, observation and subsequent treatment in medical fields. Polymer nanocomposites are being actively used to deliver various bioactive substances for applications like drug delivery, wound healing, etc. Most utilized polymeric matrices used for biomedical fields are synthetic polymers like poly(ε-caprolactone), polylactides, acrylic and tyrosine-based polymers; natural polymers like cellulose, alginate, chitosan etc.; or combination of both. These nanocomposites can be easily processed into various forms depending upon use. Usually nanocomposite hydrogels, films and membranes are highly recommended for biomedical use. Out of these, nanocomposite hydrogel has grabbed the attention of material scientists for its versatile applications and unparalleled applicability and adjustable biochemical and physical attributes [83]. The use of nanocomposite hydrogels in some of the important biomedical fields has been discussed below.

5.4.1 Tissue Engineering

Hydrogel, in general, is a rather interesting candidate for generating tissue engineering scaffolds attributed to its high-water absorption affinity, porous morphology, non-toxicity, and most essentially its biomimicking ability [84, 85]. But the virgin matrix hydrogels are limited by lack of mechanical and optical properties [86, 87]. Incorporation of nanoparticles overcomes the limitations of composite hydrogels along with introduction of sites of adhesion [88], and providing growth factors to assist cellular growth [89]. Moreover, introduction of nanoparticles further improves the self-healing nature, thermal properties, supports stem cell differentiation [85], and hence, allows formulation of implantable scaffolds for tissue engineering.

5.4.2 Drug Delivery

Stimuli-responsive drug delivery is defined as a controlled form of drug delivery triggered by stimulus either internal or external [90]. But, without tuning the properties of the drug carrier hydrogels, the drug release may occur in an unregulated manner [91]. Incorporation of nanoparticles offers a greater control over the rate of crosslinking, porosity, and hence enhances its applicability in drug administration. Nanoparticles may also have direct effect on the hydrogel's response to external stimuli like electromagnetic radiations and hence allows the orthogonal drug administration [92] and used for effective antibacterial application [93]. Dai et al. designed a series of poly(lactic-co-glycolic acid) (PLGA) nanoparticles incorporated chitosan/glycerophosphate (CS/GP)-based thermosensitive hydrogels for the purpose of release of the drug interferon α-2 b. Incorporation of the polymeric nanoparticles enhanced the mechanical strength of the hydrogels as well as elongated the release time up to 1.5 to 3 times as compared to virgin matrix hydrogels [94].

5.4.3 Wound Healing

Conventional wound dressing materials usually requires antibiotics to inhibit microbial infection [94]. The incorporation of nanoparticles has rendered angiogenic, antipathogenic, and bioadhesive properties to the hydrogels [95, 96]. As reported in the earlier section, nanoparticles addition can accentuate the mechanical strength and sites of cell adhesion, which may enable good resolution prints and post removal cell viability of the nanocomposite hydrogels [97]. Recently, Khan et al. reported PVA/PVP hydrogels embedded with Ag@ZnO core–shell nanocomposites for wound healing applications. The synthesized hydrogels showed improved antibacterial behaviour as well as wound healing efficacy in mouse models [98].

5.5 Packaging Industry

Nanocomposite films have gained popularity in packaging industry, especially for the purpose of food packaging. Polymeric nanocomposite films have offered an effective alternative to the traditionally used plastic-based packaging materials. These new age packaging materials exhibit improved gas barrier properties, water resistance, mechanical and thermal stability, as well as biodegradability which ensures safe and secure packaging of the consumables as well as increases its shelf life. Commonly used polymers for this purpose are synthetic polymers like polypropylene, polycaprolactone, polylactic acid (PLA), poly(butylene adipate-*co*-terephthalate) (PBAT),biopolymers like cellulose, chitosan, alginates or their combination to form biodegradable packaging materials. Further incorporation of nanodimensional fillers enhances the above said properties concerned for the bioprinting application [99]. Xie et al. fabricated PBAT/organoclay nanocomposites to study its packaging efficiency and reported improved optical, thermal, mechanical, barrier properties [100]. The material exhibited a 41.9% elevation in tensile strength as compared to virgin PBAT films.

6 Conclusion

Nanohybrid materials are new class of smart materials with potential applications because of their versatile properties. Out of different applications the designing of hybrid materials for water purification has achieved a special attention of researchers with a challenge for removal of contaminants from drinking waters. Based on this issue present chapter gives a concluding remark towards history, classifications and possible synthesis of new age hybrid materails.

References

1. Eddy MD (2008) The language of mineralogy: John Walker, chemistry and the edinburgh medical school. 1750–1800. ISBN0754663329, 9780754663324
2. Feynman RP (1960) There's plenty of room at the bottom. Eng Sci 23:22–36
3. Schmidt D, Shah D, Giannelis EP (2002) New advances in polymer/ layered silicate nanocomposites. Curr Opin Solid State Mater Sci 6(3):205–212
4. Kumar SK, Jouault N, Benicewicz B, Neely T (2013) Nanocomposites with polymer grafted nanoparticles. Macromolecules 46:3199–3214
5. Qi W, Zhang X, Wang H (2018) Self-assembled polymer nanocomposites for biomedical application. Curr Opin Colloid Interf Sci 35:36–41
6. Ko KH (2016) Origins of human intelligence: the chain of tool-making and brain evolution. Anthropol Notebooks 22(1):5–22
7. Sarunas M (2002) European prehistory: a survey. Springer. ISBN 0306467933, 9780306467936
8. He L (1996) Chinese ceramics: the new standard guide. Thames and Hudson, London. ISBN 0500237271, 9780500237274
9. Zhu C, Du D, Eychmuller A, Lin Y (2015) Engineering ordered and nonordered porous noble metal nanostructures: synthesis, assembly, and their applications in electrochemistry. Chem Rev 115(16):8896–8943
10. Schmidt G (2003) Properties of polymer–nanoparticle composites. Curr Opin Colloid Interf Sci 8:145–155
11. Liu K, Cao M, Fujishima A, Jiang L (2014) Bio-inspired titanium dioxide materials with special wettability and their applications. Chem Rev 114(19):10044–10094
12. Gacitua W, Ballerini A, Zhang J (2005) Polymer nanocomposites: synthetic and natural fillers a review. Maderas Cienc y Tecnol 7(3):159–178
13. Tjong SC, Ma ZY (2000) Microstructural and mechanical characteristics of in situ metal matrix matrix composites. Mater Sci Eng 29:49–113
14. Torralba JM, da Costa CE, Velasco F (2003) P/M aluminum matrix composites: an overview. J Mater Process Technol 133:203–206
15. Yazdani B, Porwal H, Xia Y, Yan H, Reece MJ, Zhu Y (2015) Role of synthesis method on microstructure and mechanical properties of graphene/carbon nanotube toughened Al_2O_3 nanocomposites. Ceram Int 41(8):9813–9822

16. Niihara K (1991) New design concept of structural ceramics—Ceramic nanocomposites. J Am Ceram Soc 99(10):974–982
17. Porwal H, Grasso S, Reece M (2013) Review of graphene-ceramic matrix composites. Adv Appl Ceram 112(8):443
18. Pokropivny V, Skorokhod V (2007) Classification of nanostructures by dimensionality and concept of surface forms engineering in nanomaterial science. Mater Sci Eng C 27(5–8):990–993
19. Capsal J, Dantras E, Dandurand J, Lacabanne C (2011) Molecular mobility in piezoelectric hybrid nanocomposites with 0–3 connectivity: volume fraction influence. J Noncrystl. Solids 357(19–20):3410–3415
20. Wang Z (2006) Piezoelectric nanogenerators based on zinc oxide nanowire arrays. Science 312(5771):242–246
21. She G, Zhang X, Shi W, Fan X (2007) Chang J Electrochemical/chemical synthesis of highly-oriented single-crystal ZnO nanotube arrays on transparent conductive substrates. Electrochem Commun 9(12):2784–2788
22. Yamano A, Takata K, Kozuka H (2012) Ferroelectric domain structures of 0.4-μm-thick $Pb(Zr,Ti)O_3$ films prepared by polyvinylpyrrolidone-assisted Sol–Gel method. J Appl Phys 111(5):054109
23. Qi Y, Jafferis N, Lyons K, Lee C, Ahmad H, McAlpine M (2010) Piezoelectric ribbons printed onto rubber for flexible energy conversion. Nano Lett 10(2):524–528
24. Liu JL, Bashir S (2015) Advanced nanomaterials and their applications in renewable energy. Elsevier, Amsterdam, The Netherlands. ISBN 9780128015285
25. Zhang Q, Deng Y, Hu Z, Liu Y, Yao M, Liu P (2014) Seaurchin-like hierarchical $NiCo_2O_4$@$NiMoO_4$core–shell nanomaterials for high performance supercapacitors. Chem Phys 16(42):23451–23460
26. Wang M, Yan C, Ma L (2012) Graphene nanocomposites. InTech, London, UK
27. Potts JR, Dreyer DR, Bielawski CW, Ruoff RS (2011) Graphene-based polymer nanocomposites. Polymer 52:5–25
28. Manafi M, Manafi P, Agarwal S, Bharti AK, Asif M, Gupta VK (2017) Synthesis of nanocomposites from polyacrylamide and graphene oxide: application as flocculants for water purification. J Colloid Interf Sci 490:505–510
29. Li B, Weng X, Wu G, Zhang Y, Lv X, Gu G (2017) Synthesis of Fe_3O_4/polypyrrole/polyaniline nanocomposites by in-situ method and their electromagnetic absorbing properties. J Saudi Chem Soc 21(4):466–472
30. Platnieks O, Gaidukovs S, Neibolts N, Barkane A, Gaidukova G, Thakur VK (2020) Poly(butylene succinate) and graphene nanoplatelet–based sustainable functional nanocomposite materials: structure-properties relationship. Mater Today Chem 18:100351
31. Mittal G, Dhand V, Rhee KY, Park SJ, Lee WR (2015) A review on carbon nanotubes and graphene as fillers in reinforced polymer nanocomposites. J Ind Eng Chem 21:11–25
32. Zhao Y, Hou J, Bai Z, Yang Y, Guo X, Cheng H, Zhao Z, Zhang X, Chen J, Shen C (2020) Facile preparation of lightweight PE/PVDF/Fe_3O_4/CNTs nanocomposite foams with high conductivity for efficient electromagnetic interference shielding. Compos A Appl Sci Manuf 139:106095
33. Choa YH, Yang JK, Kim BH, Jeong YK, Lee JS, Nakayama T, Sekino T, Niihara K (2003) Preparation and characterization of metal: ceramic nanoporous nanocomposite powders. J Magn Magn Mater 266(1–2):12–19
34. Gharbi B, Taabouche A, Brella M, Gheriani R, Bouachiba Y, Bouabellou A, Hanini F, Barouk S, Serrar H, Rahal B (2021) Spray pyrolysis synthesized and ZnO–NiO nanostructured thin films analysis with their nanocomposites for waveguiding applications. Semiconductors 55:37–43
35. Janakiraman V, Tamilnayagam V, Sundararajan RS, Suresh S, Biju CS (2021) Structural optical and morphological properties of Ta_2O_5/SnO_2 nanocomposite thin film prepared by spray pyrolysis method. J Mater Sci: Mater Electron 32:9244–9252

36. Behera A, Mallick P, Mohapatra SS (2020) Nanocoatings for anticorrosion: an introduction. 227–243
37. Binions R, Parkin IP (2011) Novel chemical vapour deposition routes to nanocomposite thin films, Advances in nanocomposites - synthesis, characterization and industrial applications, Boreddy Reddy, IntechOpen
38. Behnam R, Roghani-Mamaqani H, Salami-Kalajahi M (2019) Preparation of carbon nanotube and polyurethane-imide hybrid composites by sol–gel reaction. Polym Compos 40:E1903–E1909
39. Lee KC (2019) Preparation of PS/PVP/TiO_2 core–shell particles under various reaction factors. J Nanosci Nanotechnol 19:6334–6340
40. Nakahira A, Niihara K (1992) Strctural ceramics-ceramic nanocomposites by sintering method:roles of nano-size particles. J Ceram Soc Jpn 100(4):448–453
41. Gholipour V, Shamanian M, Ashrafi A, Maleki A (2021) Development of aluminium-nanoclay composite by using powder metallurgy and hot extrusion process. Met Mater Int 27:3681–3694
42. Burke L, Mortimer CJ, Curtis DJ, Lewis AR, Williams R, Hawkins K, Maffeis TGG, Wright CJ (2017) In-situ synthesis of magnetic iron-oxide nanoparticlenanofibre composites using electrospinning. Mater Sci Eng C 70:512–519
43. Leszczy_nska A, Njuguna J, Pielichowski K, Banerjee JR (2007) Polymer/montmorillonite nanocomposites with improved thermal properties: part I. Factors influencing thermalstability and mechanisms of thermal stability improvement. Thermochimica Acta 453 (2):75–96
44. Bel T, Mehranpour S, Şengül AV, Camtakan Z, Baydogan N (2021) Electron beam penetration of poly (methyl methacrylate)/colemanite composite irradiated at low earth orbit space radiation environment. J Appl Polym Sci 138:45
45. Baniasadi H, Trifol J, Ranta A, Seppälä J (2021) Exfoliated clay nanocomposites of renewable long-chain aliphatic polyamide through in-situ polymerization. Compos B Eng 211:108655
46. Star A, Liu Y, Grant K, Ridvan L, Stoddart JF, Steuerman DW, Diehl MR, Boukai A, Heath JR (2003) Noncovalent side wall functionalization of single-walled carbon nanotubes. Macromolecules 36(3):553–560
47. Charitos I, Georgousis G, Klonos PA, Kyritsis A, Mouzakis D, Raptis Y, Kontos A, Kontou E (2021) The synergistic effect on the thermomechanical and electrical properties of carbonaceous hybrid polymer nanocomposites. Polym Testing 95:107102
48. Sahoo NG, Rana S, Cho JW, Li L, Chan SH (2010) Polymer nanocomposites based onfunctionalized carbon nanotubes. Prog Polym Sci 35:837–867
49. Liu F, Hu N, Zhang JY, Atobe S, Weng SY, Ning HM, Liu YL, Wu LK, Zhao YX, Mo FH, Fu SY, Xu CH, Alamusi YW (2016) The interfacial mechanical properties of functionalized graphene-polymer nanocomposites. RSC Adv 6:66658–66664
50. Oishi Y, Nakaya M, Matsui E, Hotta A (2015) Structural and mechanical properties of cellulose composites made of isolated cellulose nanofibers and poly(vinyl alcohol). Compos A Appl Sci Manuf 73:72–79
51. Makarchuk OV, Dontsova TA, Astrelin IM (2016) Magnetic nanocomposites as efficient sorption materials for removing dyes from aqueous solutions. Nanoscale Res Lett 11(1):161
52. Ramesan MT, Jayakrishnan P (2017) Role of nickel oxide nanoparticles on magnetic, thermal and temperature dependent electrical conductivity of novel Poly(vinyl cinnamate) based nanocomposites: applicability of different conductivity models. J Inorg Organomet Polym Mater 27(1):143–153
53. Yan X, Wang F, Zheng B, Huang F (2012) Stimuli-responsive supramolecular polymeric materials. Chem Soc Rev 41:6042–6065
54. Thakur VK, Kessler MR (2015) Self-healing polymer nanocomposite materials: a review. Polymer 69:369–383
55. Kuhl N, Bode S, Hager MD, Schubert US (2015) Self-healing polymers based on reversible covalent bonds. In: Advances in polymer science. Springer: New York, NY, USA, 273, pp 1–58
56. Tanaka T (2005) Dielectric nanocomposites with insulating properties. IEEE Trans Dielectr Electr Insul 12:914–928

57. Chandra A K, Kumar N R (2017) Polymer nanocomposites for automobile engineering applications. In: Properties and applications of polymer nanocomposites. Springer, Berlin/Heidelberg, Germany, pp 139–172
58. Gowri S, Almeida L, Amorim T, Carneiro N, Souto AP, Esteves MF (2010) Polymer nanocomposites for multifunctional finishing of textiles—A review. Text Res J 80:1290–1306
59. Wool RP (2008) Self-healing materials: a review. Soft Matter 4:400–418
60. Blaiszik BJ, Sottos NR, White SR (2008) Nanocapsules for self-healing materials. Compos Sci Technol 68:978–986
61. Utrera-Barrios S, Verdejo R, López-Manchado M, Santana MH (2020) Evolution of self-healing elastomers, from extrinsic to combined intrinsic mechanisms: a review. Mater Horiz 7:2882–2902
62. Sharma A, Mandal T, Goswami S (2021) Fabrication of cellulose acetate nanocomposite films with lignocelluosic nanofiber filler for superior effect on thermal, mechanical and optical properties. Nano-Struct Nano-Objects 25:100642
63. Abdelrazek EM, Abdelghany AM, Badr SI, Morsi MA (2018) Structural, optical, morphological and thermal properties of PEO/ PVP blend containing different concentrations of biosynthesized Au nanoparticles. J Mater Res Technol. 7(4):419–431
64. Hou Y, Gao S (2003) Monodisperse nickel nanoparticles prepared from a monosurfactant system and their magnetic properties. J Mater Chem 13(7):1510–1512
65. Tang NJ, Chen W, Zhong W, Jiang HY, Huang SL, Du YW (2006) Highly stable carbon-coated Fe/SiO_2 composites: synthesis, structure and magnetic properties. Carbon 44(3):423–427
66. Zhang QJ, Pan BC, Chen XQ, Zhang WM, Pan BJ, Zhang QX, Lv L, Zhao XS (2008) Preparation of polymer-supported hydrated ferric oxide based on Donnan membrane effect and its application for arsenic removal. Sci China B 51(4):379–385
67. Sinsawat A, Anderson KL, Vaia RA, Farmer BL (2003) Influence of polymer matrix composition and architecture on polymer nanocomposite formation: Coarse-grained molecular dynamics simulation. J Polym Sci, Part B: Polym Phys 41(24):3272–3284
68. Lofrano G, Carotenuto M, Libralato G, Domingos RF, Markus A, Dini L, Gautam RK, Baldantoni D, Rossi M, Sharma SK, Chattopadhyaya MC, Giugni M, Meric S (2016) Polymer functionalized nanocomposites for metals removal from water and wastewater: an overview. Water Res 92:22–37
69. Trujillo-Reyes J, Peralta-Videa J, Gardea-Torresdey J (2014) Supported and unsupported nanomaterials for water and soil remediation: are they a useful solution for worldwide pollution? J Hazard Mater 280:487–503
70. Bhatkhande DS, Pangarkar VG, Beenackers AA (2002) Photocatalytic degradation for environmental applications—a review. J Chem Technol Biotechnol 77(1):102–116
71. Gupta VK, Sharma G, Pathania D, Kothiyal NC (2015) Nanocomposite pectin Zr(IV) selenotungstophosphate for adsorptional/photocatalytic remediation of methylene blue and malachite green dyes from aqueous system. J Ind Eng Chem 21:957–964
72. Zhao X, Lv L, Pan B, Zhang W, Zhang S, Zhang Q (2011) Polymer-supported nanocomposites for environmental application: a review. Chem Eng J 170(2–3):381–394
73. Jain P, Pradeep T (2005) Potential of silver nanoparticle-coated polyurethane foam as an antibacterial water filter. Biotechnol Bioeng 90:59–63
74. Reddy AV, Mohan D, Bhattacharya A, Shah V, Ghosh P (2003) Surface modificationof ultrafiltration membranes by preadsorption of a negatively charged polymer. J Membr Sci 214:211–221
75. Berber MR (2020) Current advances of polymer composites for water treatment and desalination. J Chem 2020:1–19
76. Bet-moushoul E, Mansourpanah Y, Farhadi K, Tabatabaei M (2016) TiO_2 nanocomposite based polymeric membranes: a review on performance improvement for various applications in chemical engineering processes. Chem Eng J 283:29–46
77. Zhu Y, Peng L, Fang Z, Yan C, Zhang X, Yu G (2018) Structural engineering of 2Dnanomaterials for energy storage and catalysis. Adv Mater 30:1706347

78. Liu W, Ullah B, Kuo CC, Cai X (2019) Two-dimensional nanomaterials-based polymer composites: fabrication and energy storage applications. Adv Polym Technol 2019:1–15
79. Asen P, Shahrokhian S, Zad AI (2018) Ternary nanostructures of Cr_2O_3/graphene oxide/conducting polymers for supercapacitor application. J Electroanal Chem 823:505–516
80. Yang C, Wei H, Guan L, Guo J, Wang Y, Yan X, Zhang X, Wei S, Guo Z (2015) Polymer nanocomposites for energy storage, energy saving, and anticorrosion. J Mater Chem A 3:14929–14941
81. Pacheco-Torgal F, Jalali S (2011) Cementitious building materials reinforced with vegetable fibers: a review. Constr Build Mater;25:575–581
82. Sakr MR, Bassuoni MT (2021) Silane and methyl-methacrylate based nanocomposites as coatings for concrete exposed to salt solutions and cyclic environments. Cement Concr Compos 115:103841
83. Lee KY, Ho KKC, Schlufter K, Bismarck A (2012) Hierarchical composites reinforced with robust short sisal fiber preforms utilizing bacterial cellulose as binder. Compos Sci Technol 72:1479–1486
84. Wang K, Hao Y, Wang Y, Chen Y, Mao L, Deng Y, Chen J, Yuan S, Zhang T, Ren J, Liao W (2019) Functional hydrogels and their application in drug delivery, biosensors, and tissue engineering. Int J Polym Sci 2019:3160732
85. Wang Z, Zhao J, Tang W, Hu L, Chen X, Su Y, Zou C, Wang J, Lu W, Zhen W, Zhang R, Yang D, Peng S (2019) Multifunctional nanoengineered hydrogels consisting of black phosphorus nanosheets upregulate bone formation. Small J 15(41):1901560
86. Cheng KC, Huang CF, Wei Y, Hsu S (2019) Novel chitosan–cellulose nanofiber self-healing hydrogels to correlate self-healing properties of hydrogels with neural regeneration effects. NPG Asia Mater 11:25
87. Wang N, Ma M, Luo Y, Liu T, Zhou P, Qi S, Xu Y, Chen H (2018) Mesoporous silica nanoparticles-reinforced hydrogel scaffold together with pinacidil loading to improve stem cell adhesion. Chem Nano Mat 4:631–641
88. Gaharwar A, Rivera CP, Wu CJ, Schmidt G (2011) Transparent, elastomeric and tough hydrogels from poly(ethylene glycol) and silicate nanoparticles. Acta Biomater 7:4139–4148
89. Patel R, Purwada A, Cherchietti L, Inghirami G, Melnick A, Gaharwar AK, Singh A (2014) Microscale bioadhesive hydrogel arrays for cell engineering applications. Cell Mol Bioeng 7:394–408
90. Hu X, Gao Z, Tan H, Wang H, Mao X, Pang J (2019) An Injectable hyaluronic acid-based composite hydrogel by DA click chemistry with pH sensitive nanoparticle for biomedical application. Front Chem 7:477
91. Jalani G, Naccache R, Rosenzweig DH, Haglund L, Vetrone F, Cerruti M (2016) Photocleavable hydrogel-coated upconverting nanoparticles: a multifunctional theranostic platform for NIR imaging and on-demand macromolecular delivery. J Am Chem Soc 138:1078–1083
92. Wu C, Liu J, Zhai Z, Yang L, Tang X, Zhao L, Xu K, Zhong W (2020) Double-crosslinked nanocomposite hydrogels for temporal control of drug dosing in combination therapy. Acta Biomater 106:278–288
93. Wei S, Liu X, Zhou J, Zhang J, Dong A, Huang P, Wang W, Deng L (2020) Dual-crosslinked nanocomposite hydrogels based on quaternized chitosan and clindamycin-loaded hyperbranched nanoparticles for potential antibacterial applications. Int J Biol Macromol 155:153–162
94. Dai J, Long W, Liang Z, Wen L, Yang F, Chen G (2018) A novel vehicle for local protein delivery to the inner ear: Injectable and biodegradable thermosensitive hydrogel loaded with PLGA nanoparticles. Drug Dev Ind Pharm 44:89–98
95. Rajabi N, Kharasiha M, Emadi R, Zarrabi A, Mokhtari H, Salehi S (2020) An adhesive and injectable nanocomposite hydrogel of thiolated gelatin/gelatin methacrylate/Laponite® as a potential surgical sealant. J Colloid Interf Sci 564:155–169
96. Arno M, Inam M, Weems A, Li Z, Binch A, Platt C, Richardson S, Hoyland J, Dove A, O'Reilly R (2020) Exploiting the role of nanoparticle shape in enhancing hydrogel adhesive and mechanical properties. Nat Commun 11:1–9

97. Basha SL, Ghosh S, Vinothkumar K, Ramesh B, Kumari PHP, Mohan KV, Sukumar E (2020) Fumaric acid incorporated Ag/agar-agar hybrid hydrogel: A multifunctional avenue to tackle wound healing. Mater Sci Eng C 111:110743
98. Khan MdI, Paul P, Behera SK, Jena B, Tripathy SK, Lundborg CS, Mishra A (2021) To decipher the antibacterial mechanism and promotion of wound healing activity by hydrogels embedded with biogenic Ag@ZnO core-shell nanocomposites. Chem Eng J 417:128025
99. Nadernezhad A, Caliskan OS, Topuz F, Afghah F, Erman B, Koc B (2019) Nanocomposite bioinks based on agarose and 2D nanosilicates with tunable flow properties and bioactivity for 3D bioprinting. ACS Appl Bio Mater 2:796–806
100. Xie J, Wang Z, Zhao Q, Yang Y, Xu J, Waterhouse GIN, Zhang K, Li S, Jin P, Jin G (2018) Scale-up fabrication of biodegradable poly(butylene adipate-co-terephthalate)/organophilic–clay nanocomposite films for potential packaging applications. ACS Omega 3:1187–1196

Nanocomposites for Water Treatment

Blanca L. Martínez-Vargas, De la Cruz-Marquez Galilea I., Peralta-Hernández Juan M., Durón-Torres Sergio M., and Picos-Benítez Alain R.

Abstract Water is a precious natural resource and its quality and availability is essential for the survival of living creatures on the earth. However, rapid industrialization is continuously degrading the quality of water due to the incontrolable discharge of large amounts of pollutants into the water bodies. Water pollutants have appeared as threats for the entire biosphere, so their removal has become essential. The different types of water pollutants along with their sources and impact are ellucidated. The fundamental requirements for water purification are appropriate materials with high separation capacity, low cost, porosity, and reusability. Taking this into consideration, nanotechnology provides an opportunity to develop advanced materials for effective water purification by optimizing their properties like hydrophillicity, hydrophobicity, porosity, mechanical strength, and dispersibility. Nanoparticles having high surface area, can contribute a lot in water purification but its agglomeration restricts its use. However, agglomeration can be minimized by converting nanomaterials to nanocomposites. In this chapter, different types of nanocomposites, their preparation and properties are discussed.

Keywords Water treatment · Nanohybrids · Nanocomposites

B. L. Martínez-Vargas · A. R. Picos-Benítez (✉)
Centro de Estudios Científicos y Tecnológicos No. 18, Instituto Politécnico Nacional, 98160 Zac., Zacatecas, México
e-mail: apicosb@ipn.mx

B. L. Martínez-Vargas
e-mail: bmartinezv@ipn.mx

J. M. Peralta-Hernández
Departamento de Química, División de Ciencias Naturales y Exactas, Universidad de Guanajuato, Pueblito de Rocha, Cerro de la Venada s/n, 36040 Guanajuato, México

G. I. De la Cruz-Marquez · S. M. Durón-Torres
Unidad Académica de Ciencias Químicas, Universidad Autónoma de Zacatecas, Campus Siglo XXI Carretera Zacatecas-Guadalajara Km. 6.0, 98160 Ejido la Escondida, Zacatecas, México

S. K. Swain (ed.), *Nanohybrid Materials for Water Purification*, Composites Science and Technology, https://doi.org/10.1007/978-981-19-2332-6_3

1 Introduction

1.1 Nanocomposites

In recent years, nanocomposites have attracted the attention of scientists and technologists in water purification due to improved processability, surface area, stability, tunable properties, and cost effectiveness. Nanocomposites have showed fast decontamination hability with high selectivity to remove various pollutants. This chapter provides uptodate information about the importance of nanocompostes in the removal of differents compounds such as metal ions, dyes, pharmaceutical compounds, pesticides, and microorganism (Table 1) from polluted water. Nanocomposites have a wide range of applications from toys to aircraft. It has to be mentioned that these synthetised nanomaterials are multi-phasic, in which at least one of the phases shows dimensions in the nano range (10–100 nm). Now-a-days nanocomposite materials have emerged as suitable alternatives to overcome limitations of different engineering materials. They are reported to be the materials of twenty-first century [1, 2].

Novel materials development at the nanoscale has been progressing from individual nanoparticles synthesis as well as assembled of two or more materials. Therefore, nanomaterials are combined to obtain the maximum multifunctionality, these sets are what we know as nanocomposites. There are multiple definitions of a nanocomposites among which the following can be summarized: *Nanocomposites are materials composed of more than two phases of less than 100* nm, *they can include porous media, colloids, gels and copolymers; but the combination of a bulk solid that acts as a matrix and a nano-dimensional (or more) phase (a nanocomposite) dispersed in the matrix is more common. The mechanical, electrical, thermal, optical, electrochemical and catalytic properties of the composite nanomaterial differ markedly from those of the component materials separately.*

Table 1 Different water pollutants, types, sources and effects

Type of pollutants	Sources	Effects
Inorganic compounds	Solid waste of mining and industry (Cr (VI), Ar, Pb, nitrates)	Harness, acidity
Organic compounds	Industrial waste (dyes, pesticedes, pharmaceticals, PHAs)	Mutagenicity and carcinogenecy
Microorganism	Sewage, animals, excrement (E-coli bacilus subtilis, pseudomonas, enterococcus faecalis)	Water born disease

1.2 Types of Nanocomposites Used for Water Treatment

Nanocomposite materials can be classified, according to their dispersed medium and dispersed phase materials: ceramic, metal, and polymeric matrix nanocomposites. Using the classification shown in Fig. 1, various combinations of nanocomposites can be obtained that are used mainly for water treatment.

2 Synthesis and Characterization of Nanocomposites

Nowadays, several studies have reported a number of methods that has been used for the synthesis of nanocomposites. A brief summary of different synthesis methods of nanocomposites are showed in Table 2. However, the designing of correct preparation technique is very critical to obtain a nanocomposite with the desirable properties. The most commonly used techniques for the characterizations of PNCs are thermal analysis (TGA, DTA, DSC, TMA, and DMA), microscopic (SEM, TEM, and AFM), spectroscopic (FT–IR, Raman) and X-ray diffraction techniques. Scanning electron microscopic methods provide images of surface, which are associated with sample properties like homogeneity, roughness, porosity, etc. The images are also used to get information about compatibility and lattice mismatch. The other microscopic techniques that are used for characterization of the nanocomposites are scanning probe microscopy (SPM) and scanning tunneling microscopy (STM). The AFM uses a sharp tip to scan across the sample and appropriate to evaluate roughness of morphology [3, 4]. The transmission electron microscopy (TEM) allows qualitative

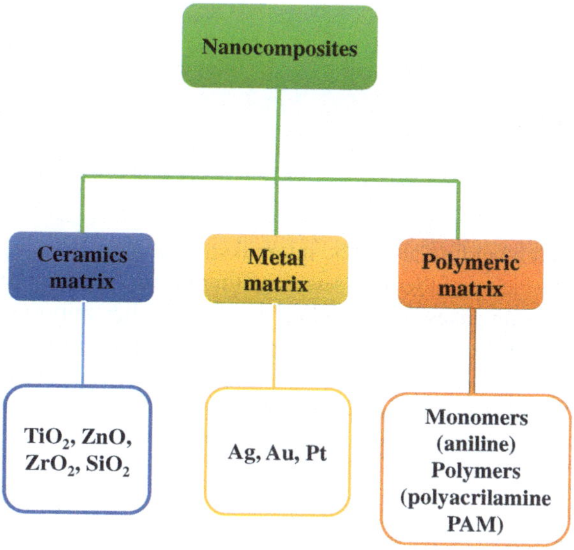

Fig. 1 Classification of nanocomposites

Table 2 Different methods used for synthesis of nanocomposites

Methods	Type nanocomposite	Characteristics	References
Sol.gel	Al_2O_3–SiO_2, Al_2O_3–TiO_2, Al_2O_3–MgO, Zeolite-inorganic compound	Sol–gel nanocomposites are further subdivided into six categories: (1) compositionally different (2) structurally different (3) both compositionally and structurallydifferent (4) nanocomposites of gels with precipitated phases; (5) nanocomposites of xerogels with metal phases; (6) nanocomposites of inorganic gels and organic molecules	[6, 7]
In-situ polymerization process	Graphene Oxide-Polyamine (GO-PA), thin film	Obtain nanosheets with high water permeability; 0.198 L/m^2h psi	[8]
Solvothermal process	BiOBr-TiO_2	Excellent photocalytic activity in the photdegradaton of rhodamine B Size 11.6 nm (XRD) Band gap-2.66 eV Surface area-184 m^2/g	[9]
Hydrothermal	Reducen graphene oxide-TiO_2	Excellent structure Great electric and optical properties High surface area	[10]
Electrochemicall synthesis	ZnO–WO_3	Size 30–40 nm Excellent photcatalyst Decolorization of methylene blue (MB)	[11]

understanding about the internal structure, spatial distribution of the various phases, and views of the defective structure through direct visualization of nanocompuesos, in some cases of individual atoms. Another important technique is wide angle X-ray diffraction pattern, which provides information about the crystallinity and latice structure of nanocomposite. Small-angle X-ray scattering (SAXS) is typically used to observe structures of the order of 10 Å or larger [5].

3 Nanocomposites for Water Purification

3.1 Heterogeneous Photocatalytic Method of Purification

Nowadays, purification of water is a hard task, since this matrix can contain a large number of hazardous pollutants, so it is important to find rapid and effective methods to purificate water sourcesthat contain several pollutants. The effects in human populations and their dangerouness of some of these pollutants is not well known. This kind of contaminants are known as Emerging Micropollutants (EMPs). These pollutants of concern are derivated by the continuous improvement of human life quality and lifestyle and all that implies. In this catergory we can find pollutants like pharmaceutical and personal care products, solvents, surfactants, pesticides, hormones and industrial chemicals. All this EMPs are found at very low concentrations in water bodies or in other natural matrix (such soil), that has been associated with numerous negative effects, as chronic toxicity, endocrine disruption effects and the continuous increased resistance to antibiotics from pathogenic bacteria [12]. Synthetic dyes are part of this group of EMP, we can find this synthetic compounds, present in wastewaters streams of the textile, food, tannery, cosmetics, plastics and pharmaceutical industries [3, 13]. The depollution of these compounds is effectively achieved by the use of highly oxidative and non selective radicals, which are mainly produced in Advance Oxidation Processes.

Advanced Oxidation Processes (AOPs) are one of the best options to remove this kind of pollutants, because they are based on the production of the highly oxidative hydroxyl radical, which is one of the strongest oxidants, with an oxidation potential of 2.80 V [14]. Among the AOPs, we canf find heterogenous photocatalysis, one of the AOPs with the most expectations.

Heterogenous photocatalysis is a photochemical process that consist in putting an organic or inorganic molecule in a liquid or a gaseous solution, in contact with a semiconductor and lately irradiate it with light, whose energy must be equivalent or higher than the band gap of the semiconductor for its excitation, thus, causing the promotion of an electron from the valence band to the conduction band, forming the so-called electron–hole pairs (Reaction 1), and the posterior creation, by oxidation-reduction reactions, ofhighly oxidizing species, like hydroxyl radicals (•OH), with the purpouse of the total pollutants remotion by oxidation, this mechanism can be seen in Fig. 2 [15, 16]

$$\text{Semiconductor} + hv \rightarrow e^-_{CB} + h^+_{VB} \qquad (1)$$

The well known electron–hole pairs or $e^-_{CB} - h^+_{VB}$, are formed when the energy of the photon, provided by an energy source (usually UV or visible light), is absorbed by the semiconductor, the irradiated energy has to be, equal or higher than the energy of the forbidden band of the semiconductor, causing the promotion of the electron from the valence band to the conduction band, generating the hole-pairs, due to the

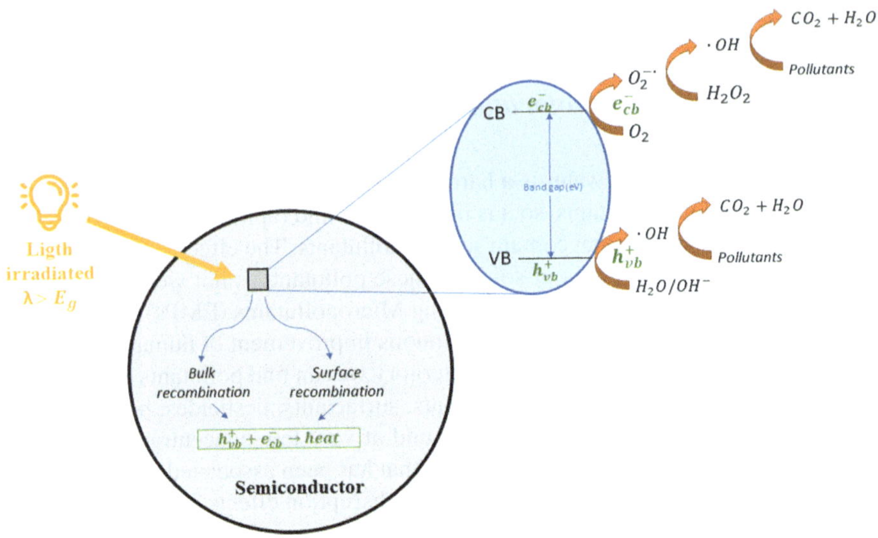

Fig. 2 Mechanism of photocatalysis process on the semiconductor surface under light irradiation

described migration of electrons, and thus creating charge carriers. The photogenerated hole allows the formation of strong oxidative and non selective radicals, while the photoexcited electron produces reducing agents. These charge carriers migrate to the surface of the photocatalyst to start redox reactions with the molecules adsorbed on the surface of the catalyst (Fig. 2). The target pollutant, can be immediately oxidized in the generated holes (Reaction 2), or by reactive oxygen species (ROS), that are generated in water [17]. The $h_{VB}{}^+$ holes in the valence band can react with hydroxyl ions or water molecules adsorbed on the surface of the catalyst forming •OH (Reactions 3 and 4). Meanwhile, the $e_{CB}{}^-$ can react with dissolved oxygen O_2 to form the superoxide radical $\bullet O_2^-$, which is represented in Reaction 5. Another way to obtain •OH can be representated by Eqs. 6–8, Finally, both oxidizing species mineralize the pollutants according to Reaction 9. [18].

$$h_{VB} + M_s \rightarrow M_s^+ \quad (2)$$

$$h_{VB}^+ + H_2O \rightarrow \bullet OH + H^+ \quad (3)$$

$$h_{VB}^+ + OH^- \rightarrow \bullet OH \quad (4)$$

$$e_{CB}^- + O_2 \rightarrow \bullet O_2^- \quad (5)$$

$$\bullet O_2^- + H_2O \rightarrow \bullet HO_2^- + \bullet OH^- \quad (6)$$

$$\bullet OH_2^- + \bullet OH^- \rightarrow H_2O_2 \quad (7)$$

$$H_2O_2 \rightarrow 2 \bullet OH \quad (8)$$

$$\bullet OH + \bullet O_2^- + \text{pollutant} \rightarrow CO_2 + H_2O \quad (9)$$

However, according to Eq. 10, the recombination process in which the electrons return to the valence band significantly reduces the quantum efficiency of the photocatalyst and occurs in a few nanoseconds [15, 16].

$$e_{CB}^- + h_{VB}^+ \rightarrow \text{heat} \quad (10)$$

So many materials has been proved as semiconductor in heterogenous photocatalysis processes.The desired characteristics in the new sinthetized photocatalysts materialsare:non toxicity, high availability, low cost, facile synthesis, chemical, thermal and mechanical stability, high values of specific superficial area, high values of electronic mobility and narrow band gap. However, non of the existing materials have all of the atributes, before mentioned alone or in pristine form. There is always the necessityof combine materials with others, in order to optimize and improve the characteristics of the involved materials and thus obtain a feature set feasible and achievable. That is why the used photocatalysts are frecuently binary or tertiary heterojunctions or hybrid materials. Moreover, combined materials can form hybrid nanocomposites, which can improve the charge separation, and thus, enhancing the photocatalysis process.

TiO_2 was first used in the photocatalytic water splitting, since then, their application as a photocatalyst has increased year by year [19]. Currently, TiO_2 is one of the most commonly used nanomaterials for heterigenous photocatalysis applicationsin the depolution of water with diferent pollutants, among these, we can find synthetic dye removal. TiO2 has been used as the main material and as a nanocomposite. TiO2 has most of thedesirable properties of photocatalyst such: chemical stability, availability and low cost [20, 21]. Also, its relatively low band gap (3.2 eV)that can be narrower by combining TiO2 with other materials[22]. Recently, it has been reported the aplication of iron doped titanium dioxide nanotubes (Fe–TNT) in photocatalytic experiments for the decolourization, of the synthetic organic dye Red Congo, from a wastewater by irradiation of visible light, maximum discoloration efficency was 86% in 180 min [23]. It is common practice the dopping of TiO_2 with transition metals to improve the photocatalytic activity by decreasing the electron–hole recombination. Vaiano et al. [24] used a photocatalysts synthesised with sulfated TiO_2 to obtain an Au–TiO_2 and Pt–TiO_2 nanocomposites by using two methods: chemical reduction and photochemical deposition. This nanocomposite was usedefficiently for the removal of the synthetic dye Patent Blue V, resulting in thediscoloration of 93%, achieved by the catalyst Au-TiO_2under 180 min irradiation. Moreover, it was concluded that Au–TiO_2 had a better photocatalytic activity than Pt–TiO_2. It is well known that the Fermi energy levels affect the interfacial charge transfer and, thus, the charge separation. According to the authors, as higher the Fermi level of the noble metal, the better the photocatalytic activity, the authors found that the Fermi level of Au is higher than Pt's [24]. Recently, it was reported the green synthesis of SnO_2/TiO_2 nanoparticles with a good cristallinity and porous nature with a narrow band gap of 2.8 eV. The photocatalytic behaivor was tested in decolourization of the synthetic dye methylene green. According whit the presented results, it is possible to achieve a 96% of decolourization in an experimental time of 75 min, by visible irradiation. Is important to mention the band gap value disminution, from the original values of separately semiconductors, 3.2 and 3.6 for TiO_2 and SnO_2, respectively. That allowed the photocatalyst activation under visible light range [21].

Several works have focused their study on a well known used photocatalysist, ZnO. Due tothe large number of literature available about the TiO_2, ZnO has become a reference of comparation in the heterogenous photocatalysis research, the first semiconductor to behave almost ideally in aqueous solution was monocrystalline n-type ZnO [17]. Even if ZnO has a wide band gap, 3.2–3.33 eV, what limits its optical absorption to the ultraviolet region, its photocatalytic performance is greatly improved by adding dopants such as transition metals [3], combining it with other semiconductors [25] or causing oxygen vacancies in its crystalline structure [26]. To mention an example of above ZnO has been used in the decomposition of anthracene by heterogeneous photocatalysis under ultraviolet light using nanocomposites of ZnO and n ZnO/p–MnO, varying the content of p–MnO, reaching degradation efficiencies of the anthracene from 76 to 100% for ZnO and for the nanocomposite with the highest p–MnO content [4].

On the other hand, in recent years, ferrites have gained special attention, due to its properties and the good performance as a photocatalyst. Kalam et al. [27] used cobalt-ferrite based nanoparticles, obtained by solvothermal method,for the degradation of methylene blue under visible light, discoloration efficiencies reach 80% in 140 min. Besides oxides, other materiales like CdS [28] and g–C_3N_4 [29–31] are widely used as an efficient photocatalysts. As it has been observed, it is necessary to combine these compounds with other materials in order to form nanocomposites with enhanced photocatalytic characteristics. Ferrites, were used for the photocatalytic degradation of two different synthetic dyes such methylene blue (MB) and rhodamine B (Rh B) by using of $NiFe_2O_4$–g–C_3N_4 nanoparticles and visible and sunlight to lead the material excitation. After 225 min discoloration efficiencies were of 98 and 99% respectively [32, 33].

In recent years, it has become more important the development of efficient photocatalysts under visible, or even sun light, in order to optimize and reduce the energy used in the process, with the main objective of making the process a self- sufficient process. Recently, researchers have focused their attention in the use of graphene in its different forms due its extraordinary chemical, optical and electric propierties, like its electronic mobility (200.000 $cm^2 . V^{-1} . s^{-1}$)and specific surface area (2630 $m^2 . g^{-1}$) [34]. This features make the graphene a potential material for their use in heterogenous phocatalysis. It is common to use graphene derivatives in heterojunction with other materials in order to improve their photocatalytic performance.

4 The Use of Nanocomposites for the Removal of Harmful Components from Waters

Advance oxidation process (AOP) is widely used in water and wastewater treatment for the removal of an extensive variety of pollutants that can cause human illness. Among the pollutants that AOP can remove from water we can find heavy metals, synthetic organic dyes, organic persistent compounds, drugs, pesticides, etc. AOP produces hydroxyl radicals this radicals attack the organic compounds by a redox reaction and hydroxylation [35]. Among the AOP technologies that are reported in the literature for the treatment and depollution of water and wastewater, we can find: ozonation [36–42], processes with UV light application [43–45], Fenton and Fenton like processes and others AOP like heterogeneous photocatalysis [46–50].

In the heterogeneous photocatalysis a photocatalyst is excited to promote an electron from the valence band to the conduction band. In this case, is necessary to excite the photocatalyst with a quantity of energy almost equal to its band gap. This promotion allows the generation of an electron–hole pair generating hydroxyl radicals (•OH) and super-oxide anions ($\bullet O_2^-$) which can react with the pollutants to mineralize them to simpler components.

Several authors reported the use of nanomaterials in the application of heterogenous photocatalysis for water treatment or purification methods. A nanomaterial can be defined as those who had an external dimension which goes from 1 to 100 nm. These materials can be divided into different types such: metals (nano-Fe0 or iron oxides, Pt, Ag, Au or nano-TiO2), nonmetals (nanocarbon-based materials, nano-SiO_2), nano semiconductor materials, high molecular materials, and nanocomposites materials. Also, hybridization can introduce a new composition and also can change the chemical structure of a material [51].

4.1 *Pollutant Removal from Water Matrixes*

In the past years several pollutants have gain attention in the scientific community due the unknow effects that these can cause to human health, and also the lack of regulations for their discharge [52]. The problem of these compounds is caused by their persistent capacity, after conventional wastewater treatments and the fact that few studies focused on the potential hazard that these pollutants represent to the aquatic and/or terrestrial organisms [53, 54]. Since some of these pollutants are discharged into wastewater streams, these pollutants can reach water streams that used for human supply or consumption such rivers, lakes, dams, etc. [55, 56].

Heterogenous photocatalysis by nanocomposites materials provides a several advantages against others AOP and conventional water treatment methods such the production of highly oxidant and non-selective radicals, this allows the complete mineralization, inactivation, removal or increase of the biodegradability of a wide type of pollutants that persist after conventional water treatments methods [35, 57,

58]; nanocomposites have been used in the inactivation of harmful gram negative bacteria such Escherichia coli [59–66], its variant E. coli-K12 [67] and Pseudomonas aeruginosa [68], also, the use of nanocomposites for the inactivation of gram positive bacteria is reported by several authors [69, 70] and for the inactivation of infectious fungus that can destroy wheat production [62]. According to several authors, the main cause of inactivation of the harmful bacteria is the production of various reactive species such hydroxyl radicals (•OH), hydroxide peroxide (H_2O_2) and the generation of some superoxide radicals ($•O_2^-$) [61, 63–65], however, some other mechanisms are reported like the adsorption of the photocatalyst toward the bacterial cells which can cause the inhibition of some metabolic processes [59, 66] or the direct contact and release of silver-mediated [60].

4.2 Heavy Metal Removal from Water

Nanomaterials have been synthetized for their use in the removal of other harmful pollutants that are of concern in water supply systems like heavy metals such lead (Pb), cadmium (Cd), mercury (Hg), chromium (Cr VI) among others. Heavy metals are naturally present in the earth crust, however, some anthropogenic activities are the main cause of the discharge of this components to the environment [71], the present of these components in natural ecosystem can cause several problems to natural wildlife or human populations. Heavy metals can bioaccumulate through the trophic chain [72], can cause several chronic diseases such liver or kidney failure, affections to the reproductive system, psychologic or neurological disorders [72–75] due this, the removal of heavy metals is an issue of concern for the scientific community.

Table 3 presents a compilation of recent studies where nanocomposites are used for the removal of heavy metals from different matrix of waters. The main reported removal mechanisms of heavy metals from water are adsorption [73, 75–80] and magnetic separation [74].

4.3 Synthetic Dye Removal from Water

Emergent pollutants (EP) can be described as those pollutants that are present in natural ecosystems at very low concentrations which make them imperceptible, because of these low concentrations, no regulations exist so they are discharged indiscriminately and are not considered as an environmental hazard for natural ecosystems or human populations [52]. Many of these compounds are present in wastewater streams and the conventional wastewater treatment methods are incapable for their remotion. This can cause that these pollutants can reach natural water sources. Some synthetic organic compounds are reported as like pharmaceutical drugs like antibiotics, analgesics, antidepressants, hormones, among others [30, 35, 46, 49, 50,

Table 3 Compilation of recent studies for different heavy metal removal

Materials	Heavy metal	Removal mechanism	Removal efficiency (%)	References
Fe_3O_4@TMU-32	Hg (II) Pb (II)	Charge transfer and electrostatic interactions with functional groups such C = O, N–H and Fe–O	99.2% for Pb (II) and 85% for Hg (II)	[72]
Graphene oxide	Pb (II) Cd (II)	Transfer of electrons amine to metal center	68.93% for Pb (II) and 59.4% for Cd (II)	[80]
Hydroxyapatite and bentonite	Pb (II) and Ni (II)	Adsorption (formation of complexes for Pb2 +, and ion exchange for Ni2 +)	Adsorption of 2.05. 5.83 and 5.99 mg of Pb2 + per gr of material; 1.01, 1.26 and 3.72 mg of Ni2 + per gr of material	[76]
Graphene-oxide	Cr (III)	Charge transfer	Removal of the 52% of the initial concentration	[77]
SiO_2 and tea waste	Cd (II) Pb (II)	Electrostatic interaction between the divalent metal and SiO- and COOH + groups of the adsorbent	67.3% for Pb (II) and 77.9% for Cd (II)	[73]
Silica-Iron oxide (MCM-Fe_2O_3	Ni (II) Cd (II) Cr (III) Zn (II) Pb (II)	Electrostatic interactions with functional groups such SiO- and Fe–O	Ni (II) = 53% Cd (II) = 79% Cr (III) = 61% Zn (II) = 89% Pb (II) = 99.5%	[75]
Fe_3O_4@SiO_2–NH_2	Cd (II)	Interaction with the polar primary amine group	Removal of the 93% of the initial Cd (II) concentration	[78]

(continued)

Table 3 (continued)

Materials	Heavy metal	Removal mechanism	Removal efficiency (%)	References
CuO–ZnO	Cr (VI) Pb (II)	For Cr (VI) and Pb (II) Interaction with functional groups such (O–H, C = O and C-H and O–H, C-H, C = O, C-O and M–O respectively)	99% for Cr(VI) and 97% for Pb (II)	[79]
Magnetic graphene oxide-Functionalized Tryptophan	Cr (VI) Cd (II) Pb (II)	Interaction with functional groups such: amino, amide, hydroxyl, carboxyl, and epoxy	Close to 100% for all heavy metals	[82]
Iron oxide waste	Cd (II) Pb (II) Cu (II)	Combine of the functional groups COO-in alginate that act as ligands allowing the combination with metal ions	90% for Cd (II), up to 98% for Pb (II) and 92% for Cu (II)	[83]
ZnO and multi-walled carbon nanotubes	Cr (VI)	Electrostatic adsorption by the electrostatic interaction with the amine and carboxylated groups, also, by chemical redox reaction with functional groups containing oxygen	From 77 to 94%	[84]
Graphene oxide-Manganese ferrite (GO–MnFe2O4)	Pb (II)	Physisorption by van der Waals forces and chemisorption by reaction with -OH groups of MnFe2O4 surface and reaction with -COOH and -OH functional groups of graphene oxide	98.68%	[85]

(continued)

Table 3 (continued)

Materials	Heavy metal	Removal mechanism	Removal efficiency (%)	References
Fe_3O_4/SiO_2/Polyamidoamine dendrimer	Cu (II) Pb (II) Cd (II)	Physical adsorption, ion exchange and complexation reaction (by reaction with amine and acetate functional groups)	Cu (II) = 99.3% Pb (II) = 98% Cd (II) = 96.5%	[86]

87–89], also, some synthetic dyes are reported [48, 71, 90–98] is well know that some of these pollutants can be found in some water supply sources.

Synthetic dyes are one of the most used chemicals worldwide since their use is indispensable for production of textile products such clothes, also, some dyes are used in pharmaceutical and food industry. The use of synthetic dye is due the low cost, the wide variety of colors and their low production costs [99, 100].

Conventional wastewater treatment systems are inefficient in the degradation of these pollutants, so they represent a hazard to aquatic and/or terrestrial organisms [53]. Some of these synthetic dyes are considered as carcinogenic but some of them are catalogued as non-toxic, however, if these dyes reaches natural water bodies, can be transformed or metabolized to harmful compounds (mutagenic, carcinogenic or genotoxic) by some natural processes [53, 99, 100].

5 Removal of Contaminants by Different Water Treatment Methods

Several treatment methods have been tested to achieve the complete mineralization of synthetic dyes in water or wastewater. Biological systems are the one of the most commonly water treatment technologies used for wastewater treatment. Biological systems involve the use of microorganism consortia where bacterial are the predominant organism. The principle of this process is the oxidation of suspended or organic material by microbial metabolism to simple inorganic components such CO_2 and H_2O (and also methane in case of anaerobic processes) and the removal of colloidal solids [101]. Some other technologies are climate depended like constructed wetlands (CW). In summer, these treatment systems can reach higher removal efficiencies, however their performance can decrease significantly in winter season. When recalcitrant pollutants are present in a wastewater treated by biological process, several removal mechanisms are involved like absorption, adsorption, or co-metabolism. These processes have several limitations to achieve high removal efficiencies like temperature, hydraulic retention time, microbial concentration, pH, etc. Their efficacy against complex organic pollutants is not well defined since some authors reported good removal efficiencies and some others reported low performance against these compounds [102–107].

As mentioned before, synthetic dyes are easily mineralized from water by advanced oxidation process, because of the generation of non-selective and highly oxidative radicals. Among the advance oxidation technologies that has been used, for the treatment of water with synthetic dyes, we can find ozonation [36–42], processes with UV light application [108–117], anodic oxidation, Fenton and Fenton-like processes [49, 110, 118–137] and heterogeneous photocatalysis [71, 74, 79, 91, 93, 127, 138–141].

6 The Use of TiO_2 for Water Purification

Heterogeneous photocatalysis consist in the excitation of the photocatalyst to promote an electron from the valence band to the conduction band. For the promotion of this electron is necessary to excite the photocatalyst with a quantity of energy almost equal to its band gap. This electron promotion generates an electron–hole pair which allows the generation of highly reactive oxidant like hydroxyl radicals (•OH) and super-oxide anions ($•O_2^-$). These oxidants react with the organic compounds and decompose them to simpler and stable components. Several studies reported the use of different materials for heterogenous photocatalysis, among all these compounds we can find dioxide titanium (TiO_2) as the most used material [142–145]. Titanium dioxide (TiO_2) is one of the most used photocatalyst because holes can be formed in the valence band by promoting an electron to the conduction band under UV light irradiation. This electron promotion, allow the generation of holes that can generate the strong oxidative and non-selective hydroxyl radicals (•OH). Also, is well known, that titanium dioxide presents a high chemical stability in aqueous solution, a high resistance to photo-corrosion and can the advantage that it can be easlily supported on other materials. There are several studies that present the inmobilization of TiO_2 as a film on an anode like BBD, titanium oxide (TiO_2) is one of the most used photocatalyst because holes can be formed in the valence band by promoting an electron to the conduction band under UV light irradiation. As mentioned before, these holes can generate the strong oxidative hydroxyl radicals (•OH). Also, TiO_2 present a high chemical stability in aqueous solution, high resistance to photo-corrosion and can be easily immobilized on a substrate. In order to improve the photocatalytic efficiency of the TiO_2 can be immobilizes as a film on an anode like BDD, titanium or glass, this process is known as photo-electrocatalysis (PEC). PEC involves the combination photocatalysis and electro oxidation processes. The combination of these AOP avoid the recombination of the electron–hole pairs of the TiO_2 by coating an anode with this photocatalyst and applying a bias potential [56].

7 Alternative Nanocomposites for the Removal of Recalcitrant Pollutants from Aqueous Solutions

It has mentioned above, that TiO_2 has been widely used for the removal of recalcitrant pollutants from water, however, several studies have focused their attention in proving novel materials for the heterogeneous photocatalysis process, also, several pollutants have been removed from aqueous solutions, proving the effecitivity of these new nano-hybrids (Tables 4 and 5).

Table 4 Nanocomposite studies for removal of different pollutants from aqueous solution

Material	Pollutant	Removal mechanism	Removal efficiency (%)	References
$g\text{-}C_3N_4/NiO/ZnO/Fe_3O_4$	Esomeprazol	Migration of the photo excited electrons from the conduction band of $g\text{–}C_3N_4$ to the valence band of ZnO, which generates •OH and • O_2^- radicals, activation of fenton related phenomenon, capture of electrons from Fe + 3 ions	Degradation: 95%	[30]
$\alpha\text{-}MnO_2$ nanorods	Pharmaceutical drug: Bactrim DS	Aromatic ring rupture by non-selective radicals such •OH and $•O_2^-$	Degradation: 91%	[88]
Porphyrin/Cu(II)	4-Nitrophenol	Direct reduction of 4-nitrophenol by the photogenerated elctrons	Degradation: ≈99%	[57]
Graphene Oxide/$MnFe_2O_4$	Pb II	Physisorption and chemisorption (electrostatic atraction, ion exchange and surface complexation)	Adsorption of near a 92%	[58]
$\gamma\text{-}Fe_2O_3/SrCO_3$	Ibuprofen	Retarding the fast recombination of charge-pair allowing the production of •OH and $•O_2^-$ radicals	Degradation: ≈99%	[89]
Porphyrin/Cu(II)	4-Nitrophenol	Direct reduction of 4-nitrophenol by the photogenerated elctrons	Degradation: ≈99%	[57]

Table 5 Nanocomosites used in the depollution of synthetic dyes from water

Material	Pollutant	Removal mechanism	Removal efficiency (%)	References
$Fe2O_3$–ZrO_2	Methylene blue	Migration of the photo excited electrons from the conduction band of g–C_3N_4 to the valence band of ZnO, which generates •OH and • O_2^-radicals, activation of fenton related phenomenon, capture of electrons from Fe + 3 ions	Up to 98%	[91]
Graphene oxide and ZnO based nanocomposite	Methylene blue	Aromatic ring rupture by non-selective radicals such •OH and•O_2^-	Degradation: 89% Mineralization 71% COD and BOD removal: 96.33 and 96.12% respectively	[71]
Carbon nitride/horseradish peroxidase	Methyl orange	Direct reduction of 4-nitrophenol by the photogenerated elctrons	Degradation: ≈99%	[146]
ZnO/cGO	Methylene blue (MB) and rhodamine B (RhB)	Adsorbtion of the material in the surface of a membrane with increased porosity, the high porosity was attributed to the oxygen rich surface groups in ZnO/cGO	Removal efficiency for both synthetic dyes: ≈99%	[94]
Mercury sulphide (HgS) nanoparticles decorated single walled carbon nanotubes (SWCNT/HgS)	Methylene blue	Photogeneration of reactive oxidants such •O_2, •O_2^- and •OH	Degradation of: 99%	[96]

(continued)

Table 5 (continued)

Material	Pollutant	Removal mechanism	Removal efficiency (%)	References
$ZnFe_2O_4$/reduced graphene oxide (G–$ZnFe_2O_4$/rGO NHs)	Methylene blue	Promotion of electron from valence band to conduction band, this provokes the inhibition of the recombination of electron–hole, allowing the generation of $\bullet O_2^-$ and •OH	Degradation of: 94%	[97]
Au–TiO_2	Methylene blue and methyl orange	An excesive presence of electrons in the TiO2 conduction band, allows an enhanced production of $\bullet O_2^-$ and •OH	Degradation of: 94% and 85% respectively	[98]
TiO_2–SiO_2	Malachite green oxalate, methylene blue and methyl orange	Increased photocatalytic activity provoked by a higher surface on the photocatalyst, which allows the trapping of more synthetic dyes molecules	COD removal efficiency: 90% for all synthethic dyes	[140]
ZnO/Ag	Methylene blue	Moving of electrons from ZnO to carbon nanotubes, preventing the recombination process. Electrons react with O2 allowing the generation of $\bullet O_2^-$ and •OH	Degradation: 80%	[90]
Carbon nanotube/titania (CNT/TiO_2)	Methyl orange	A Synergyc effect of photosensitization by CNTs andthe presence surface Ti^{3+} states and oxygen vacancies	Degradation: ≈90%	[95]

References

1. Wang C, Wang F, Wang T, Bian Y, Yang X, Jiang X (2010) PAHs biodegradation potential of indigenous consortia from agricultural soil and contaminated soil in two-liquid-phase bioreactor (TLPB). J Hazard Mater 176:41–47. https://doi.org/10.1016/j.jhazmat.2009.10.123
2. Pandey N, Shukla SK, Singh NB (2017) Water purification by polymer nanocomposites: an overview. Nanocomposites. 3:47–66. https://doi.org/10.1080/20550324.2017.1329983
3. Martínez-Vargas BL, Durón-Torres SM, Bahena D, Rodríguez-López JL, Peralta-Hernández JM, Picos A (2019) One-pot synthesis of ZnO–Ag and ZnO–Co nanohybrid materials for photocatalytic applications. J Phys Chem Solids 135:109120. https://doi.org/10.1016/j.jpcs.2019.109120
4. Martínez-Vargas BL, Cruz-Ramírez M, Díaz-Real JA, Rodríguez-López JL, Bacame-Valenzuela FJ, Ortega-Borges R, Reyes-Vidal Y, Ortiz-Frade L (2019) Synthesis and characterization of n-ZnO/p-MnO nanocomposites for the photocatalytic degradation of anthracene. J Photochem Photobiol A Chem 369:85–96. https://doi.org/10.1016/j.jphotochem.2018.10.010
5. Jiang N, Du Y, Ji P, Liu S, He B, Qu J, Wang J, Sun X, Liu Y, Li H (2020) Enhanced photocatalytic activity of novel TiO2/Ag/MoS2/Ag nanocomposites for water-treatment. Ceram Int 46:4889–4896. https://doi.org/10.1016/j.ceramint.2019.10.225
6. Komarneni S (1992) Nanocomposites 2:1219–1230
7. Liu T, Chen W, Hua Y, Liu X (2017) Au/ZnO nanoarchitectures with Au as both supporter and antenna of visible-light. Appl Surf Sci 392:616–623. https://doi.org/10.1016/j.apsusc.2016.09.033
8. Yin J, Zhu G, Deng B (2016) Graphene oxide (GO) enhanced polyamide (PA) thin-film nanocomposite (TFN) membrane for water purification. Desalination 379:93–101. https://doi.org/10.1016/j.desal.2015.11.001
9. Wei XX, Cui H, Guo S, Zhao L, Li W (2013) Hybrid BiOBr-TiO2 nanocomposites with high visible light photocatalytic activity for water treatment. J Hazard Mater 263:650–658. https://doi.org/10.1016/j.jhazmat.2013.10.027
10. Wang F, Zhang K (2011) Reduced graphene oxide-TiO2 nanocomposite with high photocatalystic activity for the degradation of rhodamine B. J Mol Catal A Chem 345:101–107. https://doi.org/10.1016/j.molcata.2011.05.026
11. Goveas JJ, Shetty S, Mascarenhas NP, D'Souza RM, Gonsalves RA (2020) Electrochemical synthesis of ZnO–WO3 nanocomposites and their photocatalytic activity. J Appl Electrochem 50:501–511. https://doi.org/10.1007/s10800-020-01407-2
12. Wanda EMM, Nyoni H, Mamba BB, Msagati TAM (2018) Application of silica and germanium dioxide nanoparticles/polyethersulfone blend membranes for removal of emerging micropollutants from water. Phys Chem Earth 108:28–47. https://doi.org/10.1016/j.pce.2018.08.004
13. Mashkoor F, Nasar A (2020) Magsorbents: potential candidates in wastewater treatment technology–a review on the removal of methylene blue dye. J Magn Magn Mater 500:166408. https://doi.org/10.1016/j.jmmm.2020.166408
14. Teodosiu C, Gilca AF, Barjoveanu G, Fiore S (2018) Emerging pollutants removal through advanced drinking water treatment: a review on processes and environmental performances assessment. J Clean Prod 197:1210–1221. https://doi.org/10.1016/j.jclepro.2018.06.247
15. Ye S, Chen Y, Yao X, Zhang J (2020) Simultaneous removal of organic pollutants and heavy metals in wastewater by photoelectrocatalysis: a review. Chemosphere 273:128503. https://doi.org/10.1016/j.chemosphere.2020.128503
16. Martínez-Huitle CA, Scialdone O, Rodrigo MA (2018) Electrochemical water and wastewater treatment. Elsevier Inc. https://doi.org/10.1016/C2016-0-04297-3
17. Schneider J, Bahnemann D (2016) Photocatalysis fundamentals and perspectives. https://doi.org/10.1039/9781782622130-FP007

18. Uddin T, Hoque E, Bhoumick MC (2020) Facile one-pot synthesis of heterostructure SnO2/ZnO photocatalyst for enhanced photocatalytic degradation of organic dye. RSC Adv 10:23554–23565. https://doi.org/10.1039/d0ra03233f
19. Fujishima A, Honda K (1972) Electrochemical photolysis of water at a semiconductor electrode. Nature 238:37–38. https://doi.org/10.1038/238037a0
20. Wang Y, He J, Wu P, Luo D, Yan R, Zhang H, Jiang W (2020) Simultaneous removal of tetracycline and Cu(II) in hybrid wastewater through formic-acid-assisted TiO2Photocatalysis. Ind Eng Chem Res 59:15098–15108. https://doi.org/10.1021/acs.iecr.0c02443
21. Magdalane CM, Kanimozhi K, Arularasu MV, Ramalingam G, Kaviyarasu K (2019) Self-cleaning mechanism of synthesized SnO2/TiO2 nanostructure for photocatalytic activity application for waste water treatment. Surf Interfaces 17:100346. https://doi.org/10.1016/j.surfin.2019.100346
22. Hu Z, Li K, Wu X, Wang N, Li X, Li Q, Li L, Lv K (2019) Dramatic promotion of visible-light photoreactivity of TiO2 hollow microspheres towards NO oxidation by introduction of oxygen vacancy. Elsevier B.V. https://doi.org/10.1016/j.apcatb.2019.117860
23. Zafar Z, Fatima R, Kim J (2021) Experimental studies on water matrix and influence of textile effluents on photocatalytic degradation of organic wastewater using Fe–TiO 2 nanotubes: towards commercial application. Environ Res 197:111120. https://doi.org/10.1016/j.envres.2021.111120
24. Vaiano V, Iervolino G, Sannino D, Murcia JJ, Hidalgo MC, Ciambelli P, Navío JA (2016) Photocatalytic removal of patent blue V dye on Au-TiO2 and Pt-TiO2 catalysts. Appl Catal B Environ 188:134–146. https://doi.org/10.1016/j.apcatb.2016.02.001
25. Dammala P, Machado J, Rani B, Murali S, Devi S, Niraj Luwang M, Sahu NK (2019) Synthesis of biphasic nanomaterials based on ZnO and SnO 2 : application towards photocatalytic degradation of acid red dye. Nano-Structures Nano-Objects 18:100292. https://doi.org/10.1016/j.nanoso.2019.100292
26. Wang J, Chen R, Xiang L, Komarneni S (2018) Synthesis, properties and applications of ZnO nanomaterials with oxygen vacancies: a review. Ceram Int 44:7357–7377. https://doi.org/10.1016/j.ceramint.2018.02.013
27. Kalam A, Al-sehemi AG, Assiri M, Du G, Ahmad I, Pannipara M (2018) Research Center for Advanced Materials Science (RCAMS), King Khalid University, Abha Zhejiang Key Laboratory for Reactive Chemistry on Solid Surfaces, Institute of Physical Nanochemistry Laboratory, Department of Chemistry, Jamia Millia Islamia, New Delhi. Results Phys. https://doi.org/10.1016/j.rinp.2018.01.045
28. Xie YS, Yuan L, Zhang N, Xu YJ (2018) Light-tuned switching of charge transfer channel for simultaneously boosted photoactivity and stability. Appl Catal B Environ 238:19–26. https://doi.org/10.1016/j.apcatb.2018.07.006
29. He Z, Chen JJ, Chen Y, Makwarimba CP, Huang X, Zhang S, Chen JJ, Song S (2018) An activated carbon fiber-supported graphite carbon nitride for effective electro-Fenton process. Electrochim Acta 276:377–388. https://doi.org/10.1016/j.electacta.2018.04.195
30. Raha S, Ahmaruzzaman M (2020) Enhanced performance of a novel superparamagnetic g-C3N4/NiO/ZnO/Fe3O4 nanohybrid photocatalyst for removal of esomeprazole: Effects of reaction parameters, co-existing substances and water matrices. Chem Eng J 395:124969. https://doi.org/10.1016/j.cej.2020.124969
31. Mishra A, Mehta A, Kainth S, Basu S (2018) Effect of g-C3N4 loading on TiO2/Bentonite nanocomposites for efficient heterogeneous photocatalytic degradation of industrial dye under visible light. J Alloys Compd 764:406–415. https://doi.org/10.1016/j.jallcom.2018.06.089
32. Palanivel B, Ayappan C, Jayaraman V, Chidambaram S, Maheswaran R, Mani A (2019) Materials Science in Semiconductor Processing Inverse spinel NiFe 2 O 4 deposited g-C 3 N 4 nanosheet for enhanced visible light photocatalytic activity. Mater Sci Semicond Process 100:87–97. https://doi.org/10.1016/j.mssp.2019.04.040
33. Barman S, Basu S (2020) Complete removal of endocrine disrupting compound and toxic dye by visible light active porous g-C 3 N 4 / H-ZSM-5 nanocomposite. Chemosphere 241:124981. https://doi.org/10.1016/j.chemosphere.2019.124981

34. Ghany NAA, Elsherif SA, Handal HT (2017) Revolution of graphene for different applications: state-of-the-art. Surf Interfaces 9:93–106. https://doi.org/10.1016/j.surfin.2017.08.004
35. Nidheesh PV (2017) Graphene-based materials supported advanced oxidation processes for water and wastewater treatment: a review. Environ Sci Pollut Res 24:27047–27069. https://doi.org/10.1007/s11356-017-0481-5
36. Gomes J, Costa R, Quinta-Ferreira RM, Martins RC (2017) Application of ozonation for pharmaceuticals and personal care products removal from water. Sci Total Environ 586:265–283. https://doi.org/10.1016/j.scitotenv.2017.01.216
37. Cruz-Alcalde A, Sans C, Esplugas S (2018) Priority pesticide dichlorvos removal from water by ozonation process: reactivity, transformation products and associated toxicity. Sep Purif Technol 192:123–129. https://doi.org/10.1016/j.seppur.2017.09.069
38. Cruz-Alcalde A, Sans C, Esplugas S (2017) Priority pesticides abatement by advanced water technologies: the case of acetamiprid removal by ozonation. Sci Total Environ 599–600:1454–1461. https://doi.org/10.1016/j.scitotenv.2017.05.065
39. Sangjan S, Kumsopa C, Wijannarong S, Aroonsrimorakot S, Thavipoke P (2013) Removal of reactive dyes from textile dyeing industrial effluent by ozonation process. APCBEE Procedia 5:279–282. https://doi.org/10.1016/j.apcbee.2013.05.048
40. Boudissa F, Mirilà D, Arus VA, Terkmani T, Semaan S, Proulx M, Nistor ID, Roy R, Azzouz A (2019) Acid-treated clay catalysts for organic dye ozonation—thorough mineralization through optimum catalyst basicity and hydrophilic character. J Hazard Mater 364:356–366. https://doi.org/10.1016/j.jhazmat.2018.09.070
41. El Hassani K, Kalnina D, Turks M, Beakou BH, Anouar A (2019) Enhanced degradation of an azo dye by catalytic ozonation over Ni-containing layered double hydroxide nanocatalyst. Sep Purif Technol 210:764–774. https://doi.org/10.1016/j.seppur.2018.08.074
42. Ghuge SP, Saroha AK (2018) Ozonation of reactive orange 4 dye aqueous solution using mesoporous Cu/SBA-15 catalytic material. J Water Process Eng 23:217–229. https://doi.org/10.1016/j.jwpe.2018.04.009
43. Yang X, Sun J, Fu W, Shang C, Li Y, Chen Y, Gan W, Fang J (2016) PPCP degradation by UV/chlorine treatment and its impact on DBP formation potential in real waters. Water Res 98:309–318. https://doi.org/10.1016/j.watres.2016.04.011
44. Kim I, Yamashita N, Tanaka H (2009) Photodegradation of pharmaceuticals and personal care products during UV and UV/H2O2 treatments. Chemosphere 77:518–525. https://doi.org/10.1016/j.chemosphere.2009.07.041
45. Kim I, Tanaka H (2009) Photodegradation characteristics of PPCPs in water with UV treatment. Environ Int 35:793–802. https://doi.org/10.1016/j.envint.2009.01.003
46. Wang J, Wang S (2016) Removal of pharmaceuticals and personal care products (PPCPs) from wastewater: a review. J Environ Manage 182:620–640. https://doi.org/10.1016/j.jenvman.2016.07.049
47. Martínez-Huitle CA, Brillas E (2009) Decontamination of wastewaters containing synthetic organic dyes by electrochemical methods: a general review. Appl Catal B Environ 87:105–145
48. Nidheesh PVV, Zhou M, Oturan MA (2018) An overview on the removal of synthetic dyes from water by electrochemical advanced oxidation processes. Chemosphere 197:210–227. https://doi.org/10.1016/j.chemosphere.2017.12.195
49. Wang N, Zheng T, Zhang G, Wang P (2016) A review on Fenton-like processes for organic wastewater treatment. J Environ Chem Eng 4:762–787. https://doi.org/10.1016/j.jece.2015.12.016
50. Katheresan V, Kansedo J, Lau SY (2018) Efficiency of various recent wastewater dye removal methods: a review. J Environ Chem Eng 6:4676–4697. https://doi.org/10.1016/j.jece.2018.06.060
51. Pang Y, Yu J, Tang L, Zeng G, Zhu C, Wei X (2018) Magnetic nanohybrid materials for water-pollutant removal. Elsevier Inc. https://doi.org/10.1016/B978-0-12-814154-0.00001-3
52. Barrios-Estrada C, de Jesús Rostro-Alanis M, Muñoz-Gutiérrez BD, Iqbal HMN, Kannan S, Parra-Saldívar R (2018) Emergent contaminants: endocrine disruptors and their laccase-assisted degradation—a review. Sci Total Environ 612:1516–1531. https://doi.org/10.1016/j.scitotenv.2017.09.013

53. Gravato C, Soares AMVM, Daam MA, Sanches ALM, Meireles G, de Oliveira DP, Zanoni MVB (2018) Red disperse dyes (DR 60, DR 73 and DR 78) at environmentally realistic concentrations impact biochemical profile of early life stages of zebrafish (Danio rerio). Chem Biol Interact 292:94–100. https://doi.org/10.1016/j.cbi.2018.07.007
54. Huong PTL, Huy LT, Phan VN, Huy TQ, Nam MH, Lam VD, Le AT (2016) Application of graphene oxide-MnFe2O4 magnetic nanohybrids as magnetically separable adsorbent for highly efficient removal of arsenic from water. J Electron Mater 45:2372–2380. https://doi.org/10.1007/s11664-015-4314-3
55. Fernández-Domene RM, Sánchez-Tovar R, Lucas-granados B, Muñoz-Portero MJ, García-Antón J (2018) Elimination of pesticide atrazine by photoelectrocatalysis using a photoanode based on WO3 nanosheets. Chem Eng J 350:1114–1124. https://doi.org/10.1016/j.cej.2018.06.015
56. Espinola-Portilla F, Navarro-Mendoza R, Gutiérrez-Granados S, Morales-Muñoz U, Brillas-Coso E, Peralta-Hernández JM (2017) A simple process for the deposition of TiO2 onto BDD by electrophoresis and its application to the photoelectrocatalysis of Acid Blue 80 dye. J Electroanal Chem 802:57–63. https://doi.org/10.1016/j.jelechem.2017.08.041
57. Li Z, Wang C, Su Z, Zhang W, Wang N, Mele G, Li J (2020) New porphyrin/Cu(II) porphyrin-TiO2 nanohybrids for improved photocatalytic oxidation and reduction activities. Mater Chem Phys 252:123228. https://doi.org/10.1016/j.matchemphys.2020.123228
58. Dai K, Liu G, Xu W, Deng Z, Wu Y, Zhao C, Zhang Z (2020) Judicious fabrication of bifunctionalized graphene oxide/MnFe2O4 magnetic nanohybrids for enhanced removal of Pb(II) from water. J Colloid Interface Sci 579:815–822. https://doi.org/10.1016/j.jcis.2020.06.085
59. Peng Z, Li Y, Wang W, Li X, Lv X, Chen X, Shi Y, Zhou S (2021) Facile synthesis of magnetic Fe@Al-ZnO nanocomposite for photocatalytic bacterial inactivation under visible-light irradiation. Mater Sci Semicond Process 123:105560. https://doi.org/10.1016/j.mssp.2020.105560
60. Vladkova T, Angelov O, Stoyanova D, Gospodinova D, Gomes L, Soares A, Mergulhao F, Ivanova I (2020) Magnetron co-sputtered TiO2/SiO2/Ag nanocomposite thin coatings inhibiting bacterial adhesion and biofilm formation. Surf Coatings Technol 384. https://doi.org/10.1016/j.surfcoat.2019.125322
61. Khudhair EM, Ammar SH, Khadim HJ (2021) Phosphotungstic acid immobilized onto ZnO coated zerovalent iron (Fe@ZnO/PW) core/shell magnetic nanocomposite for enhanced photocatalytic bacterial inactivation under visible light. J Photochem Photobiol. A Chem 404. https://doi.org/10.1016/j.jphotochem.2020.112907
62. Liu B, Mu L, Zhang J, X Han, Shi H (2020) TiO2/Cu2(OH)2CO3 nanocomposite as efficient antimicrobials for inactivation of crop pathogens in agriculture. Mater Sci Eng C 107. https://doi.org/10.1016/j.msec.2019.110344
63. Peng Z, Wu D, Wang W, Tan F, Ng TW, Chen J, Qiao X, Wong PK (2017) Fabrication of magnetic Fe@ZnO0.6S0.4 nanocomposite for visible-light-driven photocatalytic inactivation of Escherichia coli. Appl Surf Sci 396:19–25. https://doi.org/10.1016/j.apsusc.2016.11.026
64. Saha D, Visconti MC, Desipio MM, Thorpe R (2020) Inactivation of antibiotic resistance gene by ternary nanocomposites of carbon nitride, reduced graphene oxide and iron oxide under visible light. Chem Eng J 382:122857. https://doi.org/10.1016/j.cej.2019.122857
65. Sharma B, Boruah PK, Yadav A, Das MR (2018) TiO2–Fe2O3 nanocomposite heterojunction for superior charge separation and the photocatalytic inactivation of pathogenic bacteria in water under direct sunlight irradiation. J Environ Chem Eng 6:134–145. https://doi.org/10.1016/j.jece.2017.11.025
66. Vimala K, Varaprasad K, Sadiku R, Ramam K, Kanny K (2014) Development of novel protein-Ag nanocomposite for drug delivery and inactivation of bacterial applications. Int J Biol Macromol 63:75–82. https://doi.org/10.1016/j.ijbiomac.2013.10.021
67. Lingamdinne LP, Lee S, Choi JS, Lebaka VR, Durbaka VRP, Koduru JR (2021) Potential of the magnetic hollow sphere nanocomposite (graphene oxide-gadolinium oxide) for arsenic removal from real field water and antimicrobial applications. J Hazard Mater 402:123882. https://doi.org/10.1016/j.jhazmat.2020.123882

68. Alayande AB, Obaid M, Kim IS (2020) Antimicrobial mechanism of reduced graphene oxide-copper oxide (rGO-CuO) nanocomposite films: The case of Pseudomonas aeruginosa PAO1. Mater Sci Eng C 109:110596. https://doi.org/10.1016/j.msec.2019.110596
69. Lee SH, Pumprueg S, Moudgil B, Sigmund W (2005) Inactivation of bacterial endospores by photocatalytic nanocomposites. Colloids Surfaces B Biointerfaces. 40:93–98. https://doi.org/10.1016/j.colsurfb.2004.05.005
70. Singh S, Barick KC, Bahadur D (2015) Inactivation of bacterial pathogens under magnetic hyperthermia using Fe3O4-ZnO nanocomposite. Powder Technol 269:513–519. https://doi.org/10.1016/j.powtec.2014.09.032
71. Sharma M, Sondhi H, Krishna R, Srivastava SK, Rajput P, Nigam S, Joshi M (2020) Assessment of GO/ZnO nanocomposite for solar-assisted photocatalytic degradation of industrial dye and textile effluent. Environ Sci Pollut Res 27:32076–32087. https://doi.org/10.1007/s11356-020-08849-3
72. bdollahiN, Akbar Razavi SA, Morsali A, Hu ML (2020) High capacity Hg(II) and Pb(II) removal using MOF-based nanocomposite: Cooperative effects of pore functionalization and surface-charge modulation. Elsevier B.V. https://doi.org/10.1016/j.jhazmat.2019.121667
73. Joshi S, Kataria N, Garg VK, Kadirvelu K (2020) Pb2+ and Cd2+ recovery from water using residual tea waste and SiO2@TW nanocomposites. Chemosphere 257:127277. https://doi.org/10.1016/j.chemosphere.2020.127277
74. Khan FSA, Mubarak NM, Tan YH, Karri RR, Khalid M, Walvekar R, Abdullah EC, Mazari SA, Nizamuddin S (2020) Magnetic nanoparticles incorporation into different substrates for dyes and heavy metals removal—A Review. Environ Sci Pollut Res 27:43526–43541. https://doi.org/10.1007/s11356-020-10482-z
75. Kalantari S, Yousefpour M, Taherian Z (2017) Synthesis of mesoporous silica/iron oxide nanocomposites and application of optimum sample as adsorbent in removal of heavy metals. Rare Met 36:942–950. https://doi.org/10.1007/s12598-016-0709-4
76. El-Nagar DA, Massoud SA, Ismail SH (2020) Removal of some heavy metals and fungicides from aqueous solutions using nano-hydroxyapatite, nano-bentonite and nanocomposite. Arab J Chem 13:7695–7706. https://doi.org/10.1016/j.arabjc.2020.09.005
77. El-Shafai NM, Abdelfatah MM, El-Khouly ME, El-Mehasseb IM, El-Shaer A, Ramadan MS, Masoud MS, El-Kemary MA (2020) Magnetite nano-spherical quantum dots decorated graphene oxide nano sheet (GO@Fe3O4): Electrochemical properties and applications for removal heavy metals, pesticide and solar cell. Appl Surf Sci 506:144896. https://doi.org/10.1016/j.apsusc.2019.144896
78. Kamari S, Shahbazi A (2020) Biocompatible Fe3O4@SiO2-NH2 nanocomposite as a green nanofiller embedded in PES–nanofiltration membrane matrix for salts, heavy metal ion and dye removal: Long–term operation and reusability tests. Chemosphere 243:125282. https://doi.org/10.1016/j.chemosphere.2019.125282
79. Sharma M, Poddar M, Gupta Y, Nigam S, Avasthi DK, Adelung R, Abolhassani R, Fiutowski J, Joshi M, Mishra YK (2020) Solar light assisted degradation of dyes and adsorption of heavy metal ions from water by CuO–ZnO tetrapodal hybrid nanocomposite. Mater. Today Chem. 17:100336. https://doi.org/10.1016/j.mtchem.2020.100336
80. Bagheri S, Esrafili A, Kermani M, Mehralipour J, Gholami M (2020)Performance evaluation of a novel rGO-Fe0/Fe3O4-PEI nanocomposite for lead and cadmium removal from aqueous solutions. Elsevier B.V. https://doi.org/10.1016/j.molliq.2020.114422
81. Kong Q, Preis S, Li L, Luo P, Hu Y, Wei C (2020) Graphene oxide-terminated hyperbranched amino polymer-carboxymethyl cellulose ternary nanocomposite for efficient removal of heavy metals from aqueous solutions. Int J Biol Macromol 149:581–592. https://doi.org/10.1016/j.ijbiomac.2020.01.185
82. Mahmoud ME, Osman MM, Abdel-Aal H, Nabil GM (2020) Microwave-assisted adsorption of Cr(VI), Cd(II) and Pb(II) in presence of magnetic graphene oxide-covalently functionalized-tryptophan nanocomposite. J Alloys Compd 823:153855. https://doi.org/10.1016/j.jallcom.2020.153855

83. Mahmoud ME, Saleh MM, Zaki MM, Nabil GM (2020) A sustainable nanocomposite for removal of heavy metals from water based on crosslinked sodium alginate with iron oxide waste material from steel industry. J Environ Chem Eng 8:104015. https://doi.org/10.1016/j.jece.2020.104015
84. Murali A, Sarswat PK, Free ML (2020) Adsorption-coupled reduction mechanism in ZnO-Functionalized MWCNTs nanocomposite for Cr (VI) removal and improved anti-photocorrosion for photocatalytic reduction. J Alloys Compd 843:155835. https://doi.org/10.1016/j.jallcom.2020.155835
85. Verma M, Kumar A, Singh KP, Kumar R, Kumar V, Srivastava CM, Rawat V, Rao G, Kumari S, Sharma P, Kim H (2020) Graphene oxide-manganese ferrite (GO-MnFe2O4) nanocomposite: One-pot hydrothermal synthesis and its use for adsorptive removal of Pb2+ ions from aqueous medium. J Mol Liq 315:113769. https://doi.org/10.1016/j.molliq.2020.113769
86. Zarei A, Saedi S, Seidi F (2018) Synthesis and Application of Fe 3 O 4 @SiO 2 @Carboxyl-Terminated PAMAM dendrimer nanocomposite for heavy metal removal. J. Inorg. Organomet Polym Mater 28:2835–2843. https://doi.org/10.1007/s10904-018-0948-y
87. Zia J, Farhat SM, Aazam ES, Riaz U (2021) Highly efficient degradation of metronidazole drug using CaFe2O4/PNA nanohybrids as metal-organic catalysts under microwave irradiation. Environ Sci Pollut Res 28:4125–4135. https://doi.org/10.1007/s11356-020-10694-3
88. Zia J, Aazam ES, Riaz U (2020) Synthesis of nanohybrids of polycarbazole with α-MnO2 derived from Brassica oleracea: a comparison of photocatalytic degradation of an antibiotic drug under microwave and UV irradiation. Environ Sci Pollut Res 27:24173–24189. https://doi.org/10.1007/s11356-020-08149-w
89. Bastami TR, Ahmadpour A (2016) Preparation of magnetic photocatalyst nanohybrid decorated by polyoxometalate for the degradation of a pharmaceutical pollutant under solar light. Environ Sci Pollut Res 23:8849–8860. https://doi.org/10.1007/s11356-015-5985-2
90. Mohammed MKA (2020) Carbon nanotubes loaded ZnO/Ag ternary nanohybrid with improved visible light photocatalytic activity and stability. Optik (Stuttg). 217:164867. https://doi.org/10.1016/j.ijleo.2020.164867
91. Siddiqui SI, Chaudhry SA (2019) Nanohybrid composite Fe 2 O 3 -ZrO 2 /BC for inhibiting the growth of bacteria and adsorptive removal of arsenic and dyes from water. J Clean Prod 223:849–868. https://doi.org/10.1016/j.jclepro.2019.03.161
92. Zia J, Aazam ES, Riaz U (2020) Highly efficient visible light driven photocatalytic activity of MnO2 and Polythiophene/MnO2 nanohybrids against mixed organic pollutants. J Mol Struct 1207. https://doi.org/10.1016/j.molstruc.2020.127790
93. Chen L, Chuang Y, Nguyen TB, Chang JH, Lam SS, Chen CW, Di Dong C (2020) Novel molybdenum disulfide heterostructure nanohybrids with enhanced visible-light-induced photocatalytic activity towards organic dyes. J Alloys Compd 848:156448. https://doi.org/10.1016/j.jallcom.2020.156448
94. Modi A, Bellare J (2019) Efficient removal of dyes from water by high flux and superior antifouling polyethersulfone hollow fiber membranes modified with ZnO/cGO nanohybrid. J. Water Process Eng. 29:100783. https://doi.org/10.1016/j.jwpe.2019.100783
95. Abega AV, Ngomo HM, Nongwe I, Mukaya HE, Kouoh Sone PMA, Yangkou Mbianda X (2019) Easy and convenient synthesis of CNT/TiO2 nanohybrid by in-surface oxidation of Ti 3+ ions and application in the photocatalytic degradation of organic contaminants in water. Synth Met 251:1–14. https://doi.org/10.1016/j.synthmet.2019.03.012
96. Saini PK, Kumar N, Chandra R, Nath M, Minocha AK (2019) Facile synthesis of novel SWCNT/HgS nanohybrid: an effective photocatalyst for degradation of methylene blue. Mater Lett 250:5–8. https://doi.org/10.1016/j.matlet.2019.04.090
97. Krishnan S, Murugesan S, Vasanthakumar V, Priyadharsan A, Alsawalha M, Alomayri T, Yuan B (2021) Facile green synthesis of ZnFe2O4/rGO nanohybrids and evaluation of its photocatalytic degradation of organic pollutant, photo antibacterial and cytotoxicity activities. Colloids Surfaces A Physicochem Eng Asp 611:125835. https://doi.org/10.1016/j.colsurfa.2020.125835

98. Singh J, Sahu K, Satpati B, Shah J, Kotnala RK, Mohapatra S (2019) Facile synthesis, structural and optical properties of Au-TiO2 plasmonic nanohybrids for photocatalytic applications. J Phys Chem Solids 135:109100. https://doi.org/10.1016/j.jpcs.2019.109100
99. Rawat D, Mishra V, Sharma RS (2016) Detoxification of azo dyes in the context of environmental processes. Chemosphere 155:591–605. https://doi.org/10.1016/j.chemosphere.2016.04.068
100. Rawat D, Sharma RS, Karmakar S, Arora LS, Mishra V (2018) Ecotoxic potential of a presumably non-toxic azo dye. Ecotoxicol Environ Saf 148:528–537. https://doi.org/10.1016/j.ecoenv.2017.10.049
101. Tchobanoglous G, Burton FL, Stensel HD (2003) Wastewater engineering: treatment and reuse. Metcalf & Eddy, Inc.
102. Reyes-Contreras C, Matamoros V, Ruiz I, Soto M, Bayona JM (2011) Evaluation of PPCPs removal in a combined anaerobic digester-constructed wetland pilot plant treating urban wastewater. Chemosphere 84:1200–1207. https://doi.org/10.1016/j.chemosphere.2011.06.003
103. Carballa M, Omil F, Ternes T, Lema JM (2007) Fate of pharmaceutical and personal care products (PPCPs) during anaerobic digestion of sewage sludge. Water Res 41:2139–2150. https://doi.org/10.1016/j.watres.2007.02.012
104. Alvarino T, Suarez S, Lema JM, Omil F (2014) Understanding the removal mechanisms of PPCPs and the influence of main technological parameters in anaerobic UASB and aerobic CAS reactors. J Hazard Mater 278:506–513. https://doi.org/10.1016/j.jhazmat.2014.06.031
105. Massot A, Estève K, Noilet P, Méoule C, Poupot C, Mietton-Peuchot M (2012) Biodegradation of phytosanitary products in biological wastewater treatment. Water Res 46:1785–1792. https://doi.org/10.1016/j.watres.2011.12.055
106. Shawaqfeh AT (2010) Removal of pesticides from water using anaerobic-aerobic biological treatment. Chinese J Chem Eng 18:672–680. https://doi.org/10.1016/s1004-9541(10)60274-1
107. Chen Y, Cheng JJ, Creamer KS (2008) Inhibition of anaerobic digestion process: a review. Bioresour Technol 99:4044–4064
108. Pacheco-Álvarez MOAMOA, Picos A, Pérez-segura T, Peralta-hernández JMJM (2019) Proposal for highly efficient electrochemical discoloration and degradation of azo dyes with parallel arrangement electrodes. J Electroanal Chem 838:195–203. https://doi.org/10.1016/j.jelechem.2019.03.004
109. dos Santos AJ, Sirés I, Martínez-Huitle CA, Brillas E (2018) Total mineralization of mixtures of Tartrazine, Ponceau SS and Direct Blue 71 azo dyes by solar photoelectro-Fenton in pre-pilot plant. Chemosphere 210:1137–1144. https://doi.org/10.1016/j.chemosphere.2018.07.116
110. Salazar R, Gallardo-Arriaza J, Vidal J, Rivera-Vera C, Toledo-Neira C, Sandoval MA, Cornejo-Ponce L, Thiam A (2019) Treatment of industrial textile wastewater by the solar photoelectro-Fenton process: Influence of solar radiation and applied current. Sol Energy 190:82–91. https://doi.org/10.1016/j.solener.2019.07.072
111. Cai C, Zhang Z, Liu J, Shan N, Zhang H, Dionysiou DD (2016) Visible light-assisted heterogeneous Fenton with ZnFe 2 O 4 for the degradation of Orange II in water. Appl Catal B Environ 182:456–468. https://doi.org/10.1016/j.apcatb.2015.09.056
112. Pliego G, Garcia-Muñoz P, Zazo JA, Casas JA, Rodriguez JJ (2016) Improving the Fenton process by visible LED irradiation. Environ Sci Pollut Res 23:23449–23455. https://doi.org/10.1007/s11356-016-7543-y
113. El-Ghenymy A, Centellas F, Rodríguez RM, Cabot PL, Garrido JA, Sirés I, Brillas E (2015) Comparative use of anodic oxidation, electro-Fenton and photoelectro-Fenton with Pt or boron-doped diamond anode to decolorize and mineralize Malachite Green oxalate dye. Electrochim Acta 182:247–256
114. Bedolla-Guzman A, Sirés I, Thiam A, Peralta-Hernández JM, Gutiérrez-Granados S, Brillas E (2016) Application of anodic oxidation, electro-Fenton and UVA photoelectro-Fenton to decolorize and mineralize acidic solutions of reactive yellow 160 azo dye. Electrochim Acta 206:307–316. https://doi.org/10.1016/j.electacta.2016.04.166

115. Medrano-Rodríguez F, Picos-Benítez A, Brillas E, Bandala ER, Pérez T, Peralta-Hernández JM (2020) Electrochemical advanced oxidation discoloration and removal of three brown diazo dyes used in the tannery industry. J Electroanal Chem 873:114360. https://doi.org/10.1016/j.jelechem.2020.114360
116. Solano AMS, Martínez-Huitle CA, Garcia-Segura S, El-Ghenymy A, Brillas E (2016) Application of electrochemical advanced oxidation processes with a boron-doped diamond anode to degrade acidic solutions of Reactive Blue 15 (Turqueoise Blue) dye. Electrochim Acta 197:210–220. https://doi.org/10.1016/j.electacta.2015.08.052
117. Shu HY, Chang MC, Fan HJ (2004) Decolorization of azo dye acid black 1 by the UV/H2O2 process and optimization of operating parameters. J Hazard Mater 113:201–208
118. Panizza M-HM (2018) Electrochemical oxidation of organic pollutants for wastewater treatment. CurrOpin Electrochem 11:62–71. https://doi.org/10.1016/j.coelec.2018.07.010
119. Cornejo OM, Murrieta MF, Castañeda LF, Nava JL (2020) Characterization of the reaction environment in flow reactors fitted with BDD electrodes for use in electrochemical advanced oxidation processes: a critical review. Electrochim Acta 331:135373. https://doi.org/10.1016/j.electacta.2019.135373
120. Du X, Mo Z, Li Z, Zhang W, Luo Y, Nie J, Wang Z, Liang H (2021) Boron-doped diamond (BDD) electro-oxidation coupled with nanofiltration for secondary wastewater treatment: antibiotics degradation and biofouling. Environ Int 146:106291. https://doi.org/10.1016/j.envint.2020.106291
121. Chiliquinga M, Espinoza-Montero PJ, Rodríguez O, Picos A, Bandala ER, Gutiérrez-Granados S, Peralta-Hernández JM (2020) Simultaneous electrochemical generation of ferrate and oxygen radicals to blue BR dye degradation. Processes 8:1–14. https://doi.org/10.3390/PR8070753
122. He Y, Lin H, Guo Z, Zhang W, Li H, Huang W (2019) Recent developments and advances in boron-doped diamond electrodes for electrochemical oxidation of organic pollutants. Sep Purif Technol 212:802–821. https://doi.org/10.1016/j.seppur.2018.11.056
123. Picos-Benítez AR, Martínez-Vargas BL, Duron-Torres SM, Brillas E, Peralta-Hernández JM (2020) The use of artificial intelligence models in the prediction of optimum operational conditions for the treatment of dye wastewaters with similar structural characteristics. Process Saf Environ Prot 143:36–44. https://doi.org/10.1016/j.psep.2020.06.020
124. Kuchtová G, Chýlková J, Váňa J, Vojs M, Dušek L (2020) Electro-oxidative decolorization and treatment of model wastewater containing Acid Blue 80 on boron doped diamond and platinum anodes. J Electroanal Chem 863. https://doi.org/10.1016/j.jelechem.2020.114036.
125. Paniagua D, Le Borgne S, Vázquez-Duhalt R, De Biotecnología-unam I, Universidad A, Cuernavaca CC (2004) Aislamiento Y Caracterización De Bacterias Halofilas Degradadoras De Hidrocarburos
126. Phetrak A, Westerhoff P, Garcia-Segura S (2020) Low energy electrochemical oxidation efficiently oxidizes a common textile dye used in Thailand. J Electroanal Chem 871:114301. https://doi.org/10.1016/j.jelechem.2020.114301
127. Mudhoo A, Paliya S, Goswami P, Singh M, Lofrano G, Carotenuto M, Carraturo F, Libralato G, Guida M, Usman M, Kumar S (2020) Fabrication, functionalization and performance of doped photocatalysts for dye degradation and mineralization: a review. Springer International Publishing. https://doi.org/10.1007/s10311-020-01045-2
128. Park J, Wang JJ, Tafti N, Delaune RD (2018) Journal of industrial and engineering chemistry removal of eriochrome black T by sulfate radical generated from Fe-impregnated biochar / persulfate in Fenton-like reaction. J Ind Eng Chem 71:201–209. https://doi.org/10.1016/j.jiec.2018.11.026
129. H. Hamad, D. Bassyouni, E.S. El-Ashtoukhy, N. Amin, M. Abd El-Latif, Electrocatalytic degradation and minimization of specific energy consumption of synthetic azo dye from wastewater by anodic oxidation process with an emphasis on enhancing economic efficiency and reaction mechanism, Ecotoxicol. Environ. Saf. 148 (2018) 501–512. doi:https://doi.org/10.1016/j.ecoenv.2017.10.061.

130. Baddouh A, Bessegato GG, Rguiti MM, El Ibrahimi B, Bazzi L, Hilali M, Zanoni MVB (2018) Electrochemical decolorization of Rhodamine B dye: influence of anode material, chloride concentration and current density. J Environ Chem Eng 6:2041–2047. https://doi.org/10.1016/j.jece.2018.03.007
131. Vasconcelos VM, Ponce-De-León C, Nava JL, Lanza MRV (2016) Electrochemical degradation of RB-5 dye by anodic oxidation, electro-Fenton and by combining anodic oxidation-electro-Fenton in a filter-press flow cell. J Electroanal Chem 765:179–187
132. Ganiyu SO, Zhou M, Martínez-Huitle CA (2018) Heterogeneous electro-Fenton and photoelectro-Fenton processes: a critical review of fundamental principles and application for water/wastewater treatment. Appl Catal B Environ 235:103–129. https://doi.org/10.1016/j.apcatb.2018.04.044
133. Wu Q, Wang H, Yi C (2018) Preparation of photo-Fenton heterogeneous catalyst (Fe-TS-1 zeolite) and its application in typical azo dye decoloration. J Photochem Photobiol A Chem 356:138–149. https://doi.org/10.1016/j.jphotochem.2017.12.041
134. Cruz-Rizo A, Gutiérrez-Granados S, Salazar R, Peralta-Hernández JM (2017) Application of electro-Fenton/BDD process for treating tannery wastewaters with industrial dyes. Sep Purif Technol 172:296–302
135. Jain B, Hashmi A, Sanwaria S, Singh AK, Susan MABH, Singh A (2020) Zinc oxide nanoparticle incorporated on graphene oxide: an efficient and stable photocatalyst for water treatment through the Fenton process. Adv Compos Hybrid Mater 3:231–242. https://doi.org/10.1007/s42114-020-00153-5
136. Zazou H, Afanga H, Akhouairi S, Ouchtak H, Addi AA, Akbour RA, Assabbane A, Douch J, Elmchaouri A, Duplay J, Jada A, Hamdani M (2019) Treatment of textile industry wastewater by electrocoagulation coupled with electrochemical advanced oxidation process. J Water Process Eng 28:214–221. https://doi.org/10.1016/j.jwpe.2019.02.006
137. Gomathi E, Balraj B, Kumaraguru K (2018) Electrochemical degradation of scarlet red dye from aqueous environment by titanium-based dimensionally stable anodes with SS electrodes. Appl Biol Chem 61:289–293. https://doi.org/10.1007/s13765-018-0357-5
138. Rafaie HA, Nor RM, Azmina MS, Ramli NIT, Mohamed R (2017) Decoration of ZnO microstructures with Ag nanoparticles enhanced the catalytic photodegradation of methylene blue dye. J Environ Chem Eng 5:3963–3972. https://doi.org/10.1016/j.jece.2017.07.070
139. Zinatloo-Ajabshir S, Salavati-Niasari M (2017) Photo-catalytic degradation of erythrosine and eriochrome black T dyes using Nd2Zr2O7nanostructures prepared by a modified Pechini approach. Sep Purif Technol 179:77–85. https://doi.org/10.1016/j.seppur.2017.01.037
140. Jafari H, Afshar S, Zabihi O, Naebe M (2016) Enhanced photocatalytic activities of TiO2-SiO2 nanohybrids immobilized on cement-based materials for dye degradation. Res Chem Intermed 42:2963–2978. https://doi.org/10.1007/s11164-015-2190-3
141. Oyewo OA, Elemike EE, Onwudiwe DC, Onyango MS (2020) Metal oxide-cellulose nanocomposites for the removal of toxic metals and dyes from wastewater. Int J Biol Macromol 164:2477–2496. https://doi.org/10.1016/j.ijbiomac.2020.08.074
142. Pacheco-Alvarez MO, Rodríguez-Narváez OM, Wrobel K, Navarro-Mendoza R, Nava-Montes de Oca JL, Peralta-Hernández JM (2018) Improvement of the degradation of Methyl Orange using a TiO2/BDD composite electrode to promote electrochemical and photoelectro-oxidation processes. Int J Electrochem Sci 13:11549–11567. https://doi.org/10.20964/2018.12.70
143. Fan Su Y, Wang GB, Kuo DTF, Ling Chang M, Hsin Shih Y (2016) Photoelectrocatalytic degradation of the antibiotic sulfamethoxazole using TiO2/Ti photoanode. Appl Catal B Environ 186:184–192. https://doi.org/10.1016/j.apcatb.2016.01.003
144. Azis MT, Nurwahidah AT, Wibowo D, Nurdin M (2017) Photoelectrocatalyst of Fe co-doped N-TiO2/Ti nanotubes: pesticide degradation of thiamethoxam under UV–visible lights. Environ Nanotechnol Monit Manag 8:103–111. https://doi.org/10.1016/j.enmm.2017.06.002
145. Turolla A, Bestetti M, Antonelli M (2018) Optimization of heterogeneous photoelectrocatalysis on nanotubular TiO2 electrodes: reactor configuration and kinetic modelling. Chem Eng Sci 182:171–179. https://doi.org/10.1016/j.ces.2018.02.041

146. Bian J, An X, Jiang W, Liu R, Hu C, Liu H (2021) Defect-enhanced activation of carbon nitride/horseradish peroxidase nanohybrids for visible-light-driven photobiocatalytic water purification. Chem Eng J 408:127231. https://doi.org/10.1016/j.cej.2020.127231
147. Zafar Z, Fatima R, Kim JO (2021) Experimental studies on water matrix and influence of textile effluents on photocatalytic degradation of organic wastewater using Fe–TiO2 nanotubes: towards commercial application. Environ Res 197:111120. https://doi.org/10.1016/j.envres.2021.111120

Nanocomposites for Dye Removal from Water

Mohamed A. Abdelwahab, Maie Mousa, and Nehal Salahuddin

Abstract The pollution of water by potentially dyes, is severe form of environmental impact. Traditional wastewater treatment are inadequate and cannot encounter the basic standards of water quality at sensible cost or processing time. Removal of dyes from aquatic surroundings has become a main alarm due to environmental problems and the possible hazards and hazards posed by them. Nowadays, the adsorption method as one of the most effective methods of eliminating pollutants has fascinated growing consideration among chemists and environmental researchers. However, one of the tasks is to design more effective adsorbents besides preparing them via greener and safer approaches. Nanocomposites are considered talented materials for the removal of potentially toxic dyes from aqueous solution through adsorption process. The present chapter deals with the utilization of nanocomposites for removal of dyes.

Keywords Dyes · Nanocomposites · Adsorpotion

1 Introduction

Nature has offered an abundance of resources for maintaining and grow life on the earth. Water is the essential source in the world, and today's world challenges serious issues in meeting the growing demand for safe drinking water, specifically in developing countries [97, 106]. A severe worry and vital goal for human life is clean drinking water. Moreover, water is a vital resource for food manufacturing and the preservation of human health as well as environmental protection. One of the most important challenges in the globe is the lack of clean drinking water [14, 35]. The dumping of industrial effluents into our waterways is a significant source of water contamination, which poses a severe environmental risk [68]. Drinking water can

M. A. Abdelwahab · M. Mousa · N. Salahuddin (✉)
Chemistry Department, Faculty of Science, Tanta University, Tanta 31527, Egypt
e-mail: nehal.attaf@science.tanta.edu.eg

M. A. Abdelwahab
e-mail: mohamed.abdelwahab@science.tanta.edu.eg

S. K. Swain (ed.), *Nanohybrid Materials for Water Purification*, Composites Science and Technology, https://doi.org/10.1007/978-981-19-2332-6_4

be contaminated from discharged agricultural, sewage and industrial waste, making it difficult to provide safe drinking water to nearby populations and ecosystems. According to statistical projections, approximately one billion persons living in dry countries will face clean water scarcity by 2025 [29]. Around 3.2 million person die every year as a result of a lack of clean water or sanitary conditions, particularly in poor nations [110]. Water is the most essential material for all species, notably humans, because water makes up around 60% of the human body. Contamination of the environment impacts the world's three essential components: air, soil and water. Due to population expansion, industrial and human activities have increased without equivalent environmental care, and hydric resources are being progressively depleted [80]. Water treatment is a theme of concern for our health and our environment. Drinking water sources must be treated to eliminate disease-causing chemicals.

1.1 Sources of Dyes and Its Classification

The scarcity of natural pigments and enhancements in dye manufacturing technology have made researchers and manufacturers progressively more dependent on synthetic dyes. Dyes are broadly utilized in productions, for example, paper, plastics, textiles and paints. Dyes can be categorized according to the source (natural and synthetic dyes), chemical structure (chromophores) (nitro and nitoso dyes, anthraquinone dyes, Azo dyes, triarylmethane dyes, phthalocyanine dyes, indigoid dyes and sulfur dyes) and application method (disperse dyes, acid and basic dyes, reactive dyes, vat dyes and direct dyes) as shown in Fig. 1 [15].

The amount of dyes engendered annually is assessed to be more than 700,000 tonnes, with 10–15% being released in wastewater [32, 60]. Synthetic dye present in the water is one of the most severe issues challenging environmental protection

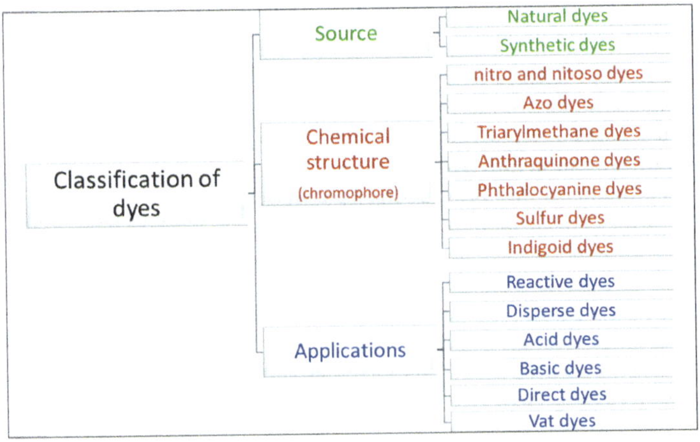

Fig. 1 Classification of dyes

strategies, owing to their difficulty being degraded into less dangerous compounds for human health and the environment. Dyes are harmful organic compounds with limited biodegradability that perform a substantial role in environmental issues such as eutrophication, visual pollution and water system disturbances [90]. Furthermore, the existence of dyes as pollutants in water resources can lower water quality and create carcinogenic and mutagenic consequences that can infect humans and wildlife. Dyes can affect a variety of difficulties when they infiltrate the environment. Because most of these organic materials are very stable toward microbial attacks, humidity, oxidizing agents and solar radiation if dismissed into rivers [11, 19], dyes can produce a decline in water properties by exposing color to the water and affecting the photosynthetic action of aquatic organisms by obstructing the access of solar light [51]. Recently, there has been much interest in eliminating coloured contaminants from wastewater [23].

1.2 Dye Removal Techniques

Nowadays, the most common processes in water treatment utilized by municipal water systems (mostly surface water treatment) involve coagulation and flocculation, sedimentation, filtration and disinfection [33]. Numerous methods have been utilized to eliminate dyes from contaminated media in both industrial wastewater and water resources such as electrochemical and membrane processes, biological treatment, coagulation, chemical oxidation, adsorption and aerobic microbial degradation (Fig. 2) [71, 89, 113]. Water treatment technology for heavy metals and dyes removal is dependent on the type of contaminant to be removed or basis for the treatment, whereas water treatment methods that are acceptable for dye removal may not be suitable for removing heavy metals. For example, chemical precipitation is used to remove heavy metals, although it is not proposed for colour removal from water. Chemical precipitation procedures are beneficial in the treatment of water and the creation of water treatment products and media [22, 55]. Treatment methods have been employed to focus on the elimination of recalcitrant dyes from wastewater and aquatic water. On the other hand, the adsorption method is widely used because of its several benefits, including economic feasibility, high efficiency in removing the dyes, and ease of processing as well as cost-effectiveness for dye removal from aquatic water and wastewater [20, 21]. The adsorption process includes a number of steps: (i) dissolving dyes in solution, (ii) exterior dye diffusion to the adsorbents' surrounds, (iii) intra/internal -particle diffusion (filling nanoparticle pores with dye materials), and (iv) desorption or adsorption on interior sites [3]. The strengths, weaknesses, opportunities and threats (SWOT) of persistent dye pollution removal based on adsorption techniques were effectively discussed on the next section [5].

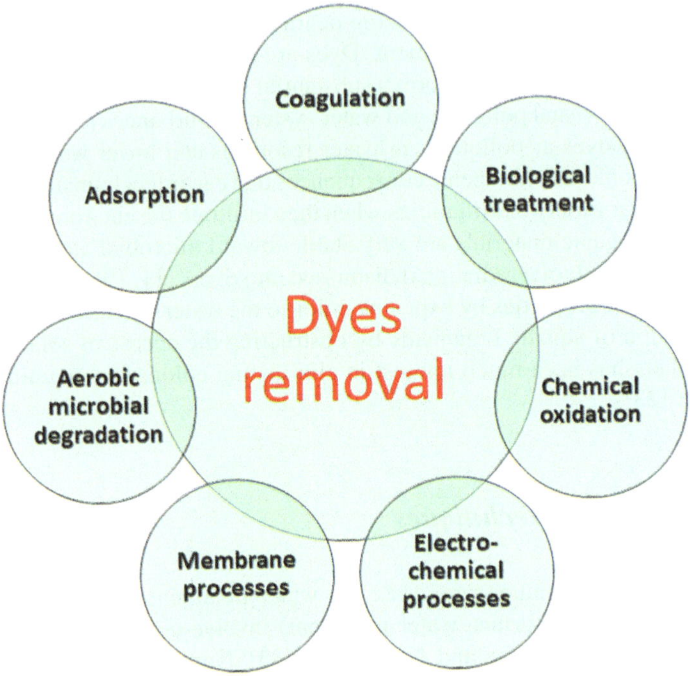

Fig. 2 Possible techniques used to remove dyes and water pollutants from contaminated media in both water resources and industrial wastewater

2 Nanocomposite Material

Nanocomposite material has expanded substantially to comprise a huge range of structures such as one, two and three-dimensional as well as amorphous materials, particularly prepared from various materials and combined together at the nanoscale. This fast-growing industry is producing a lot of new materials with unique features by mixing characteristics from different components/materials to produce a single material. There's also a chance that new qualities emerge that aren't present in the parent component materials.

2.1 *Classification of Nanocomposites*

In general, the organic–inorganic nanocomposite materials class are a rapidly expanding field of study. Effort is engrossed on the ability to get control of the

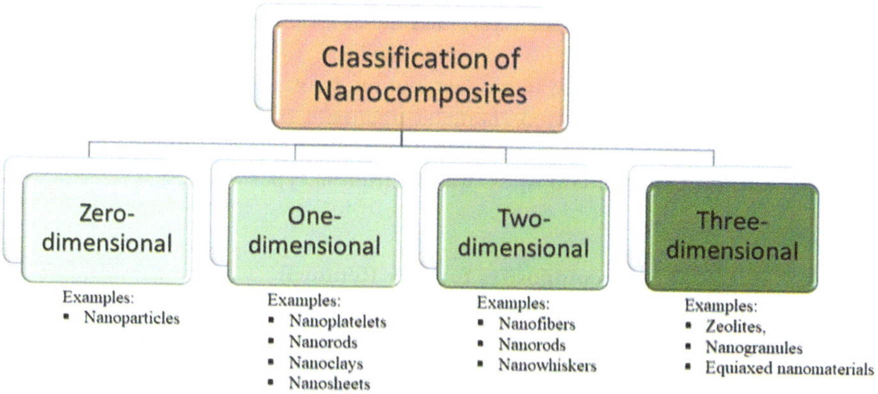

Fig. 3 Nanocomposites classification based on the dimensions

nanoscale structures via innovative synthetic approaches. The properties of nanocomposite materials depend not only on the characteristics of the matrix and the reinforcing agent/fillers, but also on their morphology and interfacial features. Nanocomposites could be classified according to the structure of nanofillers into four dimension (Fig. 3):

(a) Nanofillers with zero-dimension: Nano dimensions in any direction, for example, nanoparticles. They don't have any dimension exceed 100 nm. They might be of the amorphous, crystalline, metallic or ceramic nature.
(b) Nanofillers with one-dimension: Nano dimensions in only one direction, for example, nanotubes and nanowires (needle-type shape). Nanoclays, nanoplatelets, nanosheets and nanorods are an excellent examples of nanofillers with one-dimension.
(c) Nanofillers with two-dimension: Nano dimensions in two directions, for example, carbon nanotubes and silicates. It might also involve nanorods, nanowhiskers and nanofibers.
(d) Nanofillers with three-dimension: Nano dimensions in three directions, for example, nanogranules and zeolites.

2.2 *Efficiency of Nanocomposites for Dye Removal*

For dye removal, various organic and inorganic adsorbents (ex. Fly-ash, zeolite, saw-dust, mesoporous silica and activated carbon) are applied. These adsorbents have some drawbacks such as poor selectivity, low adsorption capacity and higher cost [78]. Adsorption using nanocomposites is the best way to remove the dye from wastewater due to its straightforward method, cheap method and efficacy to various kinds of dyes [28]. By engineering several functional groups (carboxylate, amino, acid, and hydroxyl groups) or the amalgamation of adsorbents, including metal oxides, graphene, graphene oxide (GO), and metal–organic frameworks (MOFs) in the nanocomposites, the separation capacity can be significantly improved [24, 73]. Because most dyes are resistant to biodegradation, photodegradation and oxidation, traditional biological and physico-chemical treatment techniques are unsuccessful for dye removal. Table 1 summarizes the removal efficacy and maximum adsorption capacity of different nanocomposites twards different types of dyes.

2.2.1 Clay-Based Nanocomposites for Dye Removal

Various inorganic clay has been used to remove the pollutions from the wastewater, such as laponite, sepiolite, attapulgite and montmorillonite [58]. Among these clay minerals, layered silicates montomorllonite (MMT) or nanoclay is an excellent reinforcing filler for removing the cationic and anionic dyes and can be used in different applications such as biomedical, agriculture, packaging and aerospace applications [1]. MMT has different characteristics such as high modulus, cheap material, high CEC (cation exchange capacity), non-toxicity, high surface area and ease recyclability [53]. MMT is an alumina octahedral silicates sandwiched among two tetrahedral silicate sheets [6]. By using cation exchange, the interlayer spacing of MMT was increased to enhance the adsorption capability of MMT minerals [1].

MMT has been applied in a different matrix for dye removals, such as polyvinyl alcohol and chitosan [36, 48]. The matrix of the nanocomposites also acting an crucial role in enhancing removal of the dyes. One of these matrices that was mixed with nanoclay is the hydrogel polymers (acrylic acid-acrylamide based polymers and/or copolymers) which have high adsorption capacity, oxygen barrier, thermal stability and flame retardant that can be applied in different applications (ex., agriculture and healthcare) [16, 54]. Another study utilized poly (acrylamide-co-sodium acrylate) in existence of biopolymer carrageenan and Na-MMT to enhance the absorbency of crystal violet dye (CV) from wastewater [58]. The best result has been obtained by using 10 wt.% of MMT due to enhancement of osmotic pressure. The water absorption decreased by an increasing amount of MMT, which works as a multifunctional crosslinker.

Kasgoz and Durmus [47] mixed MMT with acrylamide (AAm) and 2-acrylamido-2-methylpropanesulfonic acid sodium salt (AMPSNa) to eliminate two cationic dyes (Brilliant Cresyl Blue (BCB) and Safranine-T (ST)) from an aqueous solution. The authors found that the incorporation of MMT into the hydrogel improved the swelling capacity and diffusion capacity. However, these hydrogel composites have some drawback to the environment as the material can not be degraded which cause a "white pollution" [72].

Research is growing in the production of inexpensive, biodegradable, biocompatible and natural polymeric sorbents to overcome the disadvantages of using synthetic polymeric sorbent that are expensive and difficult to regenerate. A novel and green superabsorbent cellulose-clay nanocomposite hydrogels was produced by chemical cross-linking to remove the methylene blue dye (MB) [72]. The new materials displayed superior absorption capacity for MB solution (97% removal

Table 1 The removal efficacy and maximum adsorption capacity of different nanocomposites twards different types of dyes

Adsorbent code	The maximum removal efficiencies (%)	The maximun adsorption capacity (Q_{max})mg $g^{-1.}$	Type of Dye	References
Starch-g-PAAc/CNWs	90	2236	Methylene blue	Gomes et al. [30]
AAm-AMPSNa/clay hydrogel	–	526.3 500.0	Safranine- T Brilliant Cresyl Blue	Kaşgöz et al. [47]
GO-Cu-MOF	90	262	Methylene blue	Firouzjaei et al. [18]
G/Fe_3O_4	99.4	89.4	Fuchsine	Wang et al. [101]
Cellulose/MMT	98	1065	Methylene blue	Peng et al. [72] 2016
Chitosan/silica/ZnO	95	293.3	Methylene blue	Hassan et al [33]
Carbon/iron oxide	–	46.94	Rhodamine B	Singh et al. [95]
DTN	100	1,632	Direct blue 78	Hayati [34]
	94	1,220	Direct red 80	
Magnetic Fe@MgO		7299.2	Methyl orange	Ge et al. [26]
PES18P4CWNT0	98.7	–	Malachite green	Mousavi et al. [64]
Fe_3O_4@SiO_2-NH_2	97	–	Methyl red	Kamari et al. [44]
MCNC	99	–	Direct red16	Rafieian et al. [79]
MNC	94.21	223.82	Nylosan	Muntean et al. [65]
	85.31	114.68	Blue	
	96.50	286.91	Chromazurol S, Safranin T	

(continued)

Table 1 (continued)

Adsorbent code	The maximum removal efficiencies (%)	The maximun adsorption capacity (Q_{max})mg $g^{-1.}$	Type of Dye	References
A magnetic nanocomposite	–	113.31	Crystal violet	Singh et al. [96]
Nanocomposite superabsorbents	91.2	135.1	Crystal violet	Mahdavinia et al. [58]
CS/ZnO	96.7	–	Methylene blue	Mostafa et al. [63]
Zn0.5Co0.5Al0.5Fe1.46- La0.04O4	76	–		Ahmed et al. [7]
$CoFe_2O_4eSiO_2$	-	75.5	Malachite green	Amiri et al. [13]
ZIF-8@CNT	73.5	3300	Malachite green	Abdi et al. [2]
ZIF-8@GO	77	2034		
PANI-ZSP	–	12	Methylene blue	Gupta et al. [31]
Sm-MOF/GO	91	–	Rhodamine B	Yang et al. [108]
Chitosan/silk fibroin//hydroxyapatite	-	476	Methylene blue	Salama et al. [88]
Fe_3O_4@SiO_2-Zr-MOF	98	128	Methylene blue	Huang et al. [40]
	100	219	Methyl orange	
ZIF-67@Fe_3O_4@ESM	100	250.81	Basic Red 18	Mahmoodi et al. [59]
Ppy NF/Zn-Fe LDH	89.54	63.4	Safranin dye	Mohamed et al. [61]
G-SO_3H/Fe_3O_4	95.3	199.3	Methyl orange	Wang et al. [104]
	96.8	216.8	Brilliant yellow	
	95.5	200.6	Alizarin red	

(continued)

Table 1 (continued)

Adsorbent code	The maximum removal efficiencies (%)	The maximun adsorption capacity (Q_{max})mg $g^{-1.}$	Type of Dye	References
$Fe_3O_4 - TSC$	–	435	Malachite green	Alqadami et al. [11]
$MnFe_2O_4$/GO	94	46.08	Neutral red	Katubi et al. [45]
RHC-Fe_3O_4	–	321	Methylene Blue	Yang et al. [109]
PANI/NiO/MnO_2	97	248.4	Methyl orange	Abd Ali et al. [10]
BCA	99.5	142.8	RhodamineB	Mohammed et al. [62]
Zinc oxide	99	–	Initial dye	Nakkeeran et al. [66]
Graphene oxide intercalated montmorillonite	96	746.27	Crystal violet	Puri and Sumana [77]
(MZ/CS/AL)	–	6.14	Malachite green	Kazemiand Javanbakht [46]
PPy/Fe_3O_4/SiO_2	100	361.43	Congo red	Alzahrani et al. [12]
CuO/ZnO-T	80	-	Reactive yellow	Sharma et al. [94]
	86		Basic viole	
Cc-PEI	–	–	Methyl blue Rhodamine B	Liu et al. [57]
NiO-CMC	–	3.03	Methylene blue	Albokheet et al. [9]
(Fe_3O_4/ZA), Fe_3O_4	96.8	40.36	Methylene blue	Tran et al. [100]
$NiFe_2O_4$/GO (NGO)	–	476.19	Methylene blue	Thy et al. [99]

(continued)

Table 1 (continued)

Adsorbent code	The maximum removal efficiencies (%)	The maximun adsorption capacity (Q_{max})mg $g^{-1.}$	Type of Dye	References
CMC coated Fe_3O_4@SiO_2 MNPs	–	22.7	Methylene blue	Zirak et al. [114]
CMC/kC/AMMT	98	10.75	Methylene blue	Liu et al. [56]
CMC-AM/nHA-MS	–	–	Acid Blue 113	Varaprasad et al. (2018)
GONH	98.9	348.82	MV cationic dye	Bardajee et al. [82]
CS/CNFs	–	259.13	Cango red	Kafil et al. [43]
		553.12	Acid blue 1	
Montmorillonite-silica	84.10	172.40	Congo red	Zaccheus et al. [111]
NSG	99.5	38.46	Eriochrome Black T	Ahmed et al. [8]
Fe_3O_4@C	99.1	184	Rhodamine B	Kong et al. [49]
Z-Fe_3O_4 NC	97.5	2.57407	Methylene blue	Nyankson et al. [67]
St/Pn nanocomposite	99	3692.4	Reactive Black 5 Reactive Violet 4decolorized	Janaki et al. [41]
	98	422.40		
	87			
GO-$ZnFe_2O_4$	99	171	Methyl orange	Sadighian et al. [84]
XG-g-PAM/SiO_2	–	209.205	Congo red	Ghorai et al. [27]

(continued)

Table 1 (continued)

Adsorbent code	The maximum removal efficiencies (%)	The maximun adsorption capacity (Q_{max})mg $g^{-1.}$	Type of Dye	References
Guar gum-graft-poly (acrylamide)/silica (g-GG/SiO_2)	–	579.01	Reactive blue and Congo red	Pal et al. [69]
RGO–Ni)	90	65.31	Rhodamine B	Jinendra, U., et al. [42]
ZnS/NaAla-Gel-cl-polyAAm	97.37	-	Biebrich scarlet	Priya et al. [76]
	95.45		Crystal violet	
magnetic graphene oxide-hydrogels	–	769.23	Crystal violet	Pourjavadi et al. [74]
semi-interpenetrating network (semi-IPN)	97.58	–	Crystal violet	Sharma et al. [52]
	95.42		Eosin yellow	
XA-cl-poly(AAm)	96.82	4.7	Eosin yellow Eriochrome black-T	Sharma et al. [92]
	98.73	8.0		
BFO/Mrcl-poly(AAm)-IPN-poly(AA)	97.49	2.01	Auramine-O	Sharma et al. [93]
	95.39	3.06	Malachite green	
	94.56	7.55	Crystal violet	
PPy@CS/MMT/GO	99	76.4	Biebrich scarlet	Salahuddin et al. [87]
	97	26.7	Eosin Y	
	95		neutral red	
	66		Safranin O	
	85		Titan yellow	
PPy/CS/GO	63	0.12	Ponceau 4 R	Salahuddinet al. [86]
(M-PPy/CS/GO)	64.5	6.799	Ponceau 4 R	Salahuddin et al.[85]

efficacy for 10 mg.L^{-1} MB). Another study was applied for removing MB from wastewater by using carboxymethyl cellulose (CMC), biodegradable polysaccharide kappa-carrageenan (obtained from red seaweed) and activated MMT (A-MMT) [56]. A-MMT was prepared by an acid treatment process using sulfuric acid. The authors showed that the maximum MB removal was 98% (12.25 mg.g^{-1}) with using 0.4% A-MMT, and the adsorption still excellent after five consecutive cycles. Recently, multi-functional nanohybrid consisting of polypyrrole/chitosan, grapheneoxide and montmorillonite nanohybrid displayed high efficacy for the removal of neutral, anionic, and cationic dyes (Titan yellow, Safranin O, Neutral red, Eosiny, Biebrich scarlet) from aqueous solutio [87].

2.2.2 Carbon-Based Nanocomposites for Dye Removal

A novel adventure to improve the removal of the dye from aquatic media is directed to use carbon nano adsorbents graphene (discovered in 2004), which is an allotrope of carbon (SP^2 bonded carbon atom sheet) and has a two-dimensional hexagonal structure. Graphene powder has many different characteristics such as low density, superior surface area, chemically stable and superior thermal, electrical, mechanical and optical properties. Graphene has been used as an adsorbent material for removing the dye from aquatic water. Sulfonation favors the surface of the graphene Zeta potential, thus improve the interaction of sulfonated graphene with aromatic pollutants and thus promises the high adsorption of organic dyes on sulfonated graphene [25, 81, 91]. In addition, the sulfonated graphene nanosheets ($G–SO_3H$) have the ability to increase the dispersion in an aqueous solution [112].

Recently, blending graphene with a magnetic nanoparticles recorded unique chemical, physical characteristics and adsorption capacity compared to graphene material [4, 83]. Magnetic graphene nanocomposites have the ability to respond to a magnet, facilitating the separation of graphene from dispersion rapidly and effectively [25]. Wang et al. [101] found that a magnetic/graphene (Fe_3O_4/G) nanocomposite can effectively adsorb fuchsine dye. In another report, reduced GO-Nickel nanocomposite was prepared to remove Rhodamine B (RhB) dye from aquatic solution [42]. The results displayed that the pH influence the dye removal and showed higher removal at pH 8. According to the results of the kinetic investigation, RhB dye adsorption follows a pseudo-second-order kinetic model. Consequently, magnetic $Fe_3O_4/G\text{-}SO_3H$ nanocomposite was synthesized and applied for removal of anionic dyes (methylene orange MO, alizarin red (AR) and brilliant yellow (BY)) and cationic dyes (victoria blue (VB), safranine T (ST) and neutral red (NR)) from aqueous medium as shown in Fig. 4a, b [105]. At pH 6, the new nanocomposite material showed higher adsorption capacity in cationic dyes than anionic dyes (Fig. 4c). The maximum adsorption capacities of $G\text{-}SO_3H/Fe_3O_4$ for ST, NR, and VB dyes were 199.3, 216.8, and 200.6 mg g^{-1} respectively [105].

Graphene oxide, an electron-rich, hydrophobic nanomaterial, with large specific area has been served as an amazingly brilliant adsorbent. Graphene oxide (GO)

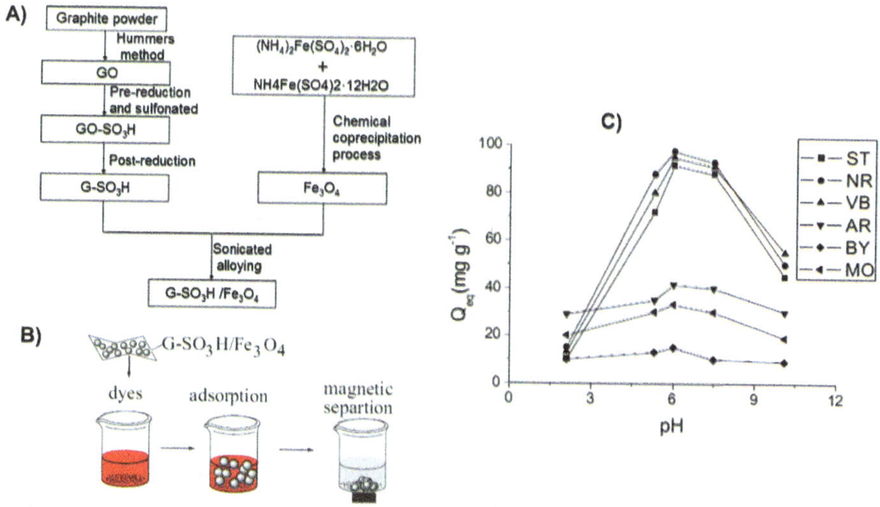

Fig. 4 **A** preparation of nanocomposite material from sulfonated graphene ($G\text{-}SO_3H$) and magnetic Fe_3O_4 nanomaterials; **B** Dye removal mechanism from aqueous solution using G- $G\text{-}SO_3H/Fe_3O_4$; **C** Effect of pH on the adsorption capacity of dye removal (MO, AR and BY) and cationic dyes (VB, ST and NR) (Reprinted with permission from Wang et al. [104])

is equivalent to graphene but with a high adsorption capacity and high surface-to-volume ratio and a functionalized group (hydroxyl, carboxyl and epoxy group) that allowed dispersion in the water and compatibility with hydrophilic dyes. Various researchers have explored the use of GO as an adsorbent for dye removals such as CV, MB, methylene orange and direct dye [23]. Puri and Sumana [77] formed a nanocomposite by the incorporation of GO into layered silicates MMT. The new nanocomposites have been used in the removal of CV dye from wastewater with a high adsorption capacity of ~746 mg/g as well as efficiently ~96%.

However, it was discovered that after the adsorption process, GO was very hard to eliminate from the aqueous medium by the standard separation method. As a result, it may raise the cost of industrial application while also polluting the treated water [17]. To solve the problem, GO hydrogel nanocomposites based on poly(acrylic acid)-g-salep polysaccharides biopolymer was applied for removing the cationic methyl violet (MV) dye in aqueous media [82]. By increasing the pH and temperature of the solution, the MV adsorption capacity on nanocomposites was enhanced, and adsorption equilibrium was achieved after 30 min.

In addition, it was reported that GO membranes are unstable in aqueous solutions due to poor adhesion between adjacent GO layers, limiting their application in wastewater. For this reason, a new nanocomposites membrane composed of samarium metal–organic framework (MOF) and GO was synthesized [108]. The Sm-MOF/GO (mass ratio of MOF: mass ratio of the reactant = 0.31) nanocomposite showed higher permeance (26 $m^{-2}\,h^{-1}\,bar^{-1}$) and high rejections (~91%) to MB and RhB. A porous crystalline 3D MOFs was made by coordination bonds from metal ions and organic ligands. Firouzjaei et al. observed the adsorption of MB by using copper-based MOF, immobilized on GO and reported that the adsorption capacity improved from 106 to 142 mg/g by increasing the temperature from 25 to 65 °C. Existence of GO in the MOF network enhanced the adsorption capability of capturing the dye [18]. Another MOF zeolitic imidazolate framework (ZIF-8) was incorporated with CNT and GO to prepare a new nanocomposite [2]. The new material was used to remove a malachite green (MG) dye from aquatic medium. Hybrid nanocomposite material showed a higher rate than sole MOF. The ZIF-GO nanocomposite material exhibited a higher adsorption capacity (3300 mg/g) than ZIF-8-CNT nanocomposites (2034 mg/g) at 20 °C.

A novel nanocomposite based on Chitosan/carbon nanoflowers was applied for adsorption of acid black1 (AB1) and cango red (CR) anionic dyes. The maximum adsorption was recorded to be 259.13 and 553.12 mg g^{-1}, respectively within 60 min [43]. In another report, the adsorption behavior of polypyrrole/chitosan/graphene oxide to Ponceau 4R (P4R) dye under different conditions such as different, salt concentrations, temperatures degrees and dye concentrations was studied and recorded high efficiency than the constituents [86].

It is worth to mention that magnetic adsorbents are characterized by high specific surface area, simple modification, structural flexibility, processing viability, mechanical strength properties and by applying an external magnetic field the adsorbents are easy recovered. The magnetic adsorbents comprise elements with magnetic properties. A new magnetic manganese ferrite ($MnFe_2SO_4$) nanoparticles was applied as magnetic support (synthesized by sol–gel route) for different adsorbents because of high magnetic and thermal characteristics, cheaper material, non-toxicity and high chemical stability. Thus, to enhance the adsorption of graphene, $MnFe_2SO_4$ was added as a support for reduced graphene adsorbent. The new nanomaterials showed high adsorption capacity to remove cationic dyes (MG and MB dye) [4]. Another study presented that GO/$MnFe_2SO_4$ has a high adsorption capacity to remove arsenic As^{5+}, As^{3+} and lead Pb^{2+} from aqueous water [50]. Katubi et al. (2021) displayed that GO/$MnFe_2O_4$ nanocomposite was efficient to remove NR dye with adsorption capacity of 46.08 mg/g. In another report, Zinc ferrite ($ZnFe_2O_4$) magnetic nanoparticles were used with GO in the removal of MO dye [84]. As well, Nickel ferrite ($NiFe_2O_4$) was utilized with GO to remove MB from an aquatic water [98]. Using 500 ppm MB at pH 10 within 150 min, the adsorption capacity achieved 343.98 mg/g.

2.2.3 Metal Oxide-Based Nanocomposites for Dye Removal

Metal oxide-based nanomaterials such as TiO_2, SiO_2, iron oxide, zinc oxide (ZnO), gallium oxide (Ga_2O_3), nickel oxide (NiO), copper oxide (CuO) [75] having exceptional functionality in their probable application to the detoxification of industrial effluents, groundwater, surface water and drinking water. However, limitations due to lose their selectivity, mechanical strength, and high capacity [70] were cured by immobilization these metal oxides into polymeric materials and designing core shell nanocomposites [38]. Mesoporous silica-gibbsite nanocomposites (NSG)

($62.34\ m^2 \cdot g^{-1}$, a pore radius of 22.717 nm, and a pore volume of $0.7081\ cm^3 \cdot g^{-1}$) recorded an efficient capacity to adsorb Eriochrome Black T dye (EBT) through an exothermic process (Inas A [8]. Magnetic nanocomposite based on cobalt ferrite-silica ($CoFe_2O_4eSiO_2$) recorded adsorbtion capacity of $75.5 \pm 1.21\ mg\ g^{-1}$ for the removal of MG dye from water. Interestinly the adsorbent was regenerated and used repeatedly and separated from water by applying a magnetic field [13].

Iron oxide (Fe_3O_4) nanoparticles have been used in designing nanocomposites due to their excellent properties such as high adsorption, superparamagnetic behavior, good compatibility, low toxicity, high surface energy, and large surface area [37, 102, 103]. The magnetic nanocomposite based on carbon/ iron oxide was applied for removal RhB from aqueous solution with maximum adsorption efficiency of 46.94 mg/g[95]. A shell of Fe_3O_4 MNP coated by a functionalized amorphous silica extracted from rice husk showed high efficiency for the removal of methyl red (MR) dye (97%) due to the presence of hydrophilic functional groups on the surface [44]. In another report, a magnetic Fe@MgO nanocomposites presented admirable adsorption properties towards methyl orange dye (MO) with the maximum adsorption capacities of $6947.9\ mg \bullet g^{-1}$ through chemical adsorption via complexing with MgO, associated with a slight mineralization of MO [26]. Consequently, magnetic nanoparticles dispersed within a matrix of activated carbon (MNC) recorded the maximum adsorption capacities of $223.82\ mg\ g^{-1}$ for Nylosan Blue (NB), $114.68\ mg\ g^{-1}$ for Chromazurol (ChS), and $286.91\ mg\ g^{-1}$ for Basic Red 2 (BR2). The adsorption of three dyes on MNC was favorable, spontaneous, and endothermic with good efficiency (greater than 65%) afterward seven adsorption—desorption cycles [65]. The efficacy of a non magnetic carbon was compared with magnetic carbon nanocomposite for adsorption of CV dye from aquous solution. The adsorption capacity was improved from 40 mg/g to 81.7 upon using magnetic nanocomposite. The maximum adsorption was recorded to be 113.31 mg/g under the optimum conditions (concentration 240 mg/l; temperature 50 °C; pH 8.50; dose 1 g/l), which was very close to the experimental value (111.80 mg/g) [96].

A nano-sized $Fe_3O_4@SiO_2$ core (15 nm) coated with a shell of Zr-MOFs (about 5 nm) recorded high adsorbtion capacity for organic dyes [39] and remained unchanged after 6 recycles. In another report, zeolitic imidazolate framework-67 (ZIF-67) ($1403.7\ m^2/g$) grown on the surface of Fe_3O_4@ egg shell membrane ($1263.9\ m^2/g$) recorded a maximum adsorption capacity of 250.81 mg/g for BR18 (Basic Red 18) following pseudo-second order [59]. Consequently, a core shell nanocomposite designed from $Zn_{0.5}Co_{0.5}\ Al_{0.5}Fe_{1.46}La_{0.04}O_4$/PVP succeeded in purifying inked-water and dyes with higher efficiency of 90% in comparsion with pure ferrite (76%) (MA [7]. Trisodium citrate based magnetite nanocomposite (Fe_3O_4 –TSC) was applied for MG dye removal from aqueous medium [11]. The maximum adsorption capacity (qe) was calculated to be $435\ mg\ g^{-1}$ following pseudo-second-order kinetic model with greatest recovery of MG dye in 0.1 M HCl. In another publication, magnetic Fe_3O_4/zeolite NaA nanocomposite exhibited [100] a high adsorption capacity (~$40.36\ mg \cdot g^{-1}$) and removal efficiency (~96.8%) compared with the zeolite NaA (~$32.99\ mg \cdot g^{-1}$ and 79.11%, respectively) due to the interaction between active sites on the surfaces and edges of the invert spinel ferrite Fe_3O_4 nanoparticles and zeolite NaA with MB molecules.

It is wotarth to mention that multifunctional magnetite-polypyrrole/chitosan /grapheneoxide exhibited high adsorption efficiency of P4R dye contaminated water with easily magnetic separation. Moreover, it has been shown to be effective against antimicrobials (E. coli) and antifungals (Fusarium fungi). The adsorption kinetics of P4R dye follows the Langmuir isotherm equation and the pseudo- second-order kinetic model and the adsorption process was an endothermic (Nehal A [85]. In another report, modification of an Fe_3O_4/SiO_2 magnetic composite with polypyrrole (PPy) (PPy/Fe_3O_4/SiO_2) improved the maximum adsorption of (361 mg/g) CR dye with high ability to reuse the nanocomposite several times leading to reduction the overall cost of the treatment [112].

A bionanocomposite of sodium alginate, chitosan (MZ/CS/AL), infused with natural zeolite, and cross-linked with glutaraldehyde and $CaCl_2$ was used to remove methylene blue from aqueous solution [46] with high efficiency. Carboxymethyl cellulose (CMC) coated $Fe_3O_4@SiO_2$ core–shell magnetic nanoparticles (MNPs) exhibited an efficient adsorbent for the removal of cationic dye, due to the abundant adsorption site, easy separation using an external magnetic field and low cost [114].

Nano-hydroxyapatite polymeric hydrogels (Varaprasad et al., 2018) developed using waste shells (Clam and Magellan shell) carboxymethyl cellulose/acrylamide were investigated as a sorbent for Acid Blue 113 (AB) from aqueous AB solution. The diffusion coefficient value was increased with the increasing nano-hydroxyapatite content in the CMC-AM/nHA-CS (0.22353–0.27681 $cm^2\ s^{-1}$) and CMC-AM/nHA-MS (0.22377–0.29737 $cm^2\ s^{-1}$) hydrogels. Combination of conducting polymers e.g. polypyrrole with LDHs improved both the adsorption of cationic dyes as well as anionic dyes through modification of surface [61, 107].

References

1. Abdelwahab M, Agag T, Akelah A, Takeichi T (2012) Synthesis and characterization of styrene modified vinylester resin-clay nanocomposites. Polym Eng Sci 52(1):125–132. https://doi.org/10.1002/pen.22054
2. Abdi J, Vossoughi M, Mahmoodi NM, Alemzadeh I (2017) Synthesis of metal-organic framework hybrid nanocomposites based on GO and CNT with high adsorption capacity for dye removal. Chem Eng J 326:1145–1158
3. Abramian L, El-Rassy H (2009) Adsorption kinetics and thermodynamics of azo-dye Orange II onto highly porous titania aerogel. Chem Eng J 150(2–3):403–410
4. Adel M, Ahmed MA, Mohamed AA (2021) A facile and rapid removal of cationic dyes using hierarchically porous reduced graphene oxide decorated with manganese ferrite. FlatChem 26:100233. https://doi.org/10.1016/j.flatc.2021.100233
5. Adelodun B, Ajibade FO, Abdulkadir TS, Bakare HO, Choi KS (2020) SWOT analysis of agro-waste based adsorbents for persistent dye pollutants removal from wastewaters. In: Kumar V, Singh J, Kumar P (eds) Environmental degradation: causes and remediation strategies, pp 88–103

6. Agag T, Akelah A, Abdelwahab M, Takeichi T, Muto H (2010) Vinylester resin-clay hybrids using various intercalating agents. J Appl Polym Sci 115(4):2060–2068. https://doi.org/10.1002/app.31293
7. Ahmed M, Khafagy RM, Bishay ST, Saleh N (2013) Effective dye removal and water purification using the electric and magnetic Zn0. 5Co0. 5Al0. 5Fe1. 46La0. 04O4/polymer core–shell nanocomposites. J Alloy Compd 578:121–131
8. Ahmed IA, Al-Radadi NS, Hussein H, Ragab AH (2019) Environmentally friendly mesoporous nanocomposite prepared from al-dross waste with remarkable adsorption ability for toxic anionic dye. J Chem
9. Albokheet W, Gouda M, Alfaiyz Y (2021) Removal of Methylene blue dye using copper oxide/carboxymethyl cellulose nanocomposite: kinetic, equilibrium and thermodynamic studies. J Textile Sci & Fashion Technol. 8(2):1–10
10. Ali LIA, Ismail HK, Alesary HF, Aboul-Enein HY (2021) A nanocomposite based on polyaniline, nickel and manganese oxides for dye removal from aqueous solutions. Int J Environ Sci Techno. 18(7):2031–5
11. Alqadami AA, Naushad M, Abdalla MA, Khan MR, Alothman ZA (2016) Adsorptive removal of toxic dye using Fe3O4–TSC nanocomposite: equilibrium, kinetic, and thermodynamic studies. J Chem Eng Data 61(11):3806–3813
12. Alzahrani FM, Alsaiari NS, Katubi KM, Amari A, Ben Rebah F, Tahoon MA (2021) Synthesis of Polymer-Based Magnetic Nanocomposite for Multi-Pollutants Removal from Water. 13(11):1742
13. Amiri M, Salavati-Niasari M, Akbari A, Gholami T (2017) Removal of malachite green (a toxic dye) from water by cobalt ferrite silica magnetic nanocomposite: herbal and green sol-gel autocombustion synthesis. Int J Hydrogen Energy 42(39):24846–24860
14. Anderson K, Ryan B, Sonntag W, Kavvada A, Friedl L (2017) Earth observation in service of the 2030 agenda for sustainable development. Geo-Spatial Inf Sci 20(2):77–96
15. Benkhaya S, M' rabet, S., & El Harfi, A. (2020) A review on classifications, recent synthesis and applications of textile dyes. Inorg Chem Commun 115:107891. https://doi.org/10.1016/j.inoche.2020.107891
16. Bulut Y, Akçay G, Elma D, Serhatlı IE (2009) Synthesis of clay-based superabsorbent composite and its sorption capability. J Hazard Mater 171(1):717–723. https://doi.org/10.1016/j.jhazmat.2009.06.067
17. Crini G (2006) Non-conventional low-cost adsorbents for dye removal: a review. Biores Technol 97(9):1061–1085. https://doi.org/10.1016/j.biortech.2005.05.001
18. Dadashi Firouzjaei M, Akbari Afkhami F, Rabbani Esfahani M, Turner CH, Nejati S (2020) Experimental and molecular dynamics study on dye removal from water by a graphene oxide-copper-metal organic framework nanocomposite. J Water Process Eng 34:101180. https://doi.org/10.1016/j.jwpe.2020.101180
19. Das SK, Bhowal J, Das AR, Guha AK (2006) Adsorption behavior of rhodamine B on rhizopus o ryzae biomass. Langmuir 22(17):7265–7272
20. De Luca P, Chiodo A, Macario A, Siciliano C, B Nagy J, Van der Bruggen B (2021). Semi-continuous adsorption processes with multi-walled carbon nanotubes for the treatment of water contaminated by an organic textile dye. Appl Sci 11(4): 2076–3417
21. Delpiano GR, Tocco D, Medda L, Magner E, Salis A (2021) Adsorption of malachite green and alizarin red s dyes using fe-btc metal organic framework as adsorbent. Int J Mol Sci 22(2):788
22. Duan B, Zhu Z, Sun C, Zhou J, Walsh A (2020) Preparing copper catalyst by ultrasound-assisted chemical precipitation method. Ultrason Sonochem 64:105013
23. Dutta S, Gupta B, Srivastava SK, Gupta AK (2021) Recent advances on the removal of dyes from wastewater using various adsorbents: a critical review (https://doi.org/10.1039/D1MA00354B). Mater Adv 2(14):4497–4531. https://doi.org/10.1039/D1MA00354B
24. El-Ghazali S, Khatri M, Mehdi M, Kharaghani D, Tamada Y, Katagiri A et al (2021) Fabrication of poly (Ethylene-glycol 1, 4-Cyclohexane Dimethylene-Isosorbide-Terephthalate) electrospun nanofiber mats for potential infiltration of fibroblast cells. Polymers 13(8):1245

25. Fan L, Luo C, Sun M, Li X, Lu F, Qiu H (2012) Preparation of novel magnetic chitosan/graphene oxide composite as effective adsorbents toward methylene blue. Biores Technol 114:703–706
26. Ge L, Wang W, Peng Z, Tan F, Wang X, Chen J et al (2018) Facile fabrication of Fe@ MgO magnetic nanocomposites for efficient removal of heavy metal ion and dye from water. Powder Technol 326:393–401
27. Ghorai S, Sarkar AK, Panda AB, Pal S (2013) Effective removal of Congo red dye from aqueous solution using modified xanthan gum/silica hybrid nanocomposite as adsorbent. Bioresour Technol. 144:485–91
28. Gil A, Assis FCC, Albeniz S, Korili SA (2011) Removal of dyes from wastewaters by adsorption on pillared clays. Chem Eng J 168(3):1032–1040. https://doi.org/10.1016/j.cej.2011.01.078
29. Gleick PH (2000) A look at twenty-first century water resources development. Water Int 25(1):127–138. https://doi.org/10.1080/02508060008686804
30. Gomes RF, de Azevedo AC, Pereira AG, Muniz EC, Fajardo AR, Rodrigues FH (2015) Fast dye removal from water by starch-based nanocomposites. J Colloid Interface Sci. 454:200–9
31. Gupta VK, Pathania D, Kothiyal NC, Sharma G (2014) Polyaniline zirconium (IV) silicophosphate nanocomposite for remediation of methylene blue dye from waste water. J Mol Liq. 190:139–45
32. Hai FI, Yamamoto K, Fukushi K (2007) Hybrid treatment systems for dye wastewater. Crit Rev Environ Sci Technol 37(4):315–377
33. Hassan H, Salama A, El-ziaty AK, El-Sakhawy M (2019) New chitosan/silica/zinc oxide nanocomposite as adsorbent for dye removal. Int J Biol Macromol 131:520–526. https://doi.org/10.1016/j.ijbiomac.2019.03.087
34. Hayati B, Mahmoodi NM, Maleki A (2015) Dendrimer–titania nanocomposite: synthesis and dye-removal capacity. Res Chem Intermed. 41(6):3743–57
35. Hosseini J, Zare EN, Ajloo D (2019) Experimental and theoretical calculation investigation on effective adsorption of lead (II) onto poly (aniline-co-pyrrole) nanospheres. J Mol Liq 296:111789
36. Hosseinzadeh H, Zoroufi S, Mahdavinia GR (2015) Study on adsorption of cationic dye on novel kappa-carrageenan/poly(vinyl alcohol)/montmorillonite nanocomposite hydrogels. Polym Bull 72(6):1339–1363. https://doi.org/10.1007/s00289-015-1340-5
37. Hu JS, Zhong LS, Song WG, Wan LJ (2008) Synthesis of hierarchically structured metal oxides and their application in heavy metal ion removal. Adv Mater 20(15):2977–2982
38. Hua M, Zhang S, Pan B, Zhang W, Lv L, Zhang Q (2012) Heavy metal removal from water/wastewater by nanosized metal oxides: a review. J Hazard Mater 211:317–331
39. Huang L, He M, Chen B, Hu B (2018) Magnetic Zr-MOFs nanocomposites for rapid removal of heavy metal ions and dyes from water. Chemosphere 199:435–444
40. Huang L, He M, Chen B, Hu B (2018) Magnetic Zr-MOFs nanocomposites for rapid removal of heavy metal ions and dyes from water. Chemosphere 199:435–444
41. Janaki V, Vijayaraghavan K, Oh B-T, Lee K-J, Muthuchelian K, Ramasamy AK et al (2012) Starch/polyaniline nanocomposite for enhanced removal of reactive dyes from synthetic effluent. Carbohydr Polym. 90(4):1437–44
42. Jinendra U, Bilehal D, Nagabhushana BM, Kumar AP (2021) Adsorptive removal of Rhodamine B dye from aqueous solution by using graphene–based nickel nanocomposite. Heliyon 7(4):e06851. https://doi.org/10.1016/j.heliyon.2021.e06851
43. Kafil M, Nasab B, Moazed H, Jokiniemi J, Laehde A, Bhatnagar A (2019) Efficient removal of azo dyes from water with chitosan/carbon nanoflower as a novel nanocomposite synthesized by pyrolysis technique. Desalin Water Treat 142:308–320
44. Kamari S, Shahbazi A (2020) Biocompatible Fe3O4@ SiO2-NH2 nanocomposite as a green nanofiller embedded in PES–nanofiltration membrane matrix for salts, heavy metal ion and dye removal: Long–term operation and reusability tests. Chemosphere 243:125282
45. Katubi KM, Alsaiari NS, Alzahrani FM, M Siddeeg S, A Tahoon M (2021) Synthesis of Manganese ferrite/graphene oxide magnetic nanocomposite for pollutants removal from water. Processes 9(4):589. https://www.mdpi.com/2227-9717/9/4/589

46. Kazemi J, Javanbakht V (2020) Alginate beads impregnated with magnetic Chitosan@ Zeolite nanocomposite for cationic methylene blue dye removal from aqueous solution. Int J Biol Macromol 154:1426–1437
47. Kaşgöz H, Durmus A (2008) Dye removal by a novel hydrogel-clay nanocomposite with enhanced swelling properties. Polym Adv Technol 19(7):838–845. https://doi.org/10.1002/pat.1045
48. Kittinaovarat S, Kansomwan P, Jiratumnukul N (2010) Chitosan/modified montmorillonite beads and adsorption Reactive Red 120. Appl Clay Sci 48(1):87–91. https://doi.org/10.1016/j.clay.2009.12.017
49. Kong D, Liu Y, Li Y, Chi W, Huang Q, Yu C (2018) Facile preparation and dye removal properties of Fe3O4@carbon nanocomposite. 13(2):219–22
50. Kumar S, Nair RR, Pillai PB, Gupta SN, Iyengar MAR, Sood AK (2014) Graphene oxide–MnFe2O4 magnetic nanohybrids for efficient removal of lead and arsenic from water. ACS Appl Mater Interfaces 6(20):17426–17436. https://doi.org/10.1021/am504826q
51. Kumar M, Tamilarasan R (2013) Modeling studies: adsorption of aniline blue by using Prosopis Juliflora carbon/Ca/alginate polymer composite beads. Carbohyd Polym 92(2):2171–2180
52. Kumar Sharma A, Priya, Singh Kaith B, Bajaj S, Bhatia JK, Panchal S et al (2019b) Efficient capture of eosin yellow and crystal violet with high performance xanthan-acacia hybrid super-adsorbent optimized using response surface methodology. Colloids and Surfaces B: Biointerfaces. 175:314–23
53. Lawchoochaisakul S, Monvisade P, Siriphannon P (2021) Cationic starch intercalated montmorillonite nanocomposites as natural based adsorbent for dye removal. Carbohyd Polym 253:117230. https://doi.org/10.1016/j.carbpol.2020.117230
54. Li A, Wang A (2005) Synthesis and properties of clay-based superabsorbent composite. Eur Polymer J 41(7):1630–1637. https://doi.org/10.1016/j.eurpolymj.2005.01.028
55. Liu Q, Deng W, Wang Q, Lin X, Gong L, Liu C et al (2020) An efficient chemical precipitation route to fabricate 3D flower-like CuO and 2D leaf-like CuO for degradation of methylene blue. Adv Powder Technol 31(4):1391–1401
56. Liu C, Omer AM, Ouyang X-K (2018) Adsorptive removal of cationic methylene blue dye using carboxymethyl cellulose/k-carrageenan/activated montmorillonite composite beads: Isotherm and kinetic studies. Int J Biol Macromol 106:823–833. https://doi.org/10.1016/j.ijbiomac.2017.08.084
57. Liu N, Zhang Q, Qu R, Zhang W, Li H, Wei Y et al (2017) Nanocomposite Deposited Membrane for Oil-in-Water Emulsion Separation with in Situ Removal of Anionic Dyes and Surfactants. Langmuir. 33(30):7380–8
58. Mahdavinia GR, Massoumi B, Jalili K, Kiani G (2012) Effect of sodium montmorillonite nanoclay on the water absorbency and cationic dye removal of carrageenan-based nanocomposite superabsorbents. J Polym Res 19(9):9947. https://doi.org/10.1007/s10965-012-9947-9
59. Mahmoodi NM, Taghizadeh M, Taghizadeh A, Abdi J, Hayati B, Shekarchi AA (2019) Bio-based magnetic metal-organic framework nanocomposite: Ultrasound-assisted synthesis and pollutant (heavy metal and dye) removal from aqueous media. Appl Surf Sci 480:288–299
60. Mandal MK, Sharma M, Pandey S, Dubey KK (2019) Membrane technologies for the treatment of pharmaceutical industry wastewater. In: Water and wastewater treatment technologies. Springer, Berlin, pp. 103–116
61. Mohamed F, Abukhadra MR, Shaban M (2018) Removal of safranin dye from water using polypyrrole nanofiber/Zn-Fe layered double hydroxide nanocomposite (Ppy NF/Zn-Fe LDH) of enhanced adsorption and photocatalytic properties. Sci Total Environ 640:352–363
62. Mohammed MI, Baytak S (2016) Synthesis of Bentonite–Carbon Nanotube Nanocomposite and Its Adsorption of Rhodamine Dye From Water. Arabian J Sci Eng. 41(12):4775–85
63. Mostafa MH, Elsawy MA, Darwish MSA, Hussein LI, Abdaleem AH (2020) Microwave-Assisted preparation of Chitosan/ZnO nanocomposite and its application in dye removal. Mater Chem Phys. 248:122914

64. Mousavi SR, Asghari M, Mahmoodi NM (2020) Chitosan-wrapped multiwalled carbon nanotube as filler within PEBA thin film nanocomposite (TFN) membrane to improve dye removal. Carbohydr Polym. 237:116128
65. Muntean SG, Nistor MA, Muntean E, Todea A, Ianoş R, Păcurariu C (2018) Removal of colored organic pollutants from wastewaters by magnetite/carbon nanocomposites: single and binary systems. J Chem
66. Nakkeeran E, Varjani SJ, Dixit V, Kalaiselvi A (2018) Synthesis, characterization and application of zinc oxide nanocomposite for dye removal from textile industrial Waste water. Indian J Exp Biol. 56:498–503
67. Nyankson E, Adjasoo J, Efavi JK, Amedalor R, Yaya A, Manu GP et al (2019) Characterization and Evaluation of Zeolite A/Fe3O4 Nanocomposite as a Potential Adsorbent for Removal of Organic Molecules from Wastewater. J Chem. 8090756.
68. Oyewo OA, Elemike EE, Onwudiwe DC, Onyango MS (2020) Metal oxide-cellulose nanocomposites for the removal of toxic metals and dyes from wastewater. Int J Biol Macromole
69. Pal S, Patra AS, Ghorai S, Sarkar AK, Mahato V, Sarkar S et al (2015) Efficient and rapid adsorption characteristics of templating modified guar gum and silica nanocomposite toward removal of toxic reactive blue and Congo red dyes. Bioresour Technol. 191:291–9
70. Palmisano G, García-López E, Marcì G, Loddo V, Yurdakal S, Augugliaro V et al (2010) Advances in selective conversions by heterogeneous photocatalysis. Chem Commun 46(38):7074–7089
71. Pavithra KG, P, S. K., V, J., & P, S. R. (2019) Removal of colorants from wastewater: a review on sources and treatment strategies. J Ind Eng Chem 75:1–19. https://doi.org/10.1016/j.jiec.2019.02.011
72. Peng N, Hu D, Zeng J, Li Y, Liang L, Chang C (2016) Superabsorbent cellulose–clay nanocomposite hydrogels for highly efficient removal of dye in water. ACS Sustain Chem Eng 4(12):7217–7224
73. Phan D-N, Lee H, Choi D, Kang C-Y, Im SS, Kim IS (2018) Fabrication of two polyester nanofiber types containing the biobased monomer isosorbide: poly (ethylene glycol 1, 4-cyclohexane dimethylene isosorbide terephthalate) and poly (1, 4-cyclohexane dimethylene isosorbide terephthalate). Nanomaterials 8(2):56
74. Pourjavadi A, Nazari M, Hosseini SH (2015) Synthesis of magnetic graphene oxide-containing nanocomposite hydrogels for adsorption of crystal violet from aqueous solution. RSC Advances. 5(41):32263–71
75. Prasanna SS, Balaji K, Pandey S, Rana S (2019) Metal oxide based nanomaterials and their polymer nanocomposites. In: Nanomaterials and polymer nanocomposites. Elsevier, pp 123–144
76. Priya, Sharma AK, Kaith BS, Tanwar V, Bhatia JK, Sharma N et al (2019) RSM-CCD optimized sodium alginate/gelatin based ZnS-nanocomposite hydrogel for the effective removal of biebrich scarlet and crystal violet dyes. Int J Biol Macromol. 129:214–26
77. Puri C, Sumana G (2018) Highly effective adsorption of crystal violet dye from contaminated water using graphene oxide intercalated montmorillonite nanocomposite. Appl Clay Sci 166:102–112. https://doi.org/10.1016/j.clay.2018.09.012
78. Rafatullah M, Sulaiman O, Hashim R, Ahmad A (2010) Adsorption of methylene blue on low-cost adsorbents: a review. J Hazard Mater 177(1):70–80. https://doi.org/10.1016/j.jhazmat.2009.12.047
79. Rafieian F, Jonoobi M, Yu Q (2019) A novel nanocomposite membrane containing modified cellulose nanocrystals for copper ion removal and dye adsorption from water. Cellulose. 26(5):3359–73
80. Reddy DHK, Lee S-M (2013) Application of magnetic chitosan composites for the removal of toxic metal and dyes from aqueous solutions. Adv Coll Interface Sci 201:68–93
81. Rezaee R, Nasseri S, Mahvi AH, Nabizadeh R, Mousavi SA, Rashidi A et al (2015) Fabrication and characterization of a polysulfone-graphene oxide nanocomposite membrane for arsenate rejection from water. J Environ Health Sci Eng 13(1):1–11

82. Rezanejade Bardajee G, Sadat Hosseini S, Vancaeyzeele C (2019) Graphene oxide nanocomposite hydrogel based on poly(acrylic acid) grafted onto salep: an adsorbent for the removal of noxious dyes from water (https://doi.org/10.1039/C8NJ05800H). New J Chem 43(8):3572–3582. https://doi.org/10.1039/C8NJ05800H
83. Sadeghfar F, Ghaedi M, Asfaram A, Jannesar R, Javadian H, Pezeshkpour V (2018) Polyvinyl alcohol/Fe3O4@carbon nanotubes nanocomposite: Electrochemical-assisted synthesis, physicochemical characterization, optical properties, cytotoxicity effects and ultrasound-assisted treatment of aqueous based organic compound. J Ind Eng Chem 65:349–362. https://doi.org/10.1016/j.jiec.2018.05.006
84. Sadighian S, Abbasi M, Arjmandi SA, Karami H (2018) Dye removal from water by zinc ferrite-graphene oxide nanocomposite. Progress Color Colorants Coat 11(2):85–92. https://doi.org/10.30509/pccc.2018.75743
85. Salahuddin NA, EL-Daly, H. A., El Sharkawy, R. G., & Nasr, B. T. (2020) Nano-hybrid based on polypyrrole/chitosan/grapheneoxide magnetite decoration for dual function in water remediation and its application to form fashionable colored product. Adv Powder Technol 31(4):1587–1596
86. Salahuddin N, EL-Daly, H., El Sharkawy, R. G., & Nasr, B. T. (2018) Synthesis and efficacy of PPy/CS/GO nanocomposites for adsorption of ponceau 4R dye. Polymer 146:291–303
87. Salahuddin NA, EL-Daly HA, El Sharkawy RG, Nasr BT (2019) Adsorption of dyes from aqueous solutions onto multi-functional PPy/cs exfoliated nanohybrid for fashionable layered polymer nanocomposites. In: Euro-mediterranean conference for environmental integration. Springer, pp 479–486
88. Salama A, Shoueir KR, Aljohani HA (2017) Preparation of sustainable nanocomposite as new adsorbent for dyes removal. Fibers Polym. 18(9):1825–30
89. Shaban M, Abukhadra MR (2017) Geochemical evaluation and environmental application of Yemeni natural zeolite as sorbent for Cd 2+ from solution: kinetic modeling, equilibrium studies, and statistical optimization. Environ Earth Sci 76(8):310
90. Shaban M, Abukhadra MR, Shahien MG, Khan AAP (2017) Upgraded modified forms of bituminous coal for the removal of safranin-T dye from aqueous solution. Environ Sci Pollut Res 24(22):18135–18151
91. Shao D, Hu J, Chen C, Sheng G, Ren X, Wang X (2010) Polyaniline multiwalled carbon nanotube magnetic composite prepared by plasma-induced graft technique and its application for removal of aniline and phenol. J Phys Chem C 114(49):21524–21530
92. Sharma AK, Priya, Kaith BS, Panchal S, Bhatia JK, Bajaj S et al (2019c) Response surface methodology directed synthesis of luminescent nanocomposite hydrogel for trapping anionic dyes. J Environ Manage. 231:380–90
93. Sharma AK, Priya, Kaith BS, Sharma N, Bhatia JK, Tanwar V et al (2019) Selective removal of cationic dyes using response surface methodology optimized gum acacia-sodium alginate blended superadsorbent. Int J Biol Macromol. 124:331–45
94. Sharma M, Poddar M, Gupta Y, Nigam S, Avasthi DK, Adelung R et al (2020) Solar light assisted degradation of dyes and adsorption of heavy metal ions from water by CuO–ZnO tetrapodal hybrid nanocomposite. Materials Today Chemistry. 17:100336
95. Singh KP, Gupta S, Singh AK, Sinha S (2010) Experimental design and response surface modeling for optimization of Rhodamine B removal from water by magnetic nanocomposite. Chem Eng J 165(1):151–160
96. Singh KP, Gupta S, Singh AK, Sinha S (2011) Optimizing adsorption of crystal violet dye from water by magnetic nanocomposite using response surface modeling approach. J Hazard Mater 186(2–3):1462–1473
97. Sunitha S, Rao AN, Karthikeyan J (2015) Synthesis of novel cobalt doped zinc oxide/carbon nano composite for the photocatalytic degradation of acid blue 113. Orient J Chem 31:107–112
98. Thy LTM, Linh NTC, Tram NTT, Tu TH, Tai LT, Khang PT et al (2021) Fabrication and response surface methodology for the adsorption of nickel ferrite-graphene oxide nanocomposite for the removal of methylene blue from water. J Nanomater 2021:4636531. https://doi.org/10.1155/2021/4636531

99. Thy LTM, Linh NTC, Tram NTT, Tu TH, Tai LT, Khang PT et al (2021) Fabrication and Response Surface Methodology for the Adsorption of Nickel Ferrite-Graphene Oxide Nanocomposite for the Removal of Methylene Blue from Water. Journal of Nanomaterials. 2021:4636531
100. Tran NBT, Duong NB, Le NL (2021) Synthesis and characterization of magnetic Fe3O4/zeolite NaA nanocomposite for the adsorption removal of methylene blue potential in wastewater treatment. J Chem
101. Wang C, Feng C, Gao Y, Ma X, Wu Q, Wang Z (2011) Preparation of a graphene-based magnetic nanocomposite for the removal of an organic dye from aqueous solution. Chem Eng J 173(1):92–97. https://doi.org/10.1016/j.cej.2011.07.041
102. Wang X-Y, Jiang X-P, Li Y, Zeng S, Zhang Y-W (2015) Preparation Fe3O4@ chitosan magnetic particles for covalent immobilization of lipase from thermomyces lanuginosus. Int J Biol Macromol 75:44–50
103. Wang P, Shi Q, Shi Y, Clark KK, Stucky GD, Keller AA (2009) Magnetic permanently confined micelle arrays for treating hydrophobic organic compound contamination. J Am Chem Soc 131(1):182–188
104. Wang S, Wei J, Lv S, Guo Z, Jiang F (2013a) Removal of organic dyes in environmental water onto magnetic-sulfonic graphene nanocomposite. CLEAN—Soil Air Water 41(10):992–1001. https://doi.org/10.1002/clen.201200460
105. Wang S, Wei J, Lv S, Guo Z, Jiang F (2013b) Removal of organic dyes in environmental water onto magnetic-sulfonic graphene nanocomposite. CLEAN–Soil Air Water 41(10):992–1001
106. Watt E, Picard M, Maldonado B, Abdelwahab MA, Mielewski DF, Drzal LT et al (2021) Ocean plastics: environmental implications and potential routes for mitigation—a perspective (https://doi.org/10.1039/D1RA00353D). RSC Adv 11(35):21447–21462. https://doi.org/10.1039/D1RA00353D
107. Xin Q, Fu J, Chen Z, Liu S, Yan Y, Zhang J et al (2015) Polypyrrole nanofibers as a high-efficient adsorbent for the removal of methyl orange from aqueous solution. J Environ Chem Eng 3(3):1637–1647
108. Yang G, Zhang D, Zhu G, Zhou T, Song M, Qu L et al (2020) A Sm-MOF/GO nanocomposite membrane for efficient organic dye removal from wastewater. RSC Advances. 10(14):8540–7
109. Yang N, Zhu S, Zhang D, Xu S (2008) Synthesis and properties of magnetic Fe3O4-activated carbon nanocomposite particles for dye removal. Materials Letters. 62(4):645–7
110. Yaqoob AA, Mohd Noor NHB, Serrà A, Mohamad Ibrahim MN (2020) Advances and challenges in developing efficient graphene oxide-based ZnO photocatalysts for dye photo-oxidation. Nanomaterials 10(5):932. https://www.mdpi.com/2079-4991/10/5/932
111. Zaccheus S, Lamayi DW (2019) 3 TV. Montmorillonite-silica nanocomposite for dye removal from solution. Adv Appl Sci Techno. 3:1–12
112. Zhao G, Li J, Wang X (2011) Kinetic and thermodynamic study of 1-naphthol adsorption from aqueous solution to sulfonated graphene nanosheets. Chem Eng J 173(1):185–190. https://doi.org/10.1016/j.cej.2011.07.072
113. Zhao X, Zhang B, Ai K, Zhang G, Cao L, Liu X et al (2009) Monitoring catalytic degradation of dye molecules on silver-coated ZnO nanowire arrays by surface-enhanced Raman spectroscopy. J Mater Chem 19(31):5547–5553
114. Zirak M, Abdollahiyan A, Eftekhari-Sis B, Saraei M (2018) Carboxymethyl cellulose coated Fe 3 O 4@ SiO 2 core–shell magnetic nanoparticles for methylene blue removal: equilibrium, kinetic, and thermodynamic studies. Cellulose 25(1):503–515

Nanostructured Composite Materials for Treatment of Dye Contaminated Water

Soumya Ranjan Mishra, Vishal Gadore, and Md. Ahmaruzzaman

Abstract It is well known that water is an essential component for sustaining life and most human activities. The problem of water pollution has been a constant threat worldwide. Several human and industrial activities release uncountable pollutants into the water sources, most toxic and even fatal to plants and animals. Among these pollutants, dyes are the common pollutants responsible for hindered grown of plants and cancer in human beings. As scientists are aware of these threats, they proposed various methods to remove dyes from wastewater. Several physical, chemical, and biological treatment methods are proposed to remove dyes. With the development of nanocomposites and their excellent capabilities for adsorption and photocatalytic degradation, they have been proposed as suitable materials for dye removal. Nanocomposites are composed of two or more materials; one acts as a base material in which other materials are incorporated. Today, many nanocomposites have been known to display outstanding efficiencies to remove dyes from wastewater, some of which are reviewed in this work. Herein, we present the recent advances in nanocomposite materials with high efficiencies for the removal of some toxic dyes from wastewater. The mechanism of adsorption and photocatalytic degradation is presented. Furthermore, the scope of future research is also discussed.

Keywords Dyes · Pollutants · Photocatalytic degradation · Nanocomposites

1 Introduction

Water is one of the essential components for sustaining life on Earth. Many of our necessary activities are incomplete without water. Water enriched with minerals is essential for human and plant growth. Unfortunately, the development of industries has imposed a severe threat to the environment. The quality of our existing water resources is depleting day by day due to industrialization and increasing population [1–3]. Therefore, in today's scenario, water pollution has emerged as a global threat to

S. Ranjan Mishra · V. Gadore · Md. Ahmaruzzaman (✉)
Department of Chemistry, National Institute of Technology, Silchar, Assam 788010, India
e-mail: mda2002@gmail.com

S. K. Swain (ed.), *Nanohybrid Materials for Water Purification*, Composites Science and Technology, https://doi.org/10.1007/978-981-19-2332-6_5

living beings affecting several commercial activities [4–6]. Industries are discharging hundreds of various inorganic and organic contaminants into water resources continuously [7]. Many of these pollutants are highly toxic and cause severe side effects, and few are lethal to humans, being carcinogenic [8–10]. These substances are hazardous and harmful to the environment.

Many organic compounds, including Dyes, pesticides, phenols, pharmaceuticals, hydrocarbons, fertilizers, etc., are also found in wastewaters [11, 12]. These are also highly toxic and potential environmental pollutants described in various literature [3, 13, 14]. Industrial dyes are the primary contaminants that are toxic and highly carcinogenic leading to environmental pollution. Dyes are highly soluble and resistant to biological degradation, photodegradation, and chemical oxidation. Several research papers are reporting the harmful effects of dyes on plant growth [15–18]. Research also show that excess dyes in the human body lead to various harmful diseases damaging the brain, kidneys, liver, and skin [19, 20]. Due to the presence of these pollutants, ground and surface water of many parts around the globe are not fit for drinking. As the population is increasing exponentially, there will be water scarcity soon. Therefore, wastewater treatment is necessary to restore water demands and protect the environment from water pollution.

Scientists have reported several methods for wastewater treatment like reverse osmosis, microfiltration, liquid–liquid extraction, membrane filtration, photodegradation, etc. [21–24]. Aerobic and anaerobic microbial degradation is also reported but is not very common because dyes are generally toxic to most microorganisms and have high operating costs [25]. Among various dye removal techniques, adsorption and photodegradation were reported to be the best methods due to simplicity of operation and high efficiency for removal of dyes and regeneration of nanocomposites. With the development of nanotechnology, various types of nanocomposites were examined to remove dyes from wastewater. Nanocomposites are generally the combination of two or more than two different nanomaterials incorporated within the matrix of other materials. The properties of nanocomposites vary according to the synthesis technique and the composition of the materials. Thanks to the research and development, it is now possible to develop the desired nanocomposite with specific application and high efficiency. This chapter presents the recent advances in nanocomposite materials to remove some common dyes found in wastewater. The classification and characteristics of dye-containing wastewater are presented. Furthermore, the mechanisms and kinetic study of adsorption and photodegradation and the scope of future research are also discussed.

2 Classification of Dyes

The dyes can broadly be classified into ionic and non-ionic based on their charge after water dissolution or on functional groups present and their structure. Ionic dyes can further be divided into cationic and anionic dyes, and non-ionic dyes are divided

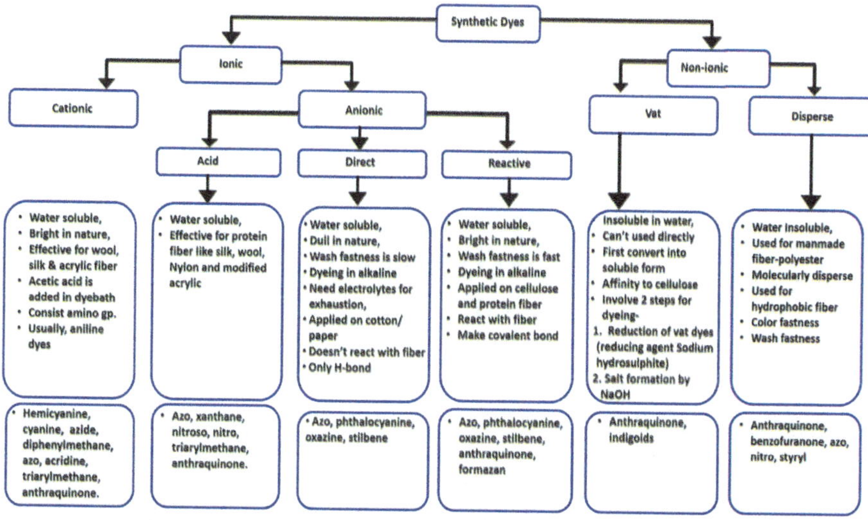

Fig. 1 Classification of dyes (Reproduced with permission from Springer [26])

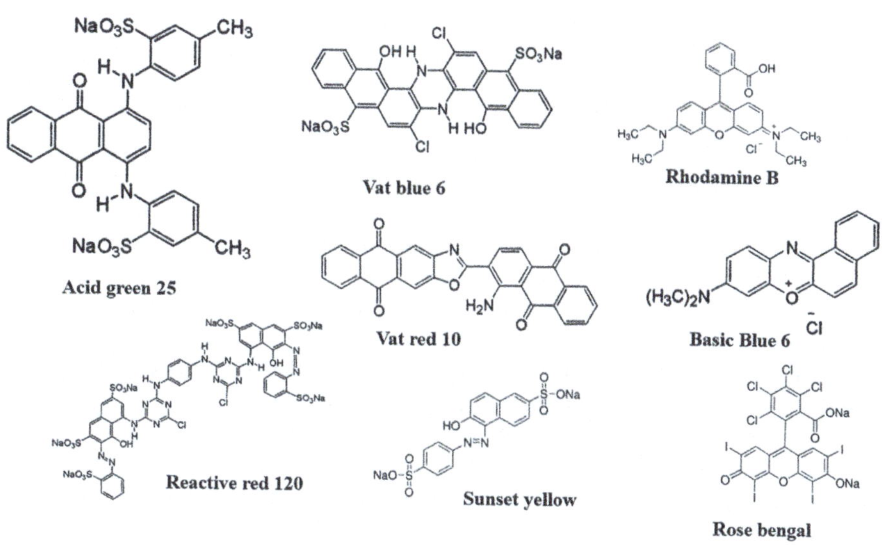

Fig. 2 Structures of some common toxic dyes found in industrial wastewater

into disperse dyes and Vat dyes (Fig. 1) [26]. Figure 2 illustrates the structure of some common dyes found in wastewater.

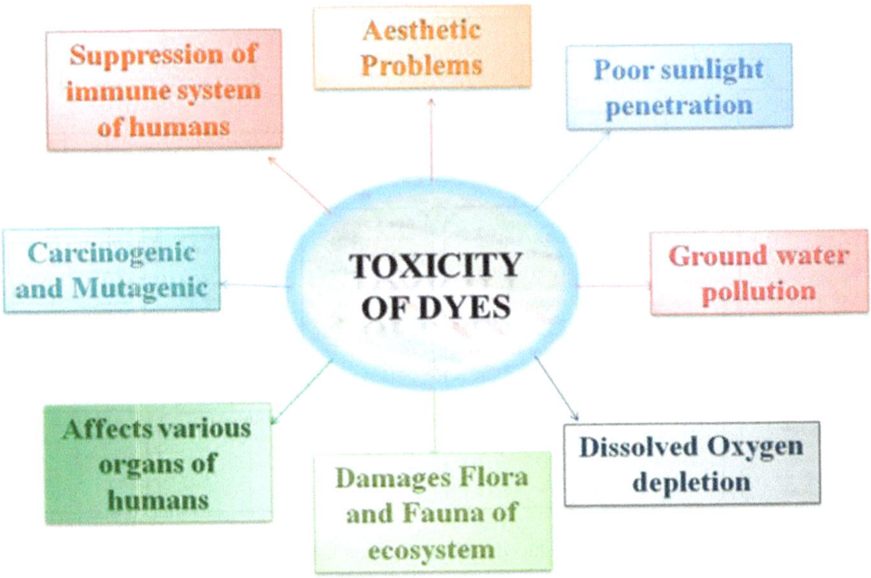

Fig. 3 Toxicity of dyes (Reproduced with permission from Elsevier [32])

3 Toxic Effects of Dyes

Dyes in water are easily detected because they are highly colored, and even a small concentration can produce intense colors [27]. The chromium-based dyes are reported to be the most toxic [28–31]. Dyes in water decrease the amount of sunlight reaching the aquatic plants and hinder their growth and development [30, 31]. Dye also has harmful effects on humans by affecting the liver, kidneys, brain, and central nervous system [32]. Figure 3 shows the toxicity of dyes [32].

Azo dyes and dyes containing anthraquinone are also very toxic and resistant to self-degradation [33]. These dyes are easily leached to the water system due to their high solubility. About 5 to 10% of the dyes get leached during the process of dying, causing environmental problems. Therefore, it is essential to remove dyes from wastewater and develop new nanocomposites with high removal efficiency.

4 Methods for Dye Removal

Various methods are reported for removing dyes, such as membrane separation, adsorption, photodegradation, and microbial degradation. Figure 4 illustrates the common methods for dye removal [32].

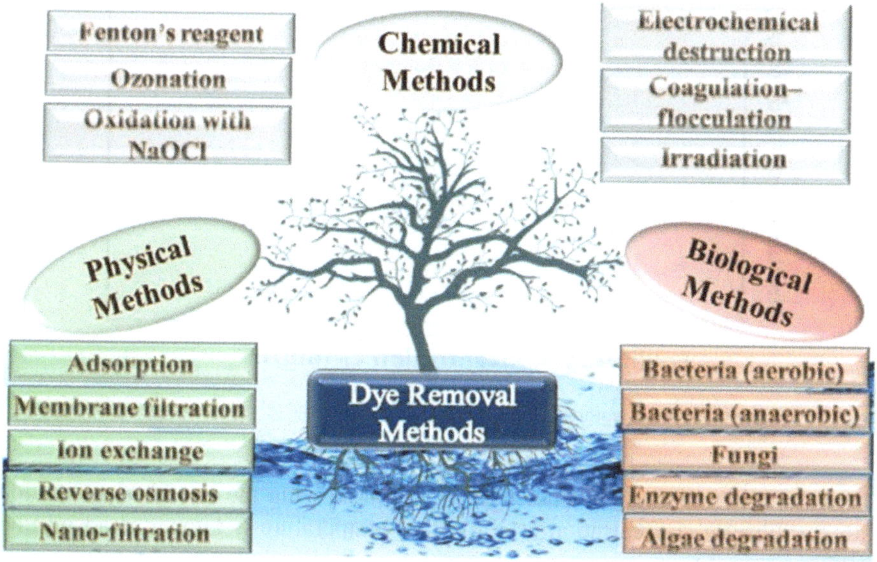

Fig. 4 Common methods for removing dyes from wastewater

4.1 Membrane Separation

Membrane separation is an advanced method for treating wastewater. The wastewater is passed through a porous semipermeable membrane. The solute particles bigger than the size of the pores are left behind, and sterile solution passes through the membrane. The solute particles are continuously removed from the membrane. The membrane separation technique includes ultrafiltration, microfiltration, nanofiltration, and reverse osmosis.

4.2 Adsorption

Adsorption is a surface phenomenon where solute particles are adsorbed on the surface of adsorbent material. Generally, physical forces help in adsorption, but chemical interactions between adsorbate and adsorbent also result in adsorption [34]. It is one of the most commonly applied techniques for removing dyes because dyes are readily adsorbed on nanocomposite surfaces and have simple operating procedures. The wastewater is missed with an adsorbent and kept for a specific period to absorb dye on the nanocomposite surface. Then the adsorbent is filtered and regenerated for further use.

Various physical and chemical factors like temperature, pressure, surface area, pH of the solution, the concentration of adsorbate, and the size of adsorbent and

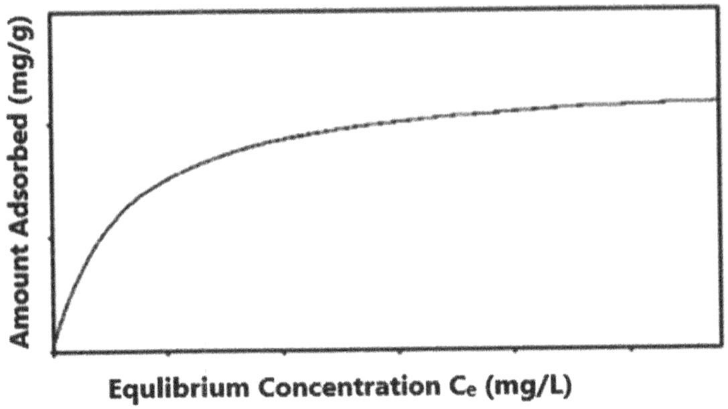

Fig. 5 A simple adsorption isotherm

adsorbate affect adsorption efficiency. The plot showing the extent of adsorption with equilibrium concentration at a constant temperature is known as adsorption isotherms (Fig. 5).

4.2.1 Adsorption Mechanism

In the case of simple solid–liquid adsorption, the transfer of solute particles to adsorbent can be characterized by two ways; external mass transfer or intraparticle diffusion or sometimes both. The mechanism of adsorption follows three steps:

- Film diffusion, where the solute particles come closer to the adsorbent surface.
- Particle diffusion, in which the solute particles movies inside the pores of the adsorbent except for a small amount remains on the surface and
- Adsorption of solute on the inner surface of the adsorbent.

The last step is very fast and occurs in equilibrium; hence it is not considered for rate constant calculations. The film diffusion and particle diffusion step control the overall adsorption process.

4.2.2 Adsorption Kinetics

Two kinetic models, pseudo-first order and pseudo-second order models, were investigated for the adsorption of pollutants in a batch system.

Pseudo-First-Order Model

Lagerengen provides the integrated for of pseudo-first-order reaction [35].

$$\frac{dq_t}{dt} = k_L(q_e - q_t) \quad (1)$$

$$\log(q_e - q_t) = \log q_e - \frac{k_t}{2.303}t \quad (2)$$

where K_l is the pseudo-first-order rate constant, the values of K_l and q_e can be calculated from the slope and intercept respectively from the plot between log (q_e-q_t) versus time.

Pseudo-Second-Order Model

The modified form of pseudo-second-order rate equation of the adsorption of pollutants can be given as

$$\frac{dq_t}{d_t} = k_2(q_e - q_1)^2 \quad (3)$$

where K_2 is the pseudo-second-order rate constant, the values of K_2 can be calculated from the plot between t/q and time. The values of the initial sorption rate (h) can be calculated from the following equation:

$$\frac{t}{q_t} = \frac{t}{k_2 q_e^2} + \frac{1}{q_e}t \quad (4)$$

4.2.3 Adsorption Isotherms

Generally, Langmuir and Freundlich's isotherms were investigated for describing the adsorption of pollutants on the adsorbent surface.

Langmuir Isotherm

This isotherm is used to describe the adsorption at homogenous sites on the surface of the adsorbent. It is successful for studying the monolayer adsorption process. For a single pollutant system, the Langmuir isotherm is given as:

$$\frac{x}{m} = \frac{V_m K C_e}{1 + K C_e} \quad (5)$$

The linear form of the above equation can be written as:

$$\frac{C_e}{x/m} = \frac{1}{KV_m} + \frac{C_e}{V_m} \tag{6}$$

where C_e is the concentration of solution at equilibrium, x/m is the amount of solute adsorbed per gram of adsorbent, m is the weight of adsorbent taken, K is the equilibrium constant, and V_m is the volume of monolayer adsorbed. The equation relating K with the heat of adsorption (q) can be written as:

$$K = K_o \exp \frac{q}{RT} \tag{7}$$

The values of K and V_m generally increase with temperature which suggests that the adsorption efficiency is enhanced at a higher temperature.

Freundlich Isotherm

This isotherm describes multilayer adsorption over a heterogeneous surface. The freundlich isotherm equation can be given as:

$$\frac{x}{m} = K_f C_e^{1/n} \tag{8}$$

The linearized form of this equation can be written as:

$$\log \frac{x}{m} = \log K_t + \log C_e^{1/n} \tag{9}$$

where K_f denotes adsorption capacity, 1/n describes adsorption intensity and other parameters have their usual meanings as in Eq. (6).

4.3 Photocatalytic Degradation

Photocatalytic degradation has been considered to be an efficient method for the degradation of dyes. It is a greener method to degrade dye by visible light irradiation. It is a highly efficient, economical, and sustainable method for wastewater treatment [36].

Several researchers reported the use of nanocomposite photocatalysts for the degradation of organic dyes from wastewater. Transition metal oxides and sulphides are known to be the best choice for photocatalytic nanocomposite materials. Titanium dioxide (TiO_2) is the most commonly used metal oxide semiconductor photocatalyst used because of its non-toxic nature and high stability. Various reactive species like superoxide radicals, hydroxide radicals, electrons and holes are generated upon irradiation of semiconductors that react with dyes in wastewater, resulting in the degradation of dyes [37].

4.3.1 Mechanism of Photodegradation

Irradiation of semiconductor photocatalyst results in the excitation of electrons from valance band (VB) to the conduction band (CB), resulting in the formation of positively charged holes (h^+) in VB. The holes react with adsorbed water molecules on the surface of the nanocomposite, generating hydroxyl radicals (OH^-). The electrons in the CB react with adsorbed oxygen to form superoxide radicals ($O_2{}^-$) which further react with H^+ ions forming hydrogen peroxide and hydroperoxyl free radicals ($\cdot HO_2$). These reactive species interact with dye molecules and initiate the degradation process [38]. The steps involved in the degradation of dyes are shown in Fig. 6 [39]. The general mechanism of photocatalytic degradation of dyes by a semiconductor photocatalyst is shown in Fig. 7 [40].

However, TiO_2 suffers from a limitation of large bandgap, making it inefficient under visible light irradiation. Therefore, scientists developed various other semiconductor photocatalysts for wastewater treatment purposes. Ceria (CeO_2), graphitic carbon nitride (g-C_3N_4) molybdenum disulphide (MoS_2) are some of the extensively investigated materials for synthesizing nanocomposites for photocatalytic applications [36, 38, 39].

4.4 *Microbial Degradation*

Scientists investigated the use of various microorganisms like bacteria, fungi, algae, and viruses to degrade dyes. Several strains of bacterias like *pseudomonas luteola, Xanthophilusazovorans, Klebsiella pneumonia, Clostridium perfringens* were investigated to degrade azo dyes from degradation wastewater. Jing et al. [41] reported about 97% degradation of 200 ppm malachite green using *Aeromonas hydrophila* LZ within 12 h. Fungi are capable of producing intracellular and extracellular enzymes which are capable of degrading dyes. Various research reported the use of white-rot fungus for the degradation of azo dyes [42]. Mycoremediation is a safe and low-cost technique for the removal of dyes [43].

Although microbial degradation is a low-cost natural method for the degradation of dyes, it is generally not a suitable and sustainable technique for real wastewater treatment applications. It is a very slow process, and only specific categories of dye could be degraded. Furthermore, wastewater contains various other toxic substances along with dyes which are fatal to microorganisms which slows down the degradation process.

Every method has its own pros and cons. Membrane separation requires expensive membranes, and additional pressure is applied to the solution to pass through the membrane. Also, it is effective towards specific size pollutants. Microbial degradation

1. **The photoexcitation of semiconductor takes place,**

$$SC + h\nu \rightarrow e^- + h^+$$

2. **Ionosorption of oxygen takes place,**

$$O_2 + e^- \rightarrow O_2^{*-}$$

3. **Followed by ionization of water,**

$$H_2O \rightarrow OH^- + H^+$$

4. **The hydroxyl ion gets oxidised,**

$$OH^- + h^+ \rightarrow {}^*OH$$

5. **The superoxide gets protonated,**

$$O_2^{*-} + H^+ \rightarrow HO_2^*$$

6. **Followed by co-scavenging of electron,**

$$HO_2^* + e^- \rightarrow HO_2^-$$

7. **Formation of hydrogen peroxide,**

$$HO_2^- + H^+ \rightarrow H_2O_2$$

8. **Followed by *OH formation,**

$$H_2O_2 + e^- \rightarrow {}^*OH + OH^-$$

9. **The pollutants get degraded by active species,**

$$\text{Pollutant} + {}^*OH \rightarrow \text{degradation product}$$

$$\text{Pollutant} + h^+ \rightarrow \text{oxidation product}$$

$$\text{Pollutant} + e^- \rightarrow \text{reduction product}$$

Fig. 6 Steps involved in the photocatalytic degradation of dyes (Reproduced with permission from Elsevier [39])

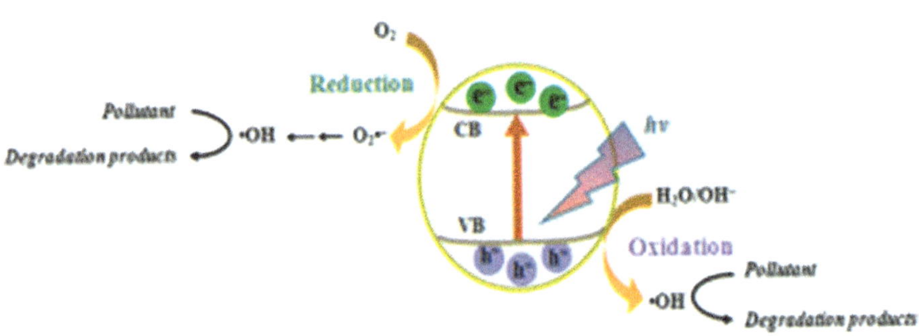

Fig. 7 Schematic illustration of a photocatalytic model of dye degradation (Reproduced with permission from Elsevier [40])

is also generally not employed for dye removal because of the toxicity of dyes towards microorganisms, and it is a time-consuming process. However, adsorption and photodegradation are the two most extensively studied methods for dye removal from wastewater because the operation cost is less, removal efficiency is high, easy operation procedure, and the process is less time-consuming.

5 Removal of Dyes Using Nanocomposites

5.1 Rhodamine B

Rhodamine B, commonly known as RhB, is a basic azo dye and belongs to the xanthene class [44]. This reddish violet-colored dye is the most toxic among all the dyes present in the xanthene class [45, 46]. RhB is a synthetic dye with high water solubility, having a molecular mass of 479 g/mol and a molecular formula of $C_{28}H_{31}N_2O_3Cl$ [47]. The structure of Rhodamine B is shown in Fig. 8 [39]. Due to its high stability, water solubility, and exceptional brightness, RhB is used in most industries [48]. However, it negatively impacts the environment as they are pretty toxic and carcinogenic [47]. It affects living organisms by irritating the respiratory tract, eyes, and skin, discontinuing their growth and development [49]. So, to remove this dye from wastewater, various nanocomposites are used and are discussed in Table 1.

Tian et al. [50] synthesized $BiIO_4$/g-C_3N_4 using the calcination process to photodegrade RhB from wastewater. The nanocomposite degraded about 90% of Rhodamine B dye in 290 min in the presence of visible light irradiation. The nanocomposite followed a pseudo-first-order reaction showing a rate constant of 0.140 min^{-1}. Using polymeric citrate and thermal exfoliation approach, Kumar et al. [51] synthesized N-doped-$SrTiO_3$/g-C_3N_4 for photodegradation of RhB. The nanocomposite showed an efficiency of nearly 100% in visible light irradiation due to high charge carrier separation, which slowed the electron–hole recombination rate.

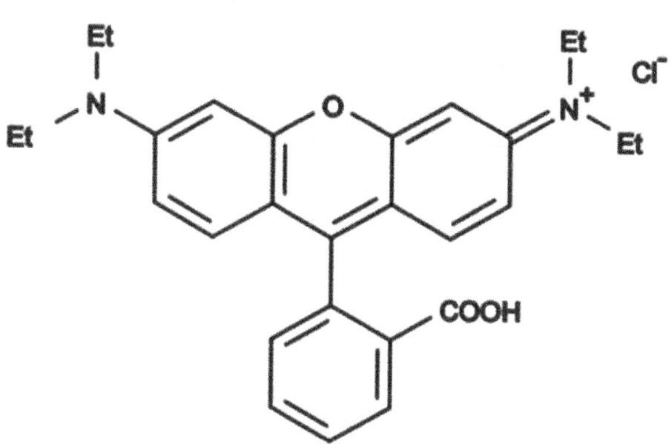

Fig. 8 Structure of Rhodamine B. "Reproduced with permission from Elsevier [39]."

Table 1 Nanocomposites for removal of Rhodamine B in aqueous medium

Nanocomposites	Dye concentration in wastewater	Nanocomposite dose (gL^{-1})	Time (min)	Efficiency	References
$BiIO_4$/g-C_3N_4	10^{-5} mol/L	1.0	290	~90%	[50]
N-doped-$SrTiO_3$/g-C_3N_4	5 mg/L	0.5	60	~100%	[51]
SiO_2/TiO_2	10^{-5} mol/L	1.5	180	~100%	[52]
Ag/Bi_2WO_6	10^{-5} mol/L	0.2	210	94.21%	[53]
Gg-cl-P(AA-co-AAm)/Fe_3O_4	100 mg/L	0.8	1440	654.87 mg/g	[54]
Co/OMC	100 mg/L	0.003	25	96%	[55]
A-rGO/Co_3O_4	100 mg/L	0.01	480	102.9 mg/g	[56]
L-Asp/Fe_3O_4	30 mg/L	1	–	10.44 mg/g	[57]

Using the one-step hetero-coagulation approach, titania-coated SiO_2 nanocomposites were produced to degrade Rhodamine B in the presence of solar light irradiation [52]. The researchers confirmed that the photodegradation followed pseudo-first-order kinetics with a rate constant of 0.00348 min^{-1} at a wavelength of 590 nm. The nanocomposite showed an efficiency of about 100% and was reusable up to 5 times. Bunluesak et al. [53] synthesized Ag/Bi_2WO_6 using the hydrothermal-photoreduction combined method. The nanocomposite showed an efficiency of 94.21% under 3.5 h for photodegradation of Rhodamine B. The increased efficiency was due to the Schottky barrier between metallic Ag nanoparticles and Bi_2WO_6 nanoplates, increasing recombination time for photogenerated electrons and holes.

Gg-cl-P(AA-co-AAm)/Fe_3O_4 is a Gum Ghatti-based magnetic Fe_3O_4 nanocomposite that removed rhodamine B by adsorption from wastewater [54]. The adsorption was relatively stable and followed pseudo-second-order kinetics. With the help of a thermodynamic study, the researchers found that the adsorption process was spontaneous and endothermic. The nanocomposite dose was 0.8 g/L, and it showed a removal efficiency of 654.87 mg/g. The nanocomposite adsorbed RhB following Langmuir isotherm and was reusable up to 3 times. Tang et al. [55] embedded Cobalt on ordered mesoporous carbon using a simple infusion method to adsorb rhodamine B from wastewater. It showed rapid adsorption of 100 mg/L of RhB within 25 min with an efficiency of 96%, and in the long run, it adsorbed 200 mg/L of RhB in 24 h with an adsorbing efficiency of 468 mg/g. The adsorption followed pseudo-second-order kinetics and the Langmuir model of the isotherm. Using the hydrothermal method, Alwan et al. [56] prepared cobalt oxide nanoparticles with graphene oxide as its support. A-rGO/Co_3O_4 adsorbed rhodamine B from wastewater with an adsorption capacity of 102.9 mg/g. The adsorption process was spontaneous and endothermic, and the equilibrium point was attained in 8 h. By taking L-Aspartic acid as a capping agent, Belachew et al. [57] synthesized L-Asp/Fe_3O_4 nanocomposite to degrade Rhodamine B from water and wastewater effectively. The adsorption followed Avrami fractional-order kinetics and Liu model isotherm. The synthesized nanocomposite showed effective removal of RhB dye with an adsorbing capacity of 10.44 mg/g.

5.2 *Methylene Blue*

Methylene blue belongs to the thiazine class, is green, basic, and water-soluble. It is famously known as MB and is used in dyeing wool, cotton, jute, textiles, and plastics [58]. It has a chemical formula of $C_{16}H_{18}ClN_3S$ and a molar mass of 319.85 g/mol. Due to the presence of tertiary amine, it is pretty toxic and is harmful to aquatic organisms [59]. The structure of MB is shown in Fig. 9 [39]. These dyes move up to human bodies via the food chain and increase heartbeat, nausea, jaundice, mental confusion, and methemoglobinemia [60, 61].

Moreover, it also causes permanent eye damage in humans as well as animals. As this dye poses a potential threat to our ecosystem, treating wastewater containing methylene blue is very important [62]. So, to remove this dye from wastewater, various nanocomposites are used and are discussed in Table 2.

To remove methylene blue from wastewater, Jiang et al. [63] synthesized Ag-CdS/g-C_3N_4 to photodegrade MB in the presence of visible light irradiation. The nanocomposite was very effective, stable, and showed no sign of photo corrosion. The removal efficiency was about 90.45% in 3 h, was reusable up to 3 times, and followed pseudo-first-order kinetics and the Langmuir–Hinshelwood model of the isotherm.

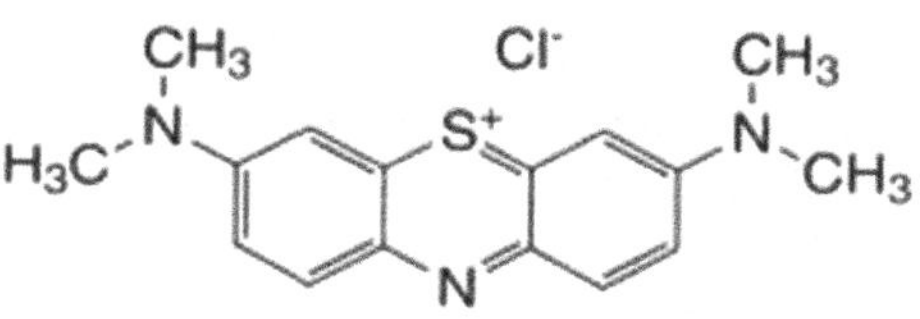

Fig. 9 Structure of Methylene blue (Reproduced with permission from Elsevier [39])

Table 2 Removal of Methylene blue from aqueous solution

Nanocomposites	Dye concentration in wastewater	Nanocomposite Dose (gL^{-1})	Time (min)	Efficiency	References
Ag-CdS/g-C_3N_4	25 mg/L	0.4	180	~92%	[63]
TiO_2/g-C_3N_4	10 mg/L	1.0	60	100%	[64]
CdS/Ag_3PO_4	2 mg/L	1	150	95%	[65]
SLGQDs/Au	100 mg/L	1	120	90%	[66]
Fe–Mn BO	50 mg/L	0.001	720	72.32 mg/g	[67]
$MgFe_2O_4$/rGO	10 mg/L	0.01	60	24.81 mg/g	[68]
Fe_3O_4@C	100 mg/L	0.1	60	18.52 mg/g	[69]
CNC/ZnO	100 mg/L	0.3	1440	64.93 mg/g	[70]

Wang et al. [64] synthesized N doped $TiO_2/g\text{-}C_3N_4$ via the in-situ microwave-assisted method. This photocatalyst was even effective at high temperatures up to 400 °C. The doped nanocomposite showed an efficiency of 100% for photodegradation of methylene blue in visible light irradiation in 60 min. Mirsalari et al. [65] produced CdS/Ag_3PO_4nanocomposite, which was used in the photodegradation of methylene blue in the presence of visible light irradiation. The removal procedure followed the first-order kinetics model and Hinselwood model of the isotherm. With the help of hydrothermal and photochemical methods, Dejpasand et al. [66] synthesized single-layer graphene quantum dots-based Au for photodegradation of methylene blue. MB was degraded under visible light irradiation with an efficiency of 90% in 2 h.

Lu et al. [67] synthesized nano-structured Fe–Mn binary oxide with the help of the co-precipitation method to adsorb MB from wastewater. The nanocomposite showed an absorbing capacity of about 72.32 mg/g and obeyed the pseudo-second-order kinetics and Langmuir isotherm models. The pH of the aqueous solution was maintained at an optimum level, and the nanocomposite was found to be reusable up to five times. To remove Methylene blue from wastewater by adsorption, Adel et al. [68] synthesized $MgFe_2O_4$/reduced graphene oxide (rGO). The nanocomposite showed excellent adsorbing progress in the first 10 min with an adsorbing capacity of 24.81 mg/g. The synthesized nanocomposite was reusable up to 4 times and followed pseudo-second-order kinetics. Xiang et al. [69] synthesized Fe_3O_4@C nanocomposite via the in-situ method for effective removal of MB. The adsorption process was exothermic, and hydrogen peroxide was added to improve the efficiency. As the nanocomposite showed an adsorbing capacity of 18.52 mg/g, it can be used for various industrial purposes. Sawdust-derived cellulose nanocrystals were embedded on ZnO nanoparticles for the efficient removal of methylene blue. The synthesized CNC/ZnO nanocomposite showed an adsorbing capacity of 64.93 mg/g and followed pseudo-second-order kinetics and the Langmuir isotherm model. The composite showed adsorbing activity when reused up to 4 times and was eco-friendly.

5.3 Methyl Orange

Methyl orange, also known as MO, belongs to the group of azo dyes. It is water-soluble and relatively acidic due to the presence of the sulphonic group [71]. Methyl orange is usually used in food coloring, dyeing of plastic, leather, and textile, and in pharma industries [72]. The chemical formula of MO is $C_{14}H_{14}N_3NaO_3S$, and its molecular mass is 327.33 g/mol. The structure of MO is shown in Fig. 10 [39]. Methyl orange is also used as a coloring agent in laboratories for determining itopride hydrochloride and determining hydrogen gas. However, methyl orange is carcinogenic and mutagenic, which is very harmful to the environment [73]. Thus, it is essential to remove this dye from the environment. Table 3 shows various nanocomposites which help in the removal of methyl orange from the aqueous solution.

Feng et al. [74] synthesized $g\text{-}C_3N_4/Bi_4O_5I_2$ by the in-situ method via thermal treatment, which enhanced the photodegradation of methyl orange in the presence

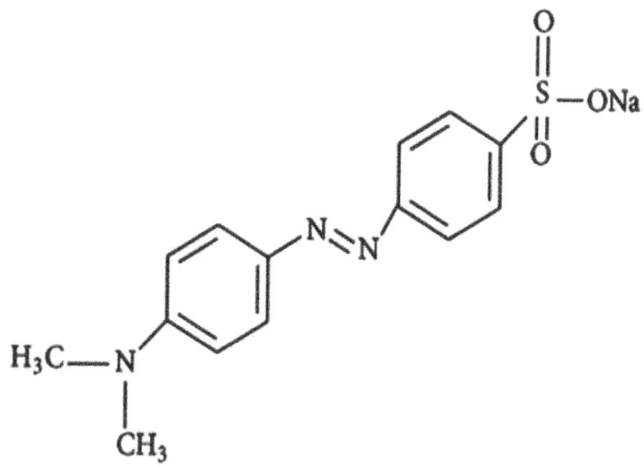

Fig. 10 Structure of Methyl orange (Reproduced with permission from Elsevier [39])

Table 3 Removal of methyl orange from aqueous solution

Nanocomposites	Dye concentration in wastewater	Nanocomposite Dose (gL^{-1})	Time (min)	Efficiency	References
$g\text{-}C_3N_4/Bi_4O_5I_2$	1 mg/L	0.010	180	0.164 min^{-1}	[74]
$g\text{-}C_3N_4/Bi_2WO_6$	2 mg/L	0.015	180	~90%	[75]
Co-ZnO	100 mg/L	0.05	130	~100%	[76]
TiO_2/MoS_2	20 mg/L	0.5	80	93%	[77]
chitosan/Al_2O_3/magnetite	100 mg/L	0.4	14	417 mg/g	[78]
CuO/NaA	100 mg/L	1	240	98%	[79]
AC/Ag	2 mg/L	–	960	72.5%	[80]
PbS/ZnO	200 mg/L	0.06	40	159 mg/g	[81]

of visible light irradiation. The rate of degradation was 82times higher than pure $g\text{-}C_3N_4$ and has a rate constant of 0.164 min^{-1}. The synthesized nanocomposite had a favorable heterostructure, which improved the charge separation, thus enhancing the rate of photodegradation. To efficiently remove methyl orange, Ge et al. [75] produced $g\text{-}C_3N_4/Bi_2WO_6$ nanocomposite, which photodegraded MO in the presence of visible light. The nanocomposite followed pseudo-first-order kinetics and the enhanced efficiency was due to the improved charge separation and stability. Adeel et al. [76] synthesized Co-ZnO for effective photodegradation of MO from wastewater. The degradation took place in visible light irradiation, and it followed the Eley–Rideal mechanism of kinetics. The presence of Co inhibits the electron–hole recombination, therefore increasing its efficiency. TiO_2/MoS_2 nanocomposite was prepared to photodegrade MO in the presence of UV–Vis light irradiation [77]. The enhanced photocatalytic activity is due to better charge carrier efficiency, higher surface area, and vacant oxygen positions. Figure 11 [7] shows the mechanism of photodegradation of methyl orange by TiO_2/MoS_2 nanocomposite.

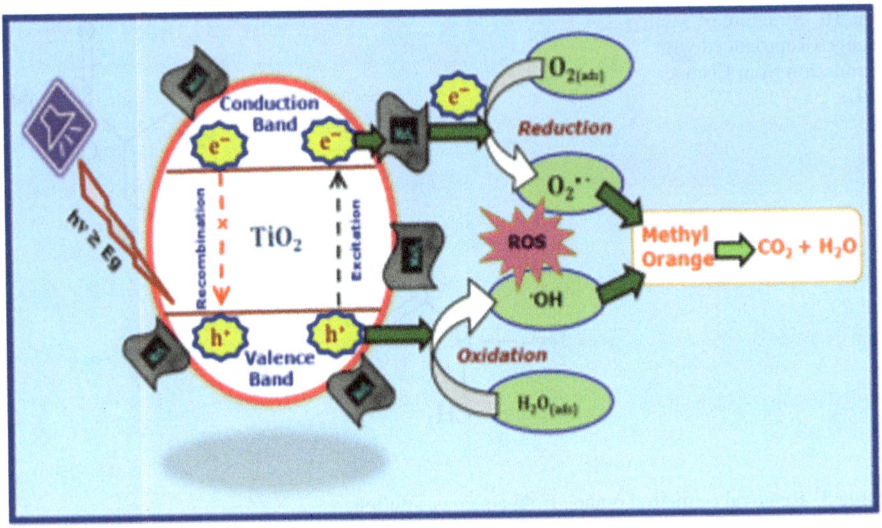

Fig. 11 Mechanism for degradation of MO by TiO_2/MoS_2 nanocomposite (Reproduced with permission from ACS publications [77])

Tanhaei et al. [78] synthesized chitosan/Al_2O_3/magnetite to remove methyl orange by adsorption. The nanocomposite showed very high adsorption capacity and followed pseudo-second-order kinetics and the Langmuir isotherm model. The optimal pH range was found to be 4–10, and the efficiency was 93%. Amokrane et al. [79] added CuO support on NaA zeolite by ion exchange method to adsorb methyl orange from aqueous solution. The researchers found that the removal was optimum at neutral pH rather than acidic or basic conditions. The adsorption of MO followed the pseudo-second-order kinetics model and the Langmuir isotherm model. Researchers synthesized activated carbon-coated silver particles to remove Methyl orange by green synthesis, adding glucose as the reducing agent and PVP as the stabilizing agent [80]. The Ac/Ag nanocomposite followed the pseudo-second-order kinetics model and obeyed both Freundlich and Langmuir isotherm. The key feature of the nanocomposite was that it was reusable even up to 10 cycles without any efficiency decrease. Liu et al. [81] synthesized PbS/ZnO nanocomposite to adsorb methyl orange by hydrothermal method. The nanocomposite is a fine example of a semiconductor heterojunction adsorbent. The removal process was spontaneous and endothermic, followed pseudo-second-order kinetics and Langmuir isotherm model.

5.4 *Malachite Green*

Malachite green is a triarylmethane dye and is used in the dyestuff and pigment industry. It has a chemical formula of $C_{46}H_{50}N_4{\cdot}2HC_2O_4{\cdot}C_2H_2O_4$ and a molecular mass of 364.91 g/mol, and its structure is shown in Fig. 12 [82]. This dye is banned in many European countries due to its toxic effect; however, some countries still use it in the fish industry, silk, dyeing, textile industry, and medicinal industry. It is very harmful to the environment as it is carcinogenic, mutagenic, shows chromosomal fractures, and causes respiratory issues. Therefore, it is crucial to remove the dye from the environment, and Table 4 shows the various nanocomposite that helps remove the toxic dye.

TiO_2/ZnO nanocomposite was synthesized by Bansal et al. [83] to degrade the carcinogenic malachite green. The ratio of TiO_2 to ZnO was 1:9, which resulted in efficient photodegradation of the dye. Fardood et al. [84] synthesized $MgFe_2O_4$ for effective photodegradation of MG in the presence of visible light irradiation. The prepared nanocomposite was magnetic, thus was easily separable from the aqueous solution, and was reusable up to 6 times. The formation of OH^* radicals on the nanocomposite surface helped oxidize malachite green, and the photodegradation mechanism fitted the first-order kinetic reaction. Diatomite embedded Ni/NiO was

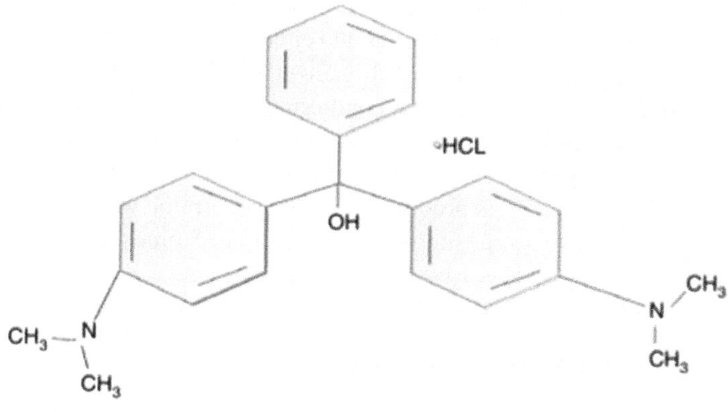

Fig. 12 Structure of Malachite green (Reproduced with permission from Degruyter.com [82])

Table 4 Removal of malachite green from aqueous solution

Nanocomposites	Dye concentration in wastewater	Nanocomposite Dose (gL^{-1})	Time (min)	Efficiency	References
TiO_2/ZnO	50 mg/L	1.0	180	72.30%	[83]
$MgFe_2O_4$	20 mg/L	0.015	60	98%	[84]
Diatomite@Ni/NiO	25 mg/L	5.0	30	100%	[85]
$ZnO\text{-}Dy_2O_3$	10 mg/L	0.01	60	99%	[86]
MWCNT-COOH	100 mg/L	1.0	10	11.73 mg/g	[87]
CuI–CuO–AC	30 mg/L	0.5	20	136.57 mg/g	[88]
MWCNT-GI	100 mg/L	0.2	60	84.7%	[89]
ZnO	2000 mg/L	0.02	120	2963 mg/g	[90]

prepared for effective photodegradation of cationic malachite green [85]. Due to the low bandgap and high surface area, the nanocomposite showed excellent photocatalytic activity. The photodegradation was controlled by electrons and holes and partially by the formed superoxide radicals. Josephine et al. [86] synthesized ZnO doped Dy_2O_3 for photodegradation of malachite green under visible light. The ZnO-Dy_2O_3 nanocomposite was prepared by precipitation method by using nitrates and sodium carbonates as the precursor. It followed pseudo-first-order kinetics with 99% of dye degradation in 1 h.

Rajabi et al. [87] synthesized carboxylate group functionalized multi-walled carbon nanotubes for adsorption of methyl orange from wastewater. The adsorption process took place for 10 min with an optimum pH of 9. The adsorption obeyed the pseudo-second-order kinetic model and the Langmuir isotherm model. The researchers found that the adsorption capacity increased by an increase in pH and temperature. To remove malachite green from the aqueous solution, researchers prepared activated carbon loaded cupric iodide-cupric oxide nanocomposite [88]. The CuI–CuO–AC nanocomposite was synthesized using the solvothermal technique and worked at an optimum pH of 7. The adsorption mechanism followed pseudo-second-order kinetics and was spontaneous and endothermic. Rong et al. [89] synthesized multi-walled carbon nanotubes functionalized with glucose for MG adsorption. The dye adsorption followed pseudo-second-order kinetics and obeyed Langmuir isotherm. ZnO nanoparticles were used to remove malachite green from wastewater efficiently [90]. The enhanced efficiency was due to strong electrostatic attraction, hydrogen bonding, and chemical precipitation. The adsorption was endothermic and spontaneous, and the nanocomposite was reusable up to 5 times.

6 Discussion and Future Prospects

Adsorption has been reported as a sustainable and efficient method for dye removal. With the development of nanomaterials and nanotechnology, scientists are continuously aiming to develop new materials for removing dyes from wastewater. However, as the properties of nanocomposites depend upon the method of synthesis, therefore, there is a need to develop facile synthesis routed for developing new nanocomposites for enhanced adsorption efficiency towards dyes. Also, most of the nano adsorbents does not show significant regeneration efficiency. The adsorption capacity decreases significantly after few cycles. Therefore, new research should aim to develop new nanocomposites that can be effectively regenerated for further use and are stable to several adsorption–desorption cycles to minimize the cost of operation further.

The disposal of used or spent adsorbents is still a challenge to overcome. Improper disposal of adsorbents can lead to secondary pollution. There is not much research data available on the life cycle of adsorbents. Studying the life cycle of adsorbents may help in understanding their disposal techniques for minimizing secondary pollution. Therefore, more research should be conducted to investigate and reduce the leaching of pollutants from nanocomposites.

Compared to several other commercially available adsorbents like activated charcoal and activated carbon, nanocomposites are not efficient in adsorption. However, the literature suggests many nanocomposites show enhanced adsorption efficiency than activated carbon, but they are costly. Hence, there is a need for new nano adsorbents with low cost, high removal efficiency, minimal secondary pollution and a facile synthesis route.

Photodegradation is also an excellent technique for dye removal from wastewater. To date, there are thousands of nanocomposites being investigated for the applications of wastewater treatment. Among them, TiO_2 has been a milestone among semiconductor photocatalytic materials. However, several metal oxides suffer from the limitation of large bandgaps, making them inactive under visible light regions. This led to a search for new materials with a lower bandgap that is active photocatalysts under visible light irradiation. But materials with low bandgap have low efficiency due to rapid charge recombination rate. Therefore, a good photocatalyst should possess a suitable bandgap, high redox ability and a wide range of light absorption properties. Such materials are very rarely found. Therefore, in search of an efficient photocatalyst for wastewater treatment under visible light irradiation, scientists came up with the idea of combining two or more than two semiconductors resulting in a heterojunction photocatalyst with improved photocatalytic properties. Most of the photocatalyst prepared are effective under UV irradiation only. Hence, new photocatalyst should be prepared, which are effective photocatalysts under visible light irradiation.

Moreover, synthesizing photocatalyst with good photocatalytic efficiency and choosing suitable materials is still a challenge for researchers. Photocatalyst doped with noble metals generally shows higher photocatalytic efficiency. It is recommended to conduct more research in developing new photocatalysts with low-cost materials with high efficiency. Sulphides of iron and molybdenum could be an excellent choice of materials for fabricating nanocomposite photocatalysts for dye removal applications.

As in the case of adsorbents, the properties of photocatalysts also depends upon the method of synthesis. Generally, hydrothermal or solvothermal techniques are recommended for the synthesis of heterojunction photocatalysts because the size of nanocomposites could be controlled by altering the concentration and temperature. Still, further research is needed to develop photocatalysts with enhanced properties by altering the synthesis techniques.

The charge transfer route and the mechanism of photocatalysis are still unclear. Two different materials with similar band positions could follow different charge transfer routes and different mechanisms. Therefore, developing techniques to capture intermediated for proper understanding of the mechanism is needed.

Photocorrosion is the critical limitation that limits the practical applications of most photocatalysts. Several researchers reported doping and nanocomposite forming to suppress photocorrosion of photocatalysts, and however, still, it is a challenge to stop the photocorrosion of the photocatalysts. More research is needed to investigate other techniques to stop the photocorrosion of nanocomposites and apply them tocurrent wastewater treatment applications.

7 Conclusion

Dyes are very harmful to the environment due to their carcinogenic and mutagenic properties. As they are highly water-soluble, they quickly get into aquatic organisms and later into the human body via the food chain. So, wastewater treatment is essential to get rid of the toxic dyes. In this literature, the classifications of the dyes along with their toxicity were discussed. The industries should treat the wastewater before dumping it into various water bodies to minimize contamination. In this chapter, we discussed some of the removal techniques applied by the industries to decrease the level of toxic dyes. The studies suggested that removing dyes by nanocomposites with the help of photodegradation and adsorption is a pretty cheap and efficient method. So, the removal of some common dyes, such as Rhodamine B, Methylene blue, Methyl orange, and Malachite green by nanocomposites, were discussed. The mechanism followed by the nanocomposite during photodegradation and adsorption of dyes was also focused. It was found that the removal of dyes by nanocomposite was exceptionally eco-friendly and cost-effective.

References

1. Nemerow NL, Dasgupta A (1991) Industrial and hazardous waste management. New York
2. Mohanta D, Mahanta A, Mishra SR, Jasimuddin S, Ahmaruzzaman M (2021) Novel SnO_2@ ZIF-8/gC_3N_4 nanohybrids for excellent electrochemical performance towards sensing of p-nitrophenol. Environ Res 197:111077
3. Ali I, Aboul-Enein HY (2004) Chiral pollutants: distribution, toxicity and analysis by chromatography and capillary electrophoresis. John Wiley & Sons
4. Metcalf L, Eddy HP, Tchobanoglous G (1991) Wastewater engineering: treatment, disposal, and reuse, vol 4. McGraw-Hill, New York
5. Droste RL, Gehr RL (2018) Theory and practice of water and wastewater treatment. John Wiley & Sons
6. Ahmaruzzaman M (2021) Biochar based nanocomposites for photocatalytic degradation of emerging organic pollutants from water and wastewater. Mater Res Bull 140:111262
7. Laws EA (2000) Aquatic pollution: an introductory text. John Wiley & Sons
8. Ali I, Aboul-Enein HY (2006) Instrumental methods in metal ion speciation. CRC Press
9. Ali I, Aboul-Enein HY, Gupta VK (2009) Nanochromatography and nanocapillary electrophoresis: pharmaceutical and environmental analyses. John Wiley & Sons
10. Vettorazzi G (1979) International regulatory aspects for pesticide chemicals. Vol I. Toxicity profiles. CRC Press Inc.
11. Meyers RA (1999) Encyclopedia of environmental pollution and cleanup. No. E5
12. Barceló D (ed) (2005) Emerging organic pollutants in waste waters and sludge. Vol 5. Springer Science & Business Media
13. World Health Organization (2010) Persistent organic pollutants: impact on child health
14. Köhler A, Hellweg S, Escher BI, Hungerbühler K (2006) Organic pollutant removal versus toxicity reduction in industrial wastewater treatment: the example of wastewater from fluorescent whitening agent production. Environ Sci Technol 40(10):3395–3401
15. Imran M, Crowley DE, Khalid A, Hussain S, Waseem Mumtaz M, Arshad M (2015) Microbial biotechnology for decolorization of textile wastewaters. Rev Environ Sci Bio/Technology 14(1):73–92

16. Mahmood F, Shahid M, Hussain S, Shahzad T, Tahir M, Ijaz M, Hussain A, Mahmood K, Imran M, Ali Khan Babar S (2017) Potential plant growth-promoting strain Bacillus sp. SR-2–1/1 decolorized azo dyes through NADH-ubiquinone: oxidoreductase activity. Bioresour Technol 235:176–184
17. Rehman K, Shahzad T, Sahar A, Hussain S, Mahmood F, Siddique MH, Siddique MA, Rashid MI (2018) Effect of reactive black 5 azo dye on soil processes related to C and N cycling. PeerJ 6:4802
18. Shafqat M, Khalid A, Mahmood T, Siddique MT, Han J-I, Habteselassie MY (2017) Evaluation of bacteria isolated from textile wastewater and rhizosphere to simultaneously degrade azo dyes and promote plant growth. J Chem Technol Biotechnol 92(10):2760–2768
19. Kadirvelu K, Kavipriya M, Karthika C, Radhika M, Vennilamani N, Pattabhi S (2003) Utilization of various agricultural wastes for activated carbon preparation and application for the removal of dyes and metal ions from aqueous solutions. Bioresour Technol 87(1):129–132
20. Horst MF, Lassalle V, Luján Ferreira M (2015) Nanosized magnetite in low cost materials for remediation of water polluted with toxic metals, azo-and antraquinonic dyes. Front Environ Sci Eng 9(5):746–769
21. Bharathiraja B, Aberna Ebenezer Selvakumari I, Iyyappan J, Varjani S (2019) Itaconic acid: an effective sorbent for removal of pollutants from dye industry effluents. Curr Opin Environ Sci Health 12: 6–17
22. Foteinis S, Maria Monteagudo J, Durán A, Chatzisymeon E (2018) Environmental sustainability of the solar photo-Fenton process for wastewater treatment and pharmaceuticals mineralization at semi-industrial scale. Sci Total Environ 612:605–612
23. Khaki MRD, Saleh Shafeeyan M, Abdul Raman AA, Ashri Wan Daud WM (2018) Evaluating the efficiency of nano-sized Cu doped TiO_2/ZnO photocatalyst under visible light irradiation. J Mol Liq 258:354–365
24. Yu A, Wang Q, Wang J, Chang C-T (2017) Rapid synthesis of colloidal silver triangular nanoprisms and their promotion of TiO_2 photocatalysis on methylene blue under visible light. Catal Commun 90:75–78
25. Padhi BS (2012) Pollution due to synthetic dyes toxicity & carcinogenicity studies and remediation. Int J Environ Sci 3(3):940–955
26. Panda SK, Aggarwal I, Kumar H, Prasad L, Kumar A, Sharma A, Vo DVN, Van Thuan D, Mishra V (2021) Magnetite nanoparticles as sorbents for dye removal: a review. Environ Chem Lett: 1–39
27. Fu Y, Viraraghavan T (2002) Removal of congo red from an aqueous solution by fungus Aspergillus niger. Adv Environ Res 7(1):239–247
28. Mittal AK, Gupta SK (1996) Biosorption of cationic dyes by dead macro fungus Fomitopsis-carnea: batch studies. Water Sci Technol 34(10):81–87
29. Fu Y, Viraraghavan T (2001) Fungal decolorization of dye wastewaters: a review. Bioresour Technol 79(3): 251–262
30. Banat IM, Nigam P, Singh D, Marchant R (1996) Microbial decolorization of textile-dyecontaining effluents: a review. Biores Technol 58(3):217–227
31. Chu HC, Chen KM (2002) Reuse of activated sludge biomass: I. Removal of basic dyes from wastewater by biomass. Process Biochem 37(6):595–600
32. Bagotia N, Sharma AK, Kumar S (2020) A review on modified sugarcane bagasse biosorbent for removal of dyes. Chemosphere:129309
33. Wang DM (2016) Environmental protection in clothing industry. In: Sustainable development: proceedings of the 2015 international conference on sustainable development (ICSD2015), pp 729–735
34. Faust,SD, Aly OM (2018) Chemistry of water treatment. CRC press
35. Gadore V, Ahmaruzzaman M (2021) Tailored fly ash materials: a recent progress of their properties and applications for remediation of organic and inorganic contaminants from water. J Water Process Eng 41:101910
36. Mishra SR, Ahmaruzzaman M (2021) Cerium oxide and its nanocomposites: structure, synthesis, and wastewater treatment applications. Mater Today Commun:102562

37. Gadore V, Ahmaruzzaman M (2021) Fly ash–based nanocomposites: a potential material for effective photocatalytic degradation/elimination of emerging organic pollutants from aqueous stream. Environ Sci Pollut Res:1–24
38. Ahmaruzzaman M, Gadore V (2021) MoS_2 based nanocomposites: an excellent material for energy and environmental applications. J Environ Chem Eng:105836
39. Ahmaruzzaman M, Ranjan Mishra S (2021) Photocatalytic performance of g-C_3N_4 based nanocomposites for effective degradation/removal of dyes from water and wastewater. Mater Res Bull:111417
40. Lam S-M a, n (2016) A review on photocatalytic application of g-C_3N_4/semiconductor (CNS) nanocomposites towards the erasure of dyeing wastewater. Mater Sci Semicond Process 47
41. Ji J, Kulshreshtha S, Kakade A, Majeed S, Li X, Liu P (2020) Bioaugmentation of membrane bioreactor with Aeromonas hydrophila LZ-MG14 for enhanced malachite green and hexavalent chromium removal in textile wastewater. Int Biodeterior Biodegrad 150:104939
42. Shah M, Banerjee A (eds) (2020) Combined application of physico-chemical & microbiological processes for industrial effluent treatment plant. Springer Nature
43. Jain A, Yadav S, Kumar Nigam V, Rani Sharma S (2017) Fungal-mediated solid waste management: a review. Mycoremed Environ Sustain:153–170
44. Ajiboye TO, Oyewo OA, Onwudiwe DC (2021) Adsorption and photocatalytic removal of Rhodamine B from wastewater using carbon-based materials. FlatChem:100277
45. Fisher P (1999) Review of using Rhodamine B as a marker for wildlife studies. Wildlife Soc Bull:318–329
46. Pourreza N, Rastegarzadeh S, Larki A (2008) Micelle-mediated cloud point extraction and spectrophotometric determination of rhodamine B using Triton X-100. Talanta 77(2):733–736
47. Jain R, Mathur M, Sikarwar S, Mittal A (2007) Removal of the hazardous dye rhodamine B through photocatalytic and adsorption treatments. J Environ Manag 85(4):956–964
48. Hu C, Thi Le A, Yong Pung S, Stevens L, Neate N, Hou X, Grant D, Xu F (2021) Efficient dye-removal via Ni-decorated graphene oxide-carbon nanotube nanocomposites. Mater Chem Phys 260:124117
49. Field MS, Wilhelm RG, Quinlan JF, Aley TJ (1995) An assessment of the potential adverse properties of fluorescent tracer dyes used for groundwater tracing. Environ Monit Assess 38(1):75–96
50. Tian N, Huang H, He Y, Guo Y, Zhang Y (2014) Novel g-C_3N_4/$BiIO_4$ heterojunction photocatalysts: synthesis, characterization and enhanced visible-light-responsive photocatalytic activity. RSC Adv 4(80):42716–42722
51. Kumar S, Tonda S, Baruah A, Kumar B, Shanker V (2014) Synthesis of novel and stable g-C_3N_4/N-doped-$SrTiO_3$ hybrid nanocomposites with improved photocurrent and photocatalytic activity under visible light irradiation. Dalton Trans 43(42):16105–16114
52. Wilhelm P, Stephan D (2007) Photodegradation of rhodamine B in aqueous solution via SiO_2@TiO_2 nano-spheres. J Photochem Photobiol, A 185(1):19–25
53. Bunluesak T, Phuruangrat A, Thongtem S, Thongtem T (2021) Visible-light-driven heterostructure Ag/Bi_2WO_6 nanocomposites synthesized by photodeposition method and used for photodegradation of rhodamine B dye. Res Chem Intermed:1–14
54. Mittal H, Mishra SB (2014) Gum ghatti and Fe_3O_4 magnetic nanoparticles based nanocomposites for the effective adsorption of rhodamine B. Carbohydr Polymers 101:1255–1264
55. Tang L, Cai Y, Yang G, Liu Y, Zeng G, Zhou Y, Li S et al (2014) Cobalt nanoparticles-embedded magnetic ordered mesoporous carbon for highly effective adsorption of rhodamine B. Appl Surf Sci 314:746–753
56. Alwan SH, Habeeb Alshamsi HA, Jasim LS (2018) Rhodamine B removal on A-rGO/cobalt oxide nanoparticles composite by adsorption from contaminated water. J Mol Struct 1161:356–365
57. Belachew N, Tadesse A, Hagos Kahsay M, ShumuyeMeshesha D, Basavaiah K (2021) Synthesis of amino acid functionalized Fe_3O_4 nanoparticles for adsorptive removal of Rhodamine B. Appl Water Sci 11(2):1–9

58. Ma H, Zhuo Q, Wang B (2009) Electro-catalytic degradation of methylene blue wastewater assisted by Fe_2O_3-modified kaolin. Chem Eng J 155(1–2):248–253
59. Zhan S, Zhou F, Huang N, Yin Y, Wang M, Yang Y, Liu Y (2015) Sonochemical synthesis of $Zn_3V_2O_7(OH)_2(H_2O)_2$ and g-C_3N_4/$Zn_3V_2O_7(OH)_2(H_2O)_2$ with high photocatalytic activities. J Mol Catal A: Chem 401:41–47
60. Yuan S, Li Z, Wang Y (2013) Effective degradation of methylene blue by a novel electrochemically driven process. Electrochem Commun 29:48–51
61. Tungprapa S, Puangparn T, Weerasombut M, Jangchud I, Fakum P, Semongkhol S, Meechaisue C, Supaphol P (2007) Electrospun cellulose acetate fibers: effect of solvent system on morphology and fiber diameter. Cellulose 14(6):563–575
62. Naeimirad M, Zadhoush A, Kotek R, EsmaeelyNeisiany R, Nouri Khorasani S, Ramakrishna S (2018) Recent advances in core/shell bicomponent fibers and nanofibers: a review. J Appl Polymer Sci 135(21):46265
63. Jiang F, Yan T, Chen H, Sun A, Xu C, Wang X (2014) Ag-C_3N_4–CdS composite catalyst with high visible-light-driven catalytic activity and photostability for methylene blue degradation. Appl Surf Sci 295:164–172
64. Wang X-J, Yang W-Y, Li F-T, Xue Y-B, Liu R-H, Hao Y-J (2013) In situ microwave-assisted synthesis of porous N-TiO_2/g-C_3N_4 heterojunctions with enhanced visible-light photocatalytic properties. Ind Eng Chem Res 52(48):17140–17150
65. Mirsalari SA, Nezamzadeh-Ejhieh A (2021) CdS–Ag_3PO_4 nano-catalyst: a brief characterization and kinetic study towards methylene blue photodegradation. Mater Sci Semicond Process 122:105455
66. Dejpasand MT, Saievar-Iranizad E, Bayat A, RahemiArdekani S (2021) Surface plasmon-induced photodegradation of methylene blue with single layer graphene quantum Dots/Au nanospheres under visible-light irradiation. J Alloys Compounds:160904
67. Lu K, Wang T, Zhai L, Wu W, Dong S, Gao S, Mao L (2019) Adsorption behavior and mechanism of Fe-Mn binary oxide nanoparticles: adsorption of methylene blue. J Colloid Interf Sci 539:553–562
68. Adel M, Ahmed MA, Mohamed AA (2021) Synthesis and characterization of magnetically separable and recyclable crumbled $MgFe_2O_4$/reduced graphene oxide nanoparticles for removal of methylene blue dye from aqueous solutions. J Phys Chem Solids 149:109760
69. Xiang H, Ren G, Zhong Y, Xu D, Zhang Z, Wang X, Yang X (2021) Fe_3O_4@C nanoparticles synthesized by in situ solid-phase method for removal of methylene blue. Nanomaterials 11(2):330
70. Oyewo OA, Adeniyi A, Bruce Sithole B, Onyango MS (2020) Sawdust-based cellulose nanocrystals incorporated with ZnO nanoparticles as efficient adsorption media in the removal of methylene blue dye. ACS Oega 5(30):18798–18807
71. Katsuda T, Ooshima H, Azuma M, Kato J (2006) New detection method for hydrogen gas for screening hydrogen-producing microorganisms using water-soluble Wilkinson's catalyst derivative. J Biosci Bioeng 102:220–226
72. Brown JP, Dietrich PS (1983) Mutagenicity of selected sulfonated azo dyes in the Salmonella/microsome assay: use of aerobic and anaerobic activation procedures. Mutat Res/Genetic Toxicol 116(3–4):305–315
73. Sun JX, Yuan YP, Qiu LG, Jiang X, Xie AJ, Shen YH, Zhu JF (2012) Fabrication of composite photocatalyst g-C_3N_4–ZnO and enhancement of photocatalytic activity under visible light. Dalton Trans 41(22):6756–6763
74. Feng Z, Zeng L, Zhang Q, Ge S, Zhao X, Lin H, He Y (2020) In situ preparation of g-C_3N_4/$Bi_4O_5I_2$ complex and its elevated photoactivity in Methyl Orange degradation under visible light. J Environ Sci 87:149–162
75. Ge L, Han C, Liu J (2011) Novel visible light-induced g-C_3N_4/Bi_2WO_6 composite photocatalysts for efficient degradation of methyl orange. Appl Catal B 108:100–107
76. Adeel M, Saeed M, Khan I, Muneer M, Akram N (2021) Synthesis and characterization of Co–ZnO and evaluation of its photocatalytic activity for photodegradation of methyl orange. ACS Omega 6(2):1426–1435

77. Kite SV, Nanaso Kadam A, Sathe DJ, Patil S, Mali SS, Hong CK, Lee SW, Garadkar KM (2021) Nanostructured TiO_2 sensitized with mos2 nanoflowers for enhanced photodegradation efficiency toward methyl orange. ACS Omega
78. Tanhaei B, Ayati A, Lahtinen M, Sillanpää M (2015) Preparation and characterization of a novel chitosan/Al_2O_3/magnetite nanoparticles composite adsorbent for kinetic, thermodynamic and isotherm studies of methyl orange adsorption. Chem Eng J 259:1–10
79. Amokrane S, Aid A, Nibou D, Trari M (2015) Adsorption of methyl orange on nanoparticles of a synthetic zeolite NaA/CuO. ComptesRendusChimie 18(3):336–344
80. Pal J, Kanti Deb M, Deshmukh DK, Verma D (2013) Removal of methyl orange by activated carbon modified by silver nanoparticles. Appl Water Sci 3(2):367–374
81. Liu S, Wang W, Cheng Y, Yao L, Han H, Zhu T, Liang Y, Fu J (2020) Methyl orange adsorption from aqueous solutions on 3D hierarchical PbS/ZnO microspheres. J Colloid Interf Sci 574:410–420
82. Tewari K, Singhal G, Arya RK (2018) Adsorption removal of malachite green dye from aqueous solution. Rev Chem Eng 34(3):427–453
83. Bansal P, Bhullar N, Sud D (2009) Studies on photodegradation of malachite green using TiO_2/ZnO photocatalyst. Desalin Water Treat 12(1–3):108–113
84. Taghavi Fardood S, Moradnia F, Mostafaei M, Afshari Z, Faramarzi V, Ganjkhanlu S (2019) Biosynthesis of $MgFe_2O_4$ magnetic nanoparticles and its application in photodegradation of malachite green dye and kinetic study. Nanochem Res 4(1):86–93
85. Liu G, Abukhadra MR, El-Sherbeeny AM, Mostafa AM, Elmeligy MA (2020) Insight into the photocatalytic properties of diatomite@ Ni/NiO composite for effective photodegradation of malachite green dye and photo-reduction of Cr (VI) under visible light. J Environ Manag 254:109799
86. Josephine GAS, Ramachandran S, Sivasamy A (2015) Nanocrystalline ZnO doped lanthanide oxide: an efficient photocatalyst for the degradation of malachite green dye under visible light irradiation. J Saudi Chem Soc 19(5):549–556
87. Rajabi M, Mirza B, Mahanpoor K, Mirjalili M, Najafi F, Moradi O, Sadegh H et al (2016) Adsorption of malachite green from aqueous solution by carboxylate group functionalized multi-walled carbon nanotubes: determination of equilibrium and kinetics parameters. J Ind Eng Chem 34:130–138
88. Nekouei F, Noorizadeh H, Nekouei S, Asif M, Tyagi I, Agarwal S, Gupta VK (2016) Removal of malachite green from aqueous solutions by cuprous iodide–cupric oxide nanocomposite loaded on activated carbon as a new sorbent for solid phase extraction: isotherm, kinetics and thermodynamic studies. J Mol Liq 213:360–368
89. Rong J, Rong X, Qiu F, Zhu X, Pan J, Zhang T, Yang D (2016) Facile preparation of glucose functionalized multi-wall carbon nanotubes and its application for the removal of cationic pollutants. Mater Lett 183:9–13
90. Zhang F, Chen X, Wu F, Ji Y (2016) High adsorption capability and selectivity of ZnO nanoparticles for dye removal. Colloids Surf A: Physicochem Eng Aspects 509:474–483

Polymer Grafted Nanocomposites for Water Decontamination

Krishna Manjari Sahu, Swapnita Patra, and Sarat K. Swain

Abstract Polymer grafted nanocomposites have received high attentions from researchers for their versatile applications and amazing properties. The process of overcoming the contamination of water is one of the important application of nanocomposites prepared from polymer grafted matrix with reinforcement of nanostructurted materials. This chapter comprises of two impressive aspects. The difference between designing a polymer nanocomposite and polymer grafted nanocomposite and the eye catching changes in properties and applications by introducing grafting technique to polymeric nanohybrid materials is the first primary attraction of the present chapter. Secondly, by accounting the important exposure towards water treatment by different polymer grafted nanohybrid materials, this chapter gives a clear purpose for designing more potential polymer grafted nanohybrid materials by reaching the needs of the book. Different grafting mechanisms, along with the photocatalytic mechanism shown by polymer grafted nanohybrid materials are discussed. Dye adsorption, removal of metal ions, desalination of salt water and the antifouling behaviour of polymer grafted nanohybrid materials and membranes are detailed.

Keyword Polymer · Grafting technique · Nanocomposites · Water decontamination · Dye Adsorption · Desalination

1 Introduction

Water is a dominant renewable natural resource and is vital for living organisms present on the earth's surface. Population growth and rapid industrialisation are the major reasons behind the pollution of water. Although water is present in copious amount, human usable fresh water reservoirs are lesser in number. In addition, contamination of water with humongous number of pollutants such as heavy metals [1], dyes [2] and microbes [3] have degraded the quality of pristine water and as such, less amount of water is left for the usage of mankind. Since the need of fresh

K. M. Sahu · S. Patra · S. K. Swain (✉)
Department of Chemistry, Veer Surendra Sai University of Technology, Burla, Odisha Sambalpur 768018, India
e-mail: skswain_chem@vssut.ac.in

S. K. Swain (ed.), *Nanohybrid Materials for Water Purification*, Composites Science and Technology, https://doi.org/10.1007/978-981-19-2332-6_6

water is increasing day by day, saline water present in huge amount on the surface of the earth can be utilized as fresh water by the desalination process [4]. Treatment, purification and decontamination of wastewater processes are not restricted to single method. The recuperation of drinking water from wastewater is a difficult and laborious process with multiple purification techniques. Commonly, there are three steps that can be employed for purification and decontamination of water, primary, secondary and tertiary water treatment. Primary water treatment process involves the removal of suspended particles and various micro pollutants through microfiltration, coagulation and chemical precipitation methods. In case of secondary wastewater treatment, naturally originated microbes play a crucial role as they have the potential to degrade chemical pollutants into eco-friendly molecules that results in lowering of negative influence on ecosystem. Tertiary water treatment process is applied to eradicate pollutants found in water even after primary and secondary purification process like monovalent ions. In order to enhance the quality of water and to make available recycled water for irrigation and industrial purposes, different materials and techniques for water purification have been proposed. Out of those, the materials which have large number of pores, high surface area and adsorptive capacity, are preferred by the researchers for advantageous outcomes [1].

1.1 Nanoparticles

Nanoparticle (NP) is an ultrafine unit with dimensions measured in nanometer and its size ranges between 1 and 100 nm [5]. The possibility to explore substance at nanoscopic range has created several opportunities to investigate NP as a principal material in water purification technique. NPs are superior to larger scale particles in various means such as high surface area, considerable electrical and magnetic properties [6]. The basic layout of nano ranged particles of a substance is mainly governed by the constituents of parent substances, atoms present within the certain area of material and the types of interaction among atoms. Based on chemical composition and chemical interaction, now-a-days, NP has been used in sundry applications such as biomedical field, food packaging, sensor and mainly in water purification treatment [7–9]. On account of distinct physical and chemical properties, metal NPs have gained much attention in recent decades in different science fields such as chemistry [10], physics [11] and material science [12]. Different NPs have been thoroughly examined and adapted in various medical applications such as wound dressing and antimicrobial coatings. Material that encloses the NPs steadily released NPs at a slower rate to provide protection against bacteria. Since NPs have considerable amount of high surface to volume ratio and high surface energy and also show excellent protection against bacteria, it can be used in fabrication of polymer composite material for the removal of microbes from sludge water [13].

1.2 Polymer Nanocomposite (PNC)

A composite is fabricated from two or more distinctive materials which are on mixing formed a stronger substance than those of individual materials. The constituent substances do not fully blend or lose their individual integrities; they mix and provide their felicitous characteristics to enhance the quality of the final product. Composites are multiphasic system, generally have two phases, one is matrix phase and another one is dispersed phase. Nanocomposite is defined as the composite consists of nanoscale material which is about 10 –100 nm at least in one phase. Nanoscale particles which are broadly used in fabrication of nanocomposite are nanotubes or lamellar nanostructure. Nanocomposites are synthesized by combining two or more components: one component must be used as dispersed phase and another is used as matrix phase. The substance which can be used as dispersed phase are allotropes of carbon, nanocluster, oxides of inorganic materials or semiconductors and organic polymers or organic and organometallic compounds whereas different bio molecules such as amino acid, protein, sugar, enzyme and sol–gel derived polymers are used as matrix phase. In agreement with the materials used as matrix, for instance ceramic, metal and polymer, nanocomposites are classified into three categories. PNCs can be engineered by the combination of a polymeric matrix and usually, inorganic nanomaterial as discontinuous phase [14]. This type of nanocomposite is also called hybrid nanocomposite. Inorganic materials such as TiO_2 [15], SiO_2 [16], ZnO [17] and Ag [18] with outstanding features are dispersed in a polymer matrix for obtaining composite. PNCs among several nanocomposites are the prominent materials of present research, investigation and development not only in industrial field but also most importantly in environmental pollution reduction and biomedical field. Different aspects of PNC such as active functionality, capable of forming film and ability to very its constituent dimension, makes PNC a foremost material to use in removal of various organic and inorganic effluents form wastewater to make water reusable. In recent times, PNC grows as a dominant material for industrial waste removal from various polluted water bodies because of its production friendly, lightweight, mechanical strength and most importantly high adsorption ability [19]. Several applications exhibited by PNCs are schematically represented in Fig. 1.

1.3 Grafting Technique

Polymer is defined as a high molecular mass compound obtained by polymerising repeating units i.e., monomers. Typically, polymers can be categorized into two different types such as homo-polymer and copolymer, based on number of repeating units used as reactants during the synthesizing process. The former type of polymer is the combination of similar type of repeating constituents whereas second type of polymer is engineered by using two dissimilar types of repeating units. According to the structure of polymer, it can also be classified into three types such as linear,

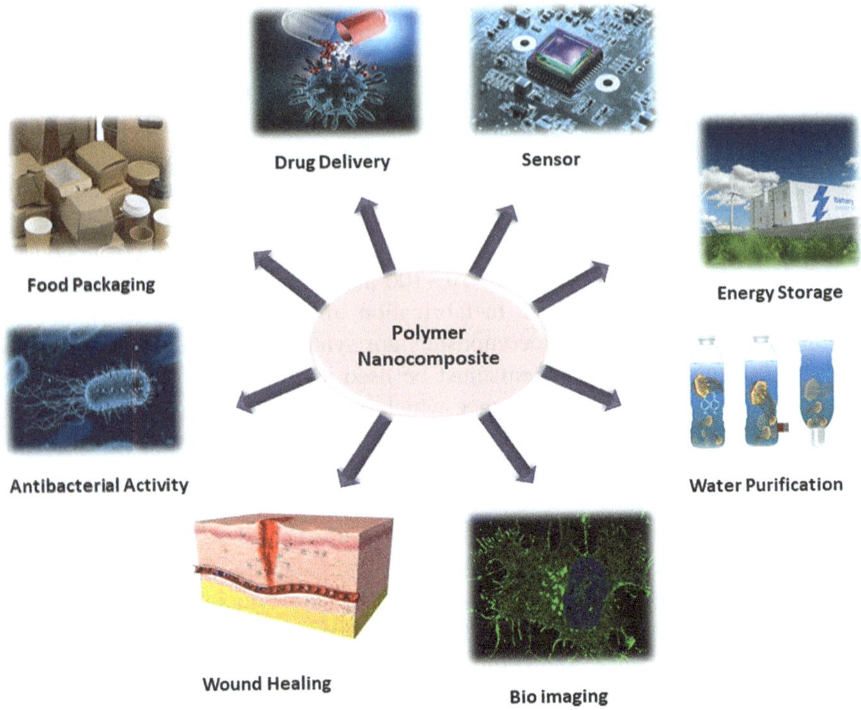

Fig. 1 Schematic representation of various applications of PNC

branched and three-dimensional network polymer. The monomers are linked together in a straight-line fashion to form linear type of polymer. Branched polymer generally consists of two parts; linear main chain and another one is branching unit that attached to the main chain through covalent bond. The polymers, in which branching fragments or parts are connected with the primary chain, are called graft polymers. To prepare graft polymers, researchers mainly follow three kinds of grafting methods such as "grafting to", "grafting from" and "grafting through" techniques. In case of "grafting to" polymerisation approach, branching polymer unit must contains a reactive end group which has to react with different functional groups present on the principal polymer chain while in case of "grafting from" approach, the formation of branching begins on the surface of primary polymer chain due to the presence of reactive initiating sites on the surface of backbone. In "grafting through" preparation method, one polymer unit containing reactive site group copolymerized with another monomer unit. Addition of branch side chain can be done during the synthesising process or after the formation of the principal chain. The property of graft polymer depends on grafting time with relation to number of graft side chains. Some specific types of surface material that can be used as backbone for the fabrication of graft polymer are manmade polymers, naturally occurring polymer and metal. Metal surface does not contain any reactive sites, so modification of metal surface is one of the mandatory steps to introduce functionally active sites. This also helps to make metal surface a favourable synthesising base for the preparation of grafted polymeric materials. The frequently used polymerisation technique for the fabrication of grafting polymer is free radical polymerisation [20].

1.4 Polymer Grafted Nanocomposite (P-g-NC)

Grafting technology is not limited to the formation of polymer only; instead, it has further applications in the fabrication of numerous advanced materials related to polymer, like PNC with extraordinary properties. The three grafting mechanisms used during the synthesis of PNCs are illustrated in Fig. 2. Generally, PNCs that are engineered through grafting methodology are considered as polymer grafted naocomposites (P-*g*-NCs). P-*g*-NCs contain polymer brushes and these brushes help nano ranged particles to disperse evenly throughout continuous phase of P-*g*-NCs. Polymer brushes have two parts: one is centre or core part that contains NP and

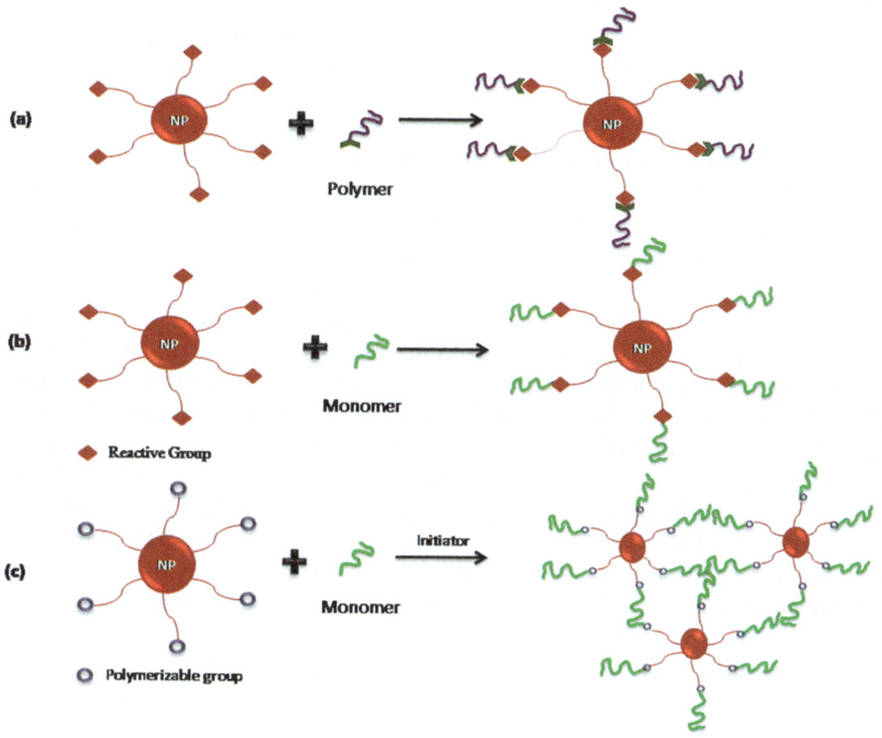

Fig. 2 Schematic representation of **a** "grafting-to" **b** "grafting-from" **c** "grafting-through" mechanism

other one is the polymers contained part that surrounds the central NP. Nanoscopic particles with unique functionality such as iron NP that has magnetic property, gold NP that has optical property or the intricate nanomaterials like virus, protein, caspid etc. can be used as central part of polymer brush. Preparation of P-*g*-NCs depends on several factors, such as types and number of functional groups present on repeating units and most importantly magnitude and uniformity of grafting on the surface of core part. Polymers used for grafting allow controlled distribution and assemblement of NPs during the P-*g*-NCs formation. Maximum outcomes have been shown by the polymer brushes that deprive of crowded grafted polymer on nanosized particle since repulsion among polymers decreases the stability of material. The particular polymer used in preparation of P-*g*-NC and its properties decides the employment of P-*g*-NCs in various application fields e.g., poly (methacrylic acid) can be used in drugs administration applications where as poly (ethylene glycol methyl ether) can be used in preparation of battery. NPs exhibit two types of charge distribution inside the atoms: one is uniformly distributed charge system and another one is direction-oriented dipole. Distribution of charged density inside the NPs governs the 2D structure of nanomaterials but this factor does not affect the entire structure of P-*g*-NCs. For example, silica NPs possess uniform charged density and have isotropic core-core interactions whereas magnetised iron NPs contain dipole, so isotropic core-core interactions are not found in iron NPs contained materials. The structure of P-*g*-NCs that have polymer brushes with same degree of polymerisation show random orientation of polymer brushes in sparsely grafted polymer brushes reign [21].

The involvement of grafting approach in the preparation process of P-*g*-NCs has significantly contributed in enhancement of properties such as absorption, photocatalytic and mechanical stability. Adsorption phenomenon follows two types of adsorption mechanism; physical and chemical adsorption. Physical adsorption is the result of various intermolecular forces such as hydrogen bonding, electrostatic interaction and Van der Waals force of attraction between adsorbent and adsorbate whereas chemical adsorption is occurred through the formation of chemical bond between the reactive sites present on the surface of both adsorbent and adsorbate. Traditional adsorbents that are used for the purpose of water treatment exhibit poor adhesion capacity since this kind of adsorbents contain only few physical active adsorption sites and also the result of chemical adsorption is not so impressive. Owing to the above facts, researchers are more interested to do fabrication and investigation related to novel P-*g*-NCs with a greater number of adsorption sites [22]. P-*g*-NCs have shown various improved properties such as mechanical and thermal stability and catalytic behaviour as compared to PNCs those are formed by techniques other than grafting method. P-*g*-NCs have notable adsorption capacity, because it comprises of NPs which have surface area more than the bulk materials. Also, the addition of nano ranged particles into the matrix of polymer through grafting technique has extremely enhanced the adhesion ability of different molecules on the surface of adsorbents since grafted sites provide additional surface area. During the purification process of water by adsorbent, the major challenge that many researchers have faced is the pollution caused by secondary pollutants. To resolve the secondary pollutants problem, recently researchers have prepared P-*g*-NCs that are able to remove different level of pollutants. Photocatalytic property of a functionalised surface is depend upon on the functional groups and its oxidative ability. P-*g*-NC contains various light responsive functional groups and the free electrons of nanosized particles help in initiation of reduction and oxidation processes to degrade numerous organic effluents into environment friendly molecules as illustrated in Fig. 3. Due to the above reasons, P-*g*-NCs exhibit excellent catalytic property [23]. Another governing character of P-*g*-NCs is mechanical stability. The degree of mechanical property of a composite material is usually determined by the mechanical strength of constituent materials and the interaction present between continuous and non-continuous phase. So, P-*g*-NC shows improved mechanical property compared to its constituent materials i.e., polymer and NP. In case of P-*g*-NCs, the mode of dispersion of nano ranged particles in matrix continuous phase plays a vital role in determining the extent of mechanical stability. The most advantageous method to acquired maximum mechanical outcome is to use grafting methodology in the preparatory process. Giovino et al. has shown the marvellous mechanical stability of P-*g*-NCs by synthesising silica-polystyrene composite. They have prepared a silica grafted brush by alternating the surface of silica using styrene polymer and then dispersed these brushes into polystyrene continuous phase system [24]. Figure 4 shows different eye-catching properties of P-*g*-NC.

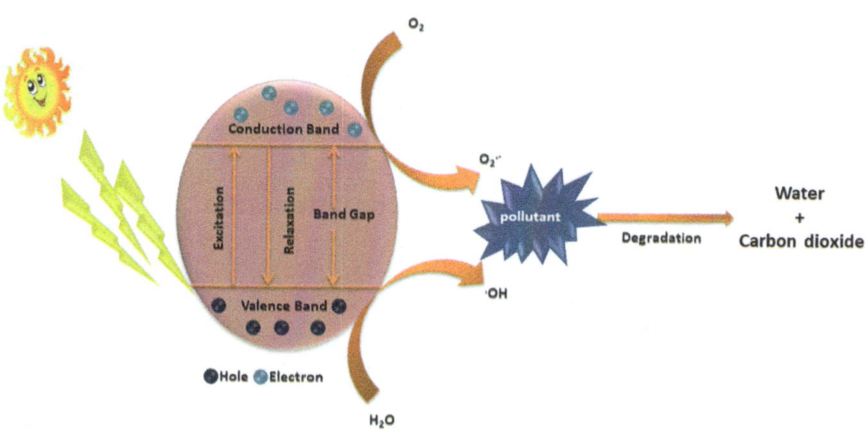

Fig. 3 Schematic illustration of photocatalytic mechanism shown by P-*g*-NCs

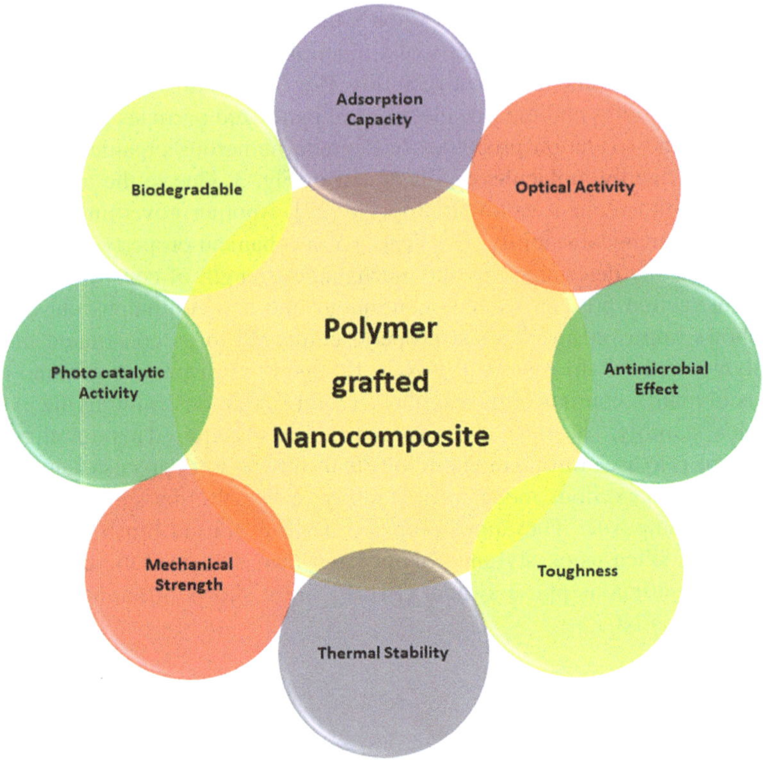

Fig. 4 Schematic representation of properties of P-*g*-NC

2 Approaches Towards Water Decontamination by P-g-NCs

2.1 Dye Adsorption

Dye has been used to give materials some astonishing shade of colours since ancient period of time. Unlike synthetic dyes, natural dyes are easily dwindling away on expose to sunlight and washing. With enhanced colour retaining property than those of natural dyes, synthetic dye is first developed by WH Perkins in 1896 to use in textile industries as well as in food industries for impartation of colour on substances. Industry in which synthetic dyes are used as raw materials releases huge amount of environmental toxifying wastewater to the nearby fresh water reservoirs. Dye contains colour showing units like auxochromes and chromophores which also give stability to the structure of synthetic dyes and make degradation of dyes more difficult. Due to the above reasons, removal of harmful dye pollutants from various water bodies is one of the most challenging tasks in water purification process.

Recently, thousands of scientists have already investigated and proposed sundry kinds of method such as biological, chemical and physical methods to extirpate dye from wastewater [25].

2.1.1 Biological Dye Removal Method

Among several types of dye removal methods, biological dye degradation method is a paramount method since the chances of release of secondary pollutants is negligible. Biological and chemical methods are mainly implemented to eradicate colour from the wastewater by detaching colour adding chromophore fragments from dyes. The standard biological methods used for this purpose are aerobic and anaerobic processes. During aerobic degradation method, oxygen hinders degradation process because of its high reduction potential value and electron accepter property. Owing to the above fact, the besmirching of coloured wastewater containing azo dye like Reactive Orange 16 is preferred anaerobic environment. Another effective biological dye degradation process is bioremediation. This process is implemented to decontaminate wastewater containing large number of organic pollutants. The wastewater release from different textile industries contains hazardous pollutants mainly azo dyes, pesticides and derivatives of phenol. The effectiveness of bioremediation process depends upon the capability of microbes to withstand in such adverse environment [26].

2.1.2 Chemical Dye Removal Method

Like biological dye removal method, chemical dye removal method is also used in the process of besmirching of organic effluents but the pathway followed for degradation mechanism is different for each case. Chemical dye degradation method includes photocatalytic process, electrochemical reduction, and reduction by Fenton's reagent. The main component for photocatalytic degradation under UV radiation of different colorants is hydroxyl free radical ($\cdot OH$). Including hydroxyl free radical, other reactive species such as superoxide (O^{2-}) and hole (h^+) promote the breaking of covalent bond to convert organic contaminants into less toxic substances through oxidation or reduction mechanism [27]. Fenton's reagent is another efficacious chemical reagent which is useful in degradation process of dye. This reagent contains hydrogen peroxide and ferrous ion which are responsible for the degradation process to treat various organic pollutants. In this dye removal process hydroxyl radical plays an important role as it has strong oxidative capacity. The reusability characteristic of this reagent makes the entire process economical [28]. Electrochemical reduction process is another notable pollutant degradation method in which metals are used as catalytic system. Different organic pollutants such as azo dyes and phenol containing pollutants have been eliminated from wastewater by using electrochemical reduction process [29].

2.1.3 Physical Dye Removal Method

The mainstay of physical dye removal method is adsorption that includes both physical and physiochemical adsorption method. This method involves the direct eradication of organic and inorganic effluents through composites or any other materials with good adsorption capability. Dyes can be of various types but for the eradication and degradation, the material that are used during purification process is depend upon the charges present on dyes. The cationic dyes require negatively charged material on the surface of adsorbent and vice-versa. Physical method is eminently convenient, since this method uses no living organism and less amount of chemical is required as compared to other methods [30]. Different biological, physical and chemical methods for dye eradication have been illustrated in Fig. 5.

Grafting technique has been introduced during the synthesis process of numerous hybrid materials to enhance the adsorption ability. Singh and Dhaliwal have fabricated AgNPs/GG/Poly(AA) based on silver NPs. They followed in situ preparation technique for the incorporation of silver NPs onto the graft copolymer made up of acrylic acid and guar gum. This nanocomposite is one of the competent adsorbent materials for effluent eradication, since the diffusion process by which dyes are adsorbed onto the material is an endothermic and spontaneous process. The NPs used in this process provide high surface area to the adsorbent for better binding ability [31]. Many materials with high adsorption potential are used in the process of

Fig. 5 Schematic illustration of biological, physical and chemical dye removal method

decontaminating water. Among those kinds of materials, hydrogel with promising adsorption capability is one of the prominent materials that have been used as adsorbent for the purification of polluted water releases from different textile industries. For the fabrication of above kind of material, biodegradable resources have attracted the attention of many researchers since these materials are less toxic and environmentally friendly in nature. Hydrogel can be synthesised by using different kinds of polymeric material. But in view of above fact, now-a-days biodegradable polymeric materials have been used in the fabrication of hydrogel nanocomposites for low production of secondary pollutants. The best known biodegradable polymeric materials are polysaccharides. Mittal et al. has engineered polysaccharide based GK-cl-P(AA-co-AAM)/SiO_2 hydrogel nanocomposite for the adsorption of methylene blue (MB) dye. This P-*g*-NC material is based on environmentally friendly polymer gum karaya (GK). At first, the matrix is prepared by grating GK onto the other two polymers: poly acrylic acid and poly acylamide. Silica nanoscopic particles have been loaded to the polymer matrices by in situ preparation method [32]. Jing et al. has investigated the effectiveness of grafting on adsorption capacity of P-*g*-NCs by preparing Lys-PGMA@Fe_3O_4. The magnitude of grafting of poly(glycidal methacrylate) (PGMA) brushes on the surface of Fe_3O_4 NPs increases with increase in polymer formation. The prepared P-*g*-NC is a pH responsive material and due to this, it is able to absorb unlike charged dye at different pH concentration levels. Experimental findings revealed that lemon yellow (LY), a negatively charged dye shows higher adsorption ability at lower pH concentration value (pH = 4.0) whereas MB, a positively charged dye has high adsorption capacity at higher pH level (pH = 10.0). The fabricated system has the adsorption capacity for LY and MB is 0.54 and 0.85 $mmol{\cdot}g^{-1}$, respectively. One of the greatest advantages of using magnetic NPs in the preparation of P-*g*-NC is that it can be easily separated out from liquid after the purification technique by the use of magnetic field [30]. Kulal and Badalamoole have investigated the eradication of dye and heavy metal by fabricating a Ggh-g-PAcM/Fe_3O_4 P-*g*-NC hydrogel materials using magnetic NPs. The adsorption capacity of the nanocomposite hydrogel is much higher than normal hydrogel since the addition of NPs, fill up the free space present within the hydrogel. As a result, the surface area of the prepared materials increases. They have also studied the adsorptivity with respect to concentration of iron NPs and they concluded that the capacity of the adsorptivity increases with increase in the concentration of NPs but up to a certain limit. The prepared materials have shown excellent property in removal of bivalent heavy metals and cationic dyes. The adsorbent also showed excellent reusability property [33].

Hydroxyapatite is a nature originated mineral and has shown its essentiality and need as constituent material to repair bone. It exhibits biocompatibilty nature since it is one of the important compositions of bone and teeth. This naturally occurring biopolymer has largely been use in the application of detoxicfication of organic effluent like dyes because of strong interaction between the hydroxyapatite and dye molecules. Graphite oxide (GO) also able to capture dyes from industrial polluted water due to the presence of functional reactive groups on the surface of GO. Hosseinzadeh and Ramin have fabricated a hydrogel i.e., starch-graft-poly(acrylamide)/grapheneoxide/hydroxyapatite for the successful eradication of malachite green (MG) organic effluent. The major reasons behind the eradication of dye are the anionic charged surface of fabricated nanocomposite hydrogel and the interactions occurred between the functional groups present on nanocomposite and the cationic charged present on MG dye [34]. Other naturally occurring mineral in nanoscale range is palygorskite, a combined form of hydrated magnesium, aluminium and silicate minerals. Surface modification by means of chemical and physical approaches shows significant change in adsorption property of palygorskite mineral. AL-Hammadi et al. has synthesized a material by grafting poly(trimesoyl chloride-melamine) on the surface of palygorskite (PTMP) for dye and heavy metal eradication from water effluents. The prepared material has shown effective removal of MO and heavy metals like nickel, arsenic, mercury, cobalt, molybdenum and chromium [35].

2.2 Metal Ion Removal

Contamination of water by metal ions is one of the major reasons that put harmful impacts on ecosystem and also the consumption of hazardous heavy metal contaminated water imbalance immune system of human body. Heavy metals are generally exposed to water reservoirs due to the extraction of metals from its respective ores, from the industries that employ heavy metals as raw material for the manufacturing of products and from agricultural wastewater. Unlike dyes, heavy metals are inorganic effluents but the techniques and materials use for eradication process are similar to organic effluents up to some extents. Adsorbents are primarily divided into two types such as traditional and non-conventional. Non-conventional materials are based on naturally originated materials and are significantly more preferable than traditional one because of their nonimmunogenic, noncarcinogenic and non-toxic nature. The functional groups present on the biopolymer have the potential to formed chelated metal ion complexes with heavy metals ion present in contaminated water. Owing to the above facts, introduction of biopolymer as one of the preparatory material for the synthesis of adsorbent has gained attention of numerous scientists who have been devoted to research area based on water purification treatment related to heavy metal elimination [36]. Sodium alginate (Alg) has the suitable characteristics to form gel by reacting with several cross linkers. Mittal et al. has studied the adsorption of Pb(II) and Cu(II) by fabricating PMMA-g-Alg/Fe_3O_4 adsorbent using oxidative free radical grafting polymerisation technique. The adsorption of Pb(II) and Cu(II) by the synthesised material is found to be 62.5 mg g^{-1} and 35.71 mg g^{-1}, respectively [37]. Hexavalent chromium ion is considered as one of the most life threatening heavy metal since it has the ability of oxidise the bio molecules present in living organisms. Among biopolymers, Chitosan is the predominant one due of the presence of hydroxyl and amine groups which play an important role in adsorption of heavy metals by generating interactions between functional groups and heavy metal ions. Sharma et al. has fabricated chitosan and N-isopropylacylamide based P-*g*-NC and investigated selective elimination of various heavy metals such as chromium(II), iron(II), lead(II) and chromium(IV) [38]. Another heavy metal that generates negative influence on environment by polluting fresh water bodies is mercury. Recently, researchers have specified the mercury metal ions as a perilous pollutant because it has the competence to create harmful effects within the organisms by entering food web. The disclosure of mercury into the surroundings is mostly because of mining, paint and pharmaceutical industries. According to HASB theory, mercury comes under the soft acid category, so it needs soft base categorized functional groups to form bonds. Since papain contains four sulfhydryl functional groups as responsive sites to interact with metal ions, papain matrix can be employed for the removal of mercury. Metin and Alver have prepared a papain chitosan/clay composite bead using graft polymerisation technique to polymerise methacrylic acid and studied the selective adsorption of mercury ions in presence of other metal ions [39].

The extensive use of non-renewable resources for energy production has dwindled the number of resources but escalated the magnitude of habitat pollution. Due to the above reason, mankind has been forced to use nuclear energy as an alternative. The major reason for disclosure of radioactive metals into the fresh water reservoirs is nuclear power plant. Magnetic NPs have been predominantly used as surface area enhancer for adsorbent species because the magnetic property helps in easy separation of NPs loaded adsorbent after the treatment of wastewater. Bayramoglu and Arica have studied the selective adsorption of U(VI) ions with other poisonous metal ions from water by synthesised a silica coated magnetic nanoranged particles grafted with polymer [40]. Elsaied et al. has fabricated a super adsorbent composite by using derivative of cellulose and magnetic iron NPs with other polymers to entrap radioactive metal ions, namely Cs^{+}, Co^{2+} and Sr^{2+} [41]. A number of P-*g*-NCs with their applications and characterization methods are summarized in Table 1.

2.3 P-g-NC as Antifoulant

Eradication of microorganisms from contaminated sewage water is the most crucial step in wastewater treatment process. Predominantly numerous filtration techniques have been employed for the detoxification and purification of municipal and biomedical wastewater. Filtration process is a result of uprising research based on technology of membrane. This method has manifested highest involvement and considerable efficacious results without adding any ancillary chemicals during the water purification processes. Various type of membranes has been grouped into two main categories i.e., polymer based membrane and inorganic based membrane. The former type of membrane is widely used as a filtration medium because it contributes substantially high thermal, mechanical properties, ease of water permeability, greater fouling property and most influentially membrane forming potential than inorganic based membrane. Polymer based membranes technologies are generally classified into four categories on the basis of recuperating filtration method and diameter of pores

Table 1 A summary of different P-gNC$_s$ investigated for dyes and heavy metals eradication along with their characterization methods

SI. No	Polymer grafted nanocomposites (P-*g*-NCs)	Characterization/Properties	Application in eradication	Pollutants	References
1	Chitosan-g-acrylic acid/SiO_2 hydrogel	FTIR, TGA, SEM, Inductively coupled plasma mass spectrometry	Heavy metal	Co(II),Cu(II), Pb(II) and Zn(II)	[42]
2	Poly(AMPS)-g-GT/GO hydrogel	FTIR,FESEM, EDS, XRD,TGA	Heavy metal	Pb(II), Cd(II), and Ag(I)	[43]
3	Fe_3O_4/PMA-g-PVA	FTIR, TGA, SEM, XRD	Heavy metal	Ag(I), Ni(II), Cd(II), Pb(II) and Co(II)	[44]
4	Poly(acrylamide)-g-chitosan@γ-Fe_2O_3	FTIR, SEM, EDS, XRD, TEM,UV–Vis	Dye	Malachite Green	[45]
5	Dext-g-PMA/SiO_2	FT-IR, ^{13}CNMR, SEM, EDAX, XRD	Dye	Methylene blue, Crystal violet and Methyl orange	[46]
6	Chitosan-grafted-polyaniline/ Co_3O_4	FTIR, FESEM, TEM, XPS, XRD, TGA, UV–Vis	Dye	Methylene blue	[47]
7	k-carrageenan-g-poly(meth acrylic acid)@ Fe_3O_4 hydrogel	FT-IR, XRD, SEM, TEM, VMS	Dye	Crystal violet	[48]
8	Gum xanthan-grafted-polyacrylic acid@ Fe_3O_4 Hydrogel	FTIR, XRD, TEM and SEM–EDS	Dye	Methyl violet	[49]
9	GA-cl-PAM/ ZnO Nanocomposite	FTIR, SEM, TGA, XRD	Dye	Malachite green	[50]
10	Gum ghatti (Ggh) and 4-acryloylmorpholine (AcM) @ Fe_3O_4	FTIR, TGA, XRD, SEM, BET, VSM	Dye and Heavy metal	Methylene blue, Rhodamine 6 G, Cu(II) and Hg(II)	[33]

present on membranes: microfiltration (MF), ultrafiltration (UF), nanofiltration (NF) and reverse osmosis (RO). Among the above mentioned membranes, MF has porosity level with pore diameter ranging from 1 to 0.1 μm while RO has the least porous structure, possessing pore size less than 0.0001 μm. Depending upon the porosity level, membranes are used in different level of water treatment process to filtrated distinct pollutants. For instance, metal ions with one valence electron are separated by NF method. The filtration process, by using above four membranes depends upon the pressure applied on polluted liquid while passing through membrane. The process by which foulants get accumulated on the upper layer or within the pores present on solid material is called fouling. One of the serious problems that membrane technology has encountered is the accumulation of pollutants on surface of membrane which first leads to slowdown and then gradually occludes liquid flow through membrane. Due to the above facts, reusability capacity of film decreases and the process become less economical. Depending upon the foulant types and characteristics, various fouling membranes are being synthesised for water purification purposes. The avoidance of fouling caused by several microbes is one of the major strenuous processes since the number of bio-pollutants greatly increased due to the nutrients present on polluted water. Except the eradication of biofoulants, all other foulants removal through adsorption process is reversible in nature. Fouling property depends upon the hydrophilic and hydrophobic nature of materials that are used in the preparation of membrane. The membrane with high hydrophobic capability has its greater fouling ability towards organic wastewater release from industries. Many of nanoranged particles like silver, titanium oxide has shown outstanding performance in the removal process of biofoulants like bacterial because of their antibacterial characteristics against both gram positive and negative bacteria. The success rate of foulants adsorption is significantly improved by using nanocomposite membrane during the purification process [51]. Incorporation of NPs onto the matrix of polymer provides greater antifouling ability. But, due to the lack of stability, the NPs start to agglomerate and the dispersion of NPs is not uniform throughout the polymer matrix. Such kind of problem has been successfully avoided by grafting the surface of nanoranged particles with suitable polymer using reversible addition/fragmentation chain transfer technique. Also, the roughness present on surface has shown great proficiency on the performance of the membrane during the filtration process of polluted water. Rafiei et al. has reported that the anti-fouling properties of PVDF/PVP-g-TiO_2 thin film P-*g*-NC is greater than polyvinylidene fuoride (PVDF) membrane because of the high surface area and roughness [52]. The bacteria, having harmful effects on living organisms are called pathogen and are mainly responsible for water borne diseases. P-*g*-NCs are also used in water disinfection process to remove pathogens. Photocatalysis is one of the impressive characteristic that a NP possesses. TiO_2 NPs hold the capacity to harm cell wall of bacteria with the help of oxygen reactive species by utilizing solar energy as a source. Wang et al. has studied the photocatalytic effect of prepared $Fe_3O_4@TiO_2$ /glycopolymer nanocomposite. Above prepared material has the capability to encapsulate and damage of E.coli bacteria by rupturing the cell membrane [53]. The rupturing and damage of a cell are demonstrated in Fig. 6. Certain specific type of materials, with required hydrophilic environment, is needed

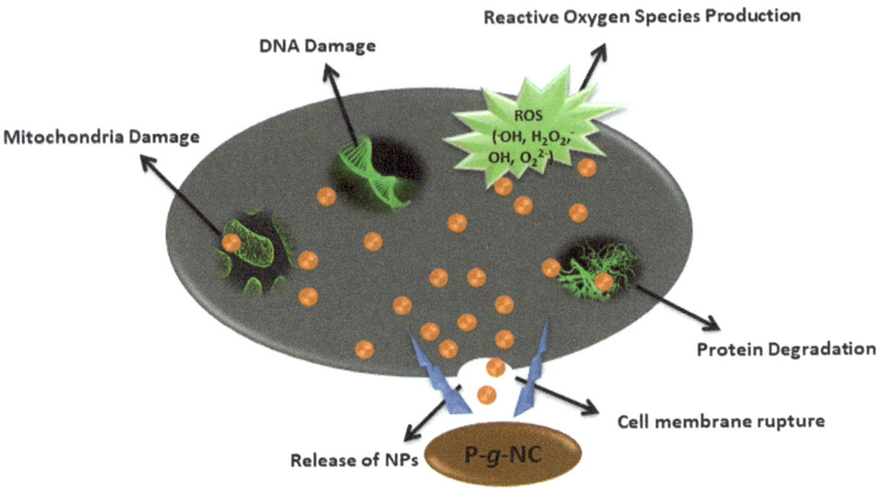

Fig. 6 Schematic representation of antibacterial activity shown by P-*g*-NCs

for the capturing of microorganisms inside the P-*g*-NCs. Hydrogel is suitable for this type of application due to the availability of pores for the entrapment of microbes. The functional groups present on GO surface have the high magnitude of adsorption capacity of pollutants by interacting with polymer. Glutamic acid (Glu) is a well-known eco-friendly and biocompatible material. Yu et al. has synthesized PVA/GO-Glu hybrid hydrogel nanocomposite beads by grafting the polyvinyl alcohol (PVA) onto the GO-Glu surface. This material shows greater loading capacity of microbes due to the enhanced size of pores present on PVA matrix [54].

2.4 Desalination

Sea and ocean hold water more than 95% of total earth's water but this kind of water has no utility in daily use of human and for industrial purposes. In view of the above reason, researchers have been focused their interest on desalinating and purifying sea water. Generally used methods for purification of saline water are distillation, multi-stage flash and electrodialysis. The extensively employed method for the purification of desalinated water is RO because this process has low cost and low energy consumption value. The most preferable material for this process is polymeric thin film composite with property enhancer nanoscopic particles incorporated within it. In last decades, researchers are more fascinated into the fabrication of membrane that capable of performing in a long-term basis and have the characteristic of low negative impact on surroundings. Forward osmosis has the cardinal importance for the purification of saline water as it overcomes certain disadvantages such as high energy consumption, less antifouling ability, less reusable capacity of membrane and expensive raw material demand [55]. β-cyclodextrin is a hydrophobic cavity containing oligosaccharide polymer. The property of β-cyclodextrin is augmented by functionalized it with GO for water treatment application. Matshetshe et al. has fabricated a thin film i.e., β-CD-f-GO/PA with greater antifouling, desalinating and antimicrobial property [56]. Bagheripour et al. has fabricated PES/PAA-Fe_3O_4 NF membrane by using immersion precipitation method and casting solution technique. Magnetic iron NPs are first grafted with poly acrylic acid to form polymer brushes and then distributed uniformly in matrices of polyethersulfone (PES). The grafted brush is used as a filler to enhance hydrophilic attraction between membrane and pollutants. They concluded that the porosity and water permeability through the prepared P-*g*-NC membrane are increasing with increase in loading of polymer brush upto 0.5%wt [57].

3 Conclusion

Treatment, purification and disinfection of wastewater are the need of hour to fulfil the demand of fresh water. Degradation of dye, eradication of heavy and radioactive metals, removal of pathogenic bacteria and desalination of brackish water are the most important applications in water purification process. By changing fabrication procedure with simple inclusion of grafting method such as "grafting to", "grafting from" and "grafting through" has shown extraordinary outcomes of P-*g*-NCs than the PNCs formed by conventional process. Out of wide varieties of materials, P-*g*-NC is the best potential candidate for this application because of its pronounced adsorption ability, notable photocatalytic behaviour, remarkable antibacterial property and exceptional mechanical and thermal stability. P-*g*-NC has been successfully implemented and investigated its predominance in various wastewater purification processes not only at laboratory level but also at commercial level.

References

1. Bolisetty S, Peydayesh M, Mezzenga R (2019) Sustainable technologies for water purification from heavy metals: review and analysis. Chem Soc Rev 48:463
2. Yadav S, Asthana A, Chakraborty R, Jain B, Singh AK, Carabineiro SAC, Susan MABH (2020) Cationic dye removal using novel magnetic/activated charcoal/β-cyclodextrin/alginate polymer nanocomposite. Nanomaterials 10(1):170
3. Vodnar DC, Mitrea L, Călinoiu LF, Szabo K, Ştefănescu BE (2020) Removal of bacteria, viruses, and other microbial entities by means of nanoparticles. Adv Nanostruct Environ Health 1:465–491
4. Lai GS, Lau WJ, Goh PS, Ismail AF, Yusof N, Tan YH (2016) Graphene oxide incorporated thin film nanocomposite nanofiltration membrane for enhanced salt removal performance. Desalination 387:14–24
5. Khan I, Saeed K, Khan I (2017) Nanoparticles: properties, applications and toxicities. Arab J Chem 12(7):908–931

6. McNamara K, Tofail SAM (2016) Nanoparticles in biomedical applications. Adv Phy: X 2(1):54–88
7. Cardoso VF, Francesko A, Ribeiro C, Bañobre-López M, Martins P, Lanceros-Mendez S (2017) Advances in magnetic nanoparticles for biomedical applications. Adv Healthc Mater 7(5):1700845
8. Kumar A, Choudhary A, Kaur H, Meheta S, Husen A (2021) Metal-based nanoparticles, sensors, and their multifaceted application in food packaging. J Nanobiotechnol 19:256
9. Zhang X, Qian J, Pan B (2016) Fabrication of novel magnetic nanoparticles of multifunctionality for water decontamination. Environ Sci Technol 50(2):881–889
10. Xu Y, Musumeci V, Aymonier C (2019) Chemistry in supercritical fluids for the synthesis of metal nanomaterials. React Chem Eng. 49:2030–2054
11. Busquets M, Espargaró A, Sabaté R, Estelrich J (2015) Magnetic nanoparticles cross the blood-brain barrier: when physics rises to a challenge. Nanomaterials 5(4):2231–2248
12. Prochowicz D, Kornowicz A, Lewiński J (2017) Interactions of native cyclodextrins with metal Ions and inorganic nanoparticles: fertile landscape for chemistry and materials science. Chem Rev 117(22):13461–13501
13. Mmelesi OK, Masunga N, Kuvarega A, Nkambule T, Mamba B, Kefeni K (2020) Cobalt ferrite nanoparticles and nanocomposites: photocatalytic, antimicrobial activity and toxicity in water treatment. Mater Sci Semicond Process 153(2010):105523
14. Rajak DK, Pagar D, Kumar R, Pruncu CI (2019) Recent progress of reinforcement materials: a comprehensive overview of composite materials. J Mater Res Technol 8(6):6354–6374
15. Ghosal K, Agatemor C, Špitálsky Z, Thomas S, Kny E (2018) Electrospinning tissue engineering and wound dressing scaffolds from polymer-titanium dioxide nanocomposites. Chem Eng J 358:1262–1278
16. Wang W, Wang P, Tang X, Elzatahry A, Wang S, Al-Dahyan D, Zhao M, Yao C, Hung C-T, Zhu X, Zhao T, Li X, Zhang F, Zhao D (2017) Facile synthesis of uniform virus-like mesoporous silica nanoparticles for enhanced cellular internalization. ACS Cent Sci 3(8):839–846
17. Buşilă M, Muşat V, Textor T, Mahltig B (2015) Synthesis and characterization of antimicrobial textile finishing based on Ag:ZnO nanoparticles/chitosan biocomposites. RSC Adv 5(28):21562–21571
18. Beyene H D, Ambaye T G (2019) Application of sustainable nanocomposites for water purification process. In: Sustainable polymer composites and nanocomposites, pp 387–412
19. Lofrano G, Carotenuto M, Libralato G, Domingos RF, Markus A, Dini L, Meric S (2016) Polymer functionalized nanocomposites for metals removal from water and wastewater: an overview. Water Res 92:22–37
20. Vega-Hernández MA, Cano-Díaz GS, Vivaldo-Lima E, Rosas-Aburto A, Hernández-Luna MG, Martinez A, Palacios-Alquisira J, Mohammadi Y, Penlidis A (2021) A review on the synthesis, characterization, and modeling of polymer grafting. Processes 9(2):375
21. Lenart WR, Hore MJA (2017) Structure–property relationships of polymer-grafted nanospheres for designing advanced nanocomposites. Nano-Struct Nano-Objects 16:428–440
22. Huber F, Berwanger J, Polesya S, Mankovsky S, Ebert H, Giessibl FJ (2019) Chemical bond formation showing a transition from physisorption to chemisorption. Science 366(6462):235–238
23. Pandey N, Shukla SK, Singh NB (2017) Water purification by polymer nanocomposites: an overview. Nanocomposites 3(2):47–66
24. Giovino M, Pribyl J, Benicewicz B, Bucinell R, Schadler L (2019) Mechanical properties of polymer grafted nanoparticle composites. Nanocomposites 4(4):244–252
25. Katheresan V, Kansedo J, Lau SY (2018) Efficiency of various recent wastewater dye removal methods: a review. J Environ Chemi Eng 6(4):4676–4697
26. Castro FD, Bassin JP, Alves TLM, Sant'Anna GL, Dezotti M (2020) Reactive Orange 16 dye degradation in anaerobic and aerobic MBBR coupled with ozonation: addressing pathways and performance. Int J Environ Sci Technol 18:1991–2010
27. Hieu Nguyen C, Lien Tran M., Tuong Van Tran T, Juang R.-S (2019) Enhanced removal of various dyes from aqueous solutions by UV and simulated solar photocatalysis over TiO_2/ZnO/rGO composites. Sep Purif Technol 23:115962

28. Shi X, Tian A, You J, Yang H, Wang Y, Xue X (2018) Degradation of organic dyes by a new heterogeneous Fenton reagent - Fe2GeS4 nanoparticle. J Hazard Mater 353:182–189
29. Popli SA, Patel P, Date M, Ruparelia J, Patel UD (2020) Rapid electro-catalytic reduction of azo dyes and phenolic compounds in the presence of metallic Palladium. Sep Purif Technol 254:117658
30. Jing S, Wang X, Tan Y (2018) Preparation of lysine-decorated polymer-brush-grafted magnetic nanocomposite for the efficient and selective adsorption of organic dye. Appl Surf Sci 441:654–662
31. Singh J, Dhaliwal AS (2020) Effective removal of methylene blue dye using silver nanoparticles containing grafted polymer of guar gum/acrylic acid as novel adsorbent. J Polym Environ 29:71–88
32. Mittal H, Maity A, Ray SS (2015) Synthesis of co-polymer-grafted gum karaya and silica hybrid organic–inorganic hydrogel nanocomposite for the highly effective removal of methylene blue. Chem Eng l 279:166–179
33. Kulal P, Badalamoole V (2020) Efficient removal of dyes and heavy metal ions from waste water using Gum ghatti – graft – poly (4-acryloylmorpholine) hydrogel incorporated with magnetite nanoparticles. J Environ Chem Eng 6:104207
34. Hosseinzadeh H, Ramin S (2018) Fabrication of starch-graft-poly (acrylamide) /graphene oxide/ hydroxyapatite nanocomposite hydrogel adsorbent for removal of malachite green dye from aqueous solution. Int J Biol Macromol 106:101–115
35. AL-Hammadi SA, Al-Absi AA, Bin-Dahman OA, Saleh TA (2018) Poly (trimesoyl chloride-melamine) grafted on palygorskite for simultaneous ultra-trace removal of methylene blue and toxic metals. J Environ Manag 226:358–364
36. Fouda-Mbanga BG, Prabakaran E, Pillay K (2021) Carbohydrate biopolymers, lignin based adsorbents for removal of heavy metals (Cd2+, Pb2+, Zn2+) from wastewater, regeneration and reuse for spent adsorbents including latent fingerprint detection: a review. Biotechnol Rep 30:e00609
37. Mittal A, Ahmad R, Hasan I (2015) Poly(methylmethacrylate)-grafted alginate/ Fe3O4 nanocomposite: synthesis and its application for the removal of heavy metal ions. Desalin Water Treat 57(42):19820–19833
38. Sharma RK, Chopra Lalita, Singh A P (2017) Sorption of Pb(II), Cu(II), Fe(II) and Cr(VI) metal ions onto cross-linked graft copolymers of chitosan with binary vinyl monomer mixtures. React Funct Polym 121:32–44
39. Metin AÜ, Alver E (2016) Fibrous polymer-grafted chitosan/clay composite beads as a carrier for immobilization of papain and its usability for mercury elimination. Bioprocess Biosyst Eng 39(7):1137–1149
40. Bayramoglu G, Arica M Y (2019) Star type polymer grafted and polyamidoxime modified silica coated-magnetic particles for adsorption of U (VI) ions from solution. Chem Eng Res Des 147:146–159
41. El-saied HA, Shahr El-Din AM, Masry BA, Ibrahim AMA (2020) Promising superabsorbent nanocomposite based on grafting biopolymer/nanomagnetite for capture of 134Cs, 85Sr and 60Co radionuclides. J Polym Environ 28(6):1749–1765
42. Pourjavadi A, Tehrani ZM, Salimi H, Banazadeh A, Abedin N (2015) Hydrogel nanocomposite based on chitosan-g-acrylic acid and modified nanosilica with high adsorption capacity for heavy metal ion removal. Iran Polym J 24:725–734
43. Sahraei R, Ghaemy M (2017) Synthesis of modified gum tragacanth/graphene oxide composite hydrogel for heavy metal ions removal and preparation of silver nanocomposite for antibacterial activity. Carbohydr Polym 157:823–833
44. Liu X, Guan J, Lai G, Xu Q, Bai X, Wang Z, Cui S (2020) Stimuli-responsive adsorption behavior toward heavy metal ions based on comb polymer functionalized magnetic nanoparticles. J Cleaner Prod 253:119915
45. Hasan I, Bhatia D,Walia S, Singh P (2020) Removal of malachite green by polyacrylamideg-chitosan γ-Fe_2O_3 nanocomposite-an application of central composite design. Groundwater Sustain Dev 100378

46. Ray J, Jana S, Mondal B, Tripathy T (2018) Enhanced and rapid adsorptive removal of toxic organic dyes from aqueous solution using a nanocomposite of saponified polymethyl acrylate grafted dextrin with embedded nanosilica. J Mol Liq 275:879–894
47. Shahabuddin S, Sarih NM, Ismail FH, Shahid MM, Huang NM (2015) Synthesis of chitosan grafted-polyaniline/Co_3O_4 nanocube nanocomposites and their photocatalytic activity toward methylene blue dye degradation. RSC Adv 5(102):83857–83867
48. Gholami M, Vardini MT, Mahdavinia GR (2016) Investigation of the effect of magnetic particles on the Crystal Violet adsorption onto a novel nanocomposite based on κ- carrageenan-g-poly(methacrylicacid). Carbohydr Polym 136772–781
49. Mittal H, Kumar V, Saruchi, Ray SS (2016) Adsorption of methyl violet from aqueous solution using gum xanthan/Fe_3O_4 based nanocomposite hydrogel. Int J Biol Macromol 891–11
50. Mittal H, Morajkar PP, Al-Alili A, Alhassan SM (2020) In-situ synthesis of ZnO nanoparticles using gum Arabic based hydrogels as a self-template for effective malachite green dye adsorption. J Polym Environ 28(6):1637–1653
51. Zahid M, Rashid A, Akram S, Rehan Z A, Razzaq W (2018) A comprehensive review on polymeric nano-composite membranes for water treatment. J Membr Sci Technol 8(1):1000179
52. Rafiei H, Abbasian M, Yegani R (2021) Polyvinylidene fluoride as a neat and the synthesized novel membranes based on PVDF/polyvinyl pyrrolidone polymer grafted with TiO_2 nanoparticles through RAFT method for water purification. Iran Polym J 30:769–780
53. Wang B, Shang C, Miao Z, Guo S, Zhang Q (2021) Lactose-containing glycopolymer grafted onto magnetic titanium dioxide nanomaterials for targeted capture and photocatalytic killing of pathogenic bacteria. Europ Polym J 142:110159
54. Yu Y, Shu Y, Ye L (2018) In situ crosslinking of poly (vinyl alcohol)/graphene oxideglutamic acid nano-composite hydrogel as microbial carrier: intercalation structure and its wastewater treatment performance. Chem Eng J 336:306–314
55. Jain H, Garg MC (2021) Fabrication of polymeric nanocomposite forward osmosis membranes for water desalination—a review. Environ Technol Innov 23:101561
56. Matshetshe K, Sikhwivhilu K, Ndlovu G, Tetyana P, Moloto N, Tetana Z (2021) Antifouling and antibacterial β-cyclodextrin decorated graphene oxide/polyamide thin-film nanocomposite reverse osmosis membranes for desalination applications. Sep Purif Technol 278:119594
57. Bagheripour E, Moghadassi AR, Hosseini SM (2017) Incorporated poly acrylic acid-co-Fe_3O_4 nanoparticles mixed matrix polyethersulfone based nanofiltration membrane in desalination process 30:821–829

Nanohybrid Polymeric Membranes for Water Purification and Separation

P. V. Chai, C. Y. Jong, S. F. Chua, and W. L. Ang

Abstract Nanohybrid based polymeric membranes technology for water purification and separation have grown over the years associated to its advancement in the past decades and remarkable impact to the membrane characteristics and performance is encouraging. In this chapter, the progress of the nanohybrid based polymeric membranes for water-based application was divided into two major sections namely membrane technology and nanohybrid polymeric membranes. The membrane technology discussed on the pressure driven membrane process, membrane configuration, membrane fabrication, membrane materials and types of membranes while the latter part discussed on graphene oxide based and non-graphene nanohybrid based polymeric membrane for water purification and separation.

Keywords Nanohybrid · Polymeric membrane · Water purification and separation · Graphene

1 Introduction

A membrane is defined as a thin sheet, film, or layer, which acts as a selective barrier between two phases and controls the passage of matter between the phases adjacent to it [1]. Membrane technologies are the processes that utilize membranes to obtain separation goals. Although membrane technologies became commercially available more than several decades ago, they are currently one of the fastest growing separation technologies for a wide range of applications.

P. V. Chai (✉) · C. Y. Jong
Department of Chemical and Petroleum Engineering, UCSI University, 56000 Cheras, Kuala Lumpur, Malaysia
e-mail: chaipv@ucsiuniversity.edu.my

S. F. Chua · W. L. Ang
Department of Chemical and Process Engineering, Universiti Kebangsaan Malaysia UKM, 43600 Bangi, Selangor, Malaysia

W. L. Ang
Centre for Sustainable Process Technology, Universiti Kebangsaan Malaysia UKM, 43600 Bangi, Selangor, Malaysia

S. K. Swain (ed.), *Nanohybrid Materials for Water Purification*, Composites Science and Technology, https://doi.org/10.1007/978-981-19-2332-6_7

In water and wastewater industries, membrane processes have been deployed to remove impurities and hazardous pollutants for the production of clean drinking water, prevention of environmental pollution from industrial effluent, and reclamation and recovery of treated water as sustainable water resource [2–4]. Tapping on its molecular-separation capability, the membrane has also been applied to control the outdoor and indoor air pollution by preventing the gaseous pollutants from being released to the atmosphere [5, 6]. Membrane technology is also widely employed in the food, pharmaceutical, and biotechnology industries, where it can separate the valuable compounds from undesired components to obtain high-purity products [7, 8]. Driven by the increasingly cost-effective membranes, with increasingly improved performance and characteristics, the applications of membranes in the industries are expected to experience rapid growth and development.

1.1 Pressure-Driven Membrane Processes

There are many different types of membrane processes available but this book chapter will focus on membrane processes related to water purification. In the water industry, pressure-driven membrane processes hold the dominant share in the market due to its maturity and ease of operation. Generally, a pressure is exerted on the solution at the feed side of the membrane to serve as a driving force to separate the stream into permeate and retentate as shown in Fig. 1 [9]. The permeate is the filtered water free from impurities or pollutants, whereas the retentate is a concentrated solution laden with rejected impurities that must be disposed of or treated by other methods. In another case, particularly for microfiltration (MF) and ultrafiltration (UF) membranes, the driving force for the membrane processes is vacuum-driven instead

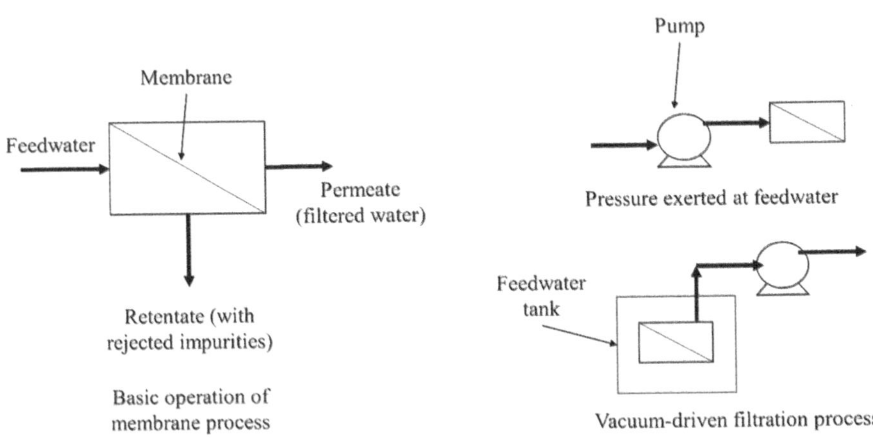

Fig. 1 Pressure-driven membrane processes

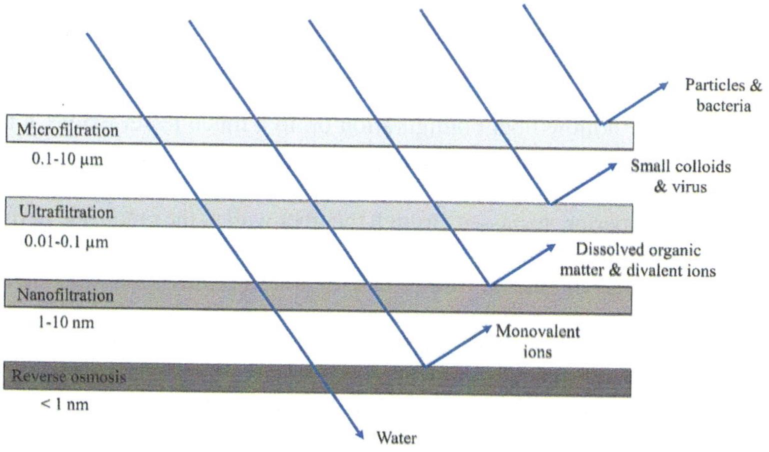

Fig. 2 Separation capability of pressure-driven membranes

of pressurized feedwater. A partial vacuum in the filtrate flow stream is created by a filtrate pump or gravity siphon, which extracts the water from the feedwater tank.

The pressure-driven membrane processes are categorized according to the membrane types, namely MF, UF, nanofiltration (NF), and reverse osmosis (RO). These membranes are permeable to water but not to components that are rejected and removed, depending on the characteristics of membranes, particularly membrane pore size [10]. As shown in Fig. 2, MF membranes possess the largest pore size and are capable of rejecting larger impurities such as particulate matter and bacteria. UF membranes with smaller pore size in between 0.01 and 0.1 µm could remove small colloids and viruses in water. The removal of dissolved organic matter and divalent ions (e.g., Ca^{2+} and Mg^{2+}) is normally conducted with NF membranes. RO membranes have the tightest and dense surface that could remove even the smallest size of monovalent species (e.g., Na^+ and Cl^-).

The selection of membrane processes relies on the quality of feed water to be treated and the desirable quality of treated water to be obtained. For instance, RO should be used for desalination of seawater since the monovalent ions could only be removed by RO membranes [11]. On the other hand, UF membranes will be used for surface water treatment since viruses and small colloids in surface water are the typical impurities to be removed [12]. The removal capability of membranes is affected by the membrane characteristics. MF and UF membranes reject the impurities through sieving action, which mainly relies on the size difference between the membrane pores and impurities size. On the other hand, NF and RO membranes possess a thin and dense permselective layer on top of the membrane surface, which separates solutes by diffusion and sieving action [13]. This also reflects that higher feed pressure is required for NF and RO processes to overcome the resistance from the permselective layer.

1.2 Membrane Configuration

Most pressure-driven membrane processes for water industry appear in the form of spiral-wound or hollow fiber configuration or, to a much lesser extent, a tubular or plate-and-frame form. Hollow fiber membranes are fibers of 0.1–1 mm diameter with a hollow space inside. Feed water is supplied either inside or outside of the membrane and the permeate passes through the fiber wall to the other side of the fiber [14]. Hollow fiber membrane module is frequently used for MF and UF membranes due to its high surface/volume ratio that could filter and produce more water in a much smaller footprint. For instance, hollow fiber UF membranes have been successfully deployed for drinking water treatment plants in China, where the membranes were capable of showing excellent rejection of particles and organic matter [15]. Nonetheless, advancement of membrane sciences has enabled the production of fiber strong enough to withstand high operating pressure. This can be seen from the reported articles on the fabrication of hollow fiber NF and RO membranes [16].

On the other hand, commercial NF and RO are normally sold in spiral-wound configurations, where flat-sheet membranes are assembled into spiral-wound elements. The spiral-wound module is particularly suitable for NF and RO membrane processes due to several features such as high-pressure durability, compactness, minimum concentration polarization, and minimum pressure drop in the permeate channel. For example, membrane desalination plants around the world deployed spiral-wound RO membranes to convert the salty water to drinkable water [17].

1.3 Membrane Fabrications

There are several methods to fabricate membranes. The synthesis conditions play a key role in altering the membrane structures and properties. Hence, understanding the membrane fabrication methods will help to control the synthesis conditions to produce membranes with desired properties. In general, the typical fabrication methods employed for the synthesis of commercial membranes are phase inversion (PI), interfacial polymerization (IP), and hollow fiber spinning.

1.3.1 Phase Inversion

PI is a process mainly used for the preparation of MF and UF membranes. It can be described as a demixing process in which the initially homogeneous polymer solution (polymer dissolved in solvent) is transformed in a controlled way from a liquid to a solid phase [18]. There are four basic techniques used to fabricate PI membranes: precipitation from vapour phase, precipitation by controlled evaporation, thermally induced phase separation, and immersion precipitation [19–21]. Out of these techniques, immersion precipitation is the most widely used method to prepare polymeric

MF and UF membranes. The polymers will be dissolved in solvent and casted on a proper supporting layer before submerged in a coagulation bath containing nonsolvent. The solvent and nonsolvent exchange will lead to the precipitation of polymers, which eventually end up as polymeric membranes [22].

1.3.2 Interfacial Polymerization

IP is the fabrication technique that revolutionises the membrane field with the production of thin film composite (TFC) RO membrane. IP involves the cross linking between two types of monomers deposited on a support layer to form a dense yet thin permselective layer. The monomers used are aqueous solution of m-phenylenediamine or piperazine and organic solution of trimesoyl chloride. The cross linking between these monomers results in the formation of polyamide layers that are responsible for the high rejection capability of NF and RO membranes towards dissolved mineral ions [23].

1.3.3 Hollow Fiber Spinning

Hollow fiber spinning is a fabrication process used to produce hollow fiber membranes. The concept is based on PIyet the polymer solution is extruded through a spinneret to form a hollow fiber shape before solidified to membrane in coagulation bath (usually water). The properties of the membranes can be altered by manipulating the dimensions of the spinneret, temperature and composition of dope (polymer) and bore (solvent) solutions, and temperature and composition of the coagulation bath [24, 25].

1.4 Membrane Materials

Intensive research has been conducted to improve the characteristics of membranes as the properties will directly affect the filtration efficiency and production of treated water. The most straightforward method to tune the membrane properties is to alter the formulation of membrane solution before casted to membrane. Generally, membranes can be synthesised from two material sources: organic and inorganic. Organic membranes are fabricated from various types of polymers such as polyethersulfone (PES), polyvinlyidenedifluoride (PVDF), polytetrafluoroethylene (PTFE), polyamide-imide, cellulose, and polyacrylonitrile (PAN) [26, 27]. Majority of the commercialised membranes (polymeric membranes) are produced from these materials due to the lower cost and ease of synthesis.

The second group of membrane materials belong to the inorganic class that includes metallic membranes, ceramic membranes, and zeolite membranes. Though the cost of inorganic membranes is higher than polymeric membranes, it outperforms the latter at other aspects such as better chemical, mechanical, and thermal resistance, which enables it to be used in harsh environments such as corrosive and high temperature environments [28, 29].

Regardless of the materials used to synthesize membranes, researchers have been actively working on studies to incorporate nanomaterials during the membrane synthesis process to further improve the characteristics and performance of the membrane.

1.5 Types of Nanocomposite Membrane

Nanocomposite membranes are membranes that could be fabricated by incorporation of nanoparticles. The techniques of nanoparticles incorporation can be divided into two main groups, which includes modification done on polymer matrix and surface modification [30]. By incorporation of nanoparticles, such as zeolite nanoparticles, carbon or nanocarbon, silica nanoparticles and zirconium dioxide, besides creating pore channel, the nanoparticles have also enhanced the membrane surface hydrophilicity and surface area at the same time showing stable performance [2, 3].

Mixed matrix membrane (MMM) is membrane prepared via polymer matrix modification. The surface modification membrane on the other hand can be divided into four main groups, which are TFC membrane, thin-film nanocomposite (TFN) membrane, TFC membrane with nanocomposite substrate and surface located nanocomposite membrane. Table 1 shows some review papers and articles regarding the advancement of different types of nanocomposite membranes in recent years.

1.5.1 Mixed Matrix Membrane (MMM)

MMM is membrane with the incorporation of a solid phase or inorganic materials into the continuous polymer matrix as shown in Fig. 3 [41]. In order to fabricate MMM, PI is one of the most commonly used method in which the nanoparticles will be blended into the dope solution before the casting and PI procedure. Besides enhancing the membrane properties and performance, the incorporation of nanoparticles such as silver (Ag) and titanium dioxide (TiO_2) nanoparticles also enable membranes to behave as multifunctional membrane by having additional functions such as antibacterial and photocatalytic properties [5–7].

1.5.2 Conventional Thin-Film Composite (TFC) Membrane

The conventional TFC membranes are thin-film membrane that could be fabricated even without the addition of nanoparticles. By referring to Fig. 4, TFC consists of a support or substrate layer and a selective layer. The substrate layer is often the

Table 1 Recent advancement of nanocomposite membrane

Title	Type of modification	Year	References
Surface modified nanocomposite membrane	Surface modification	2018	Lakhotia et al. [30]
Critical review of montmorillonite/polymer mixed-matrix filtration membranes: Possibilities and challenges	Polymer matrix modification	2018	Dlamini et al. [31]
Carbon nanotube membranes for water purification: Developments, challenges, and prospects for the future	Polymer matrix & surface modification	2018	Ihsanullah [32]
Recent advances in thin film composites membranes for brackish groundwater treatment with critical focus on Saskatchewan water sources	Surface modification	2019	Mollahosseini and Abdelrasoul [33]
Metal–organic framework membranes for wastewater treatment and water regeneration	Polymer matrix and surface modification	2019	Li et al. [34]
Challenges and opportunities in functional carbon nanotubes for membrane-based water treatment and desalination	Polymer matrix and surface modification	2019	Ali et al. [35]
Recent development of graphene oxide based forward osmosis membrane for water treatment: a critical review	Polymer matrix and surface modification	2020	Wu et al. [36]
Recent development in modification of polysulfone membrane for water treatment application	Polymer matrix and surface modification	2020	Mamah et al. [37]

(continued)

Table 1 (continued)

Title	Type of modification	Year	References
Mussel-inspired polydopamine modification of polymeric membranes for the application of water and wastewater treatment: a review	Polymer matrix and surface modification	2020	Yan et al. [38]
Layered double hydroxide-modified membranes for water treatment: recent advances and prospects	Polymer matrix and surface modification	2021	Sajid et al. [39]
Incorporation of graphene oxide-based nanocomposite in the polymeric membrane for water and wastewater treatment: a review on recent development	Polymer matrix and surface modification	2021	Ng et al. [40]

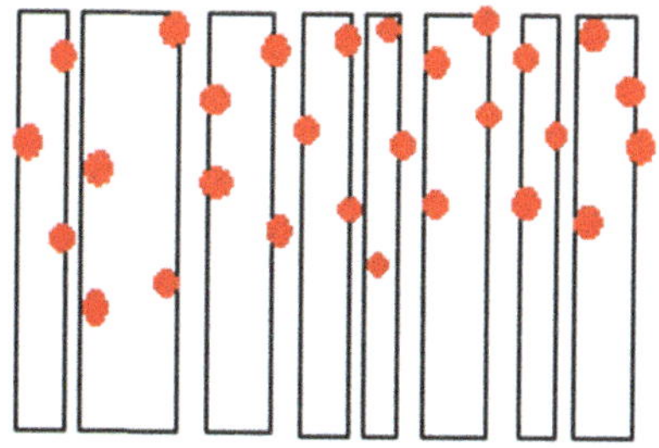

Fig. 3 Mixed matrix membrane incorporated with nanoparticles (represented by red dots)

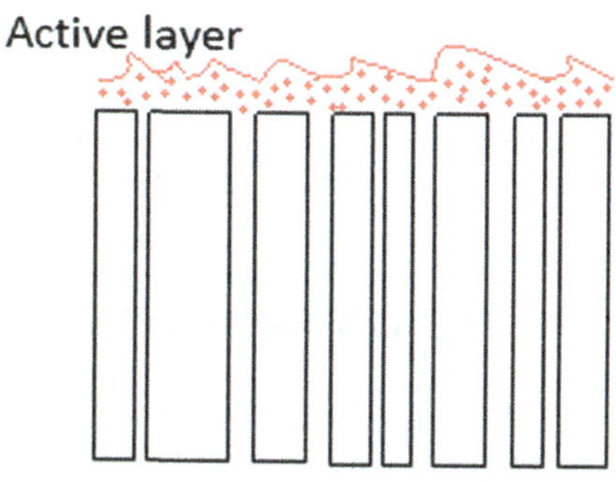

Fig. 4 TFC membrane without addition of nanoparticles

polymeric porous layer that are made of materials such as polysulfone (PSF) and PESwhich could be prepared via PI technique. On the other hand, the selective layer which also known as the active layer is a polyamide layer deposited on top of the substrate layer by using IP method. The active layer is formed by crosslinking of the aqueous and organic monomers. These thin-film membranes are mainly being utilized in NF, RO and FO process [42]. However, trade-off effect in between water permeability and solution rejection often occurs in TFC membrane. Hence, the incorporation of nanoparticles is to enhance the hydrophilicity and water permeability, without sacrificing or reducing the solute rejection.

1.5.3 Thin-Film Nanocomposite (TFN) Membrane

As shown in Fig. 5, TFN membrane has emerged from TFC membrane by the incorporation of nanoparticles. The nanoparticles could be added as nanofillers either into the aqueous or organic phase monomers. Hydrophilic nanofillers, such as the carbon-based or nanocarbon-based materials, TiO_2 and nanohybrid layered double hydroxide (LDH) are usually added into aqueous monomers phase. This is because the hydrophilic nanofillers cannot dispersed well in organic solution [43]. However, the surface of the nanofillers could be modified in order to enhance its dispersion ability in organic or non-polar solvent. Surface modification could be done on nanoparticles by using organic compound or silane compound in order to provide

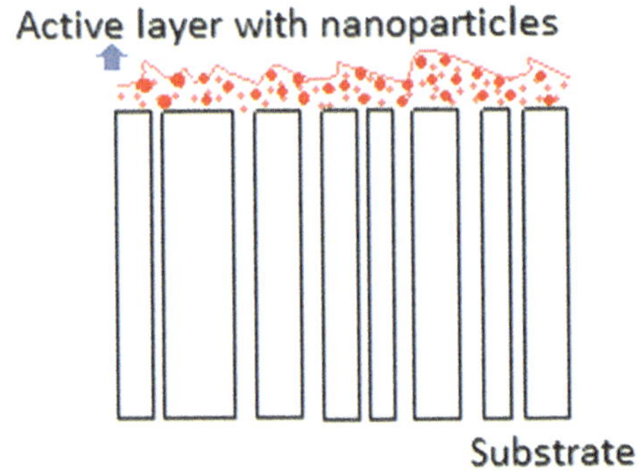

Fig. 5 TFN membrane

hydrophobic functional chains to the nanoparticles [10, 11]. These nanofillers has found to increase the hydrophilicity and reduce the fouling propensity [44].

1.5.4 TFC Membrane with Nanocomposite Substrate

For TFC membrane with substrate that is incorporated with nanoparticles (Fig. 6), the structure is of similar as TFN membrane and MMM. The difference is the position of adding the nanoparticles, in TFN nanoparticles are added to the selective layer (either aqueous or organic phase monomers), whereas for this TFC membrane, the nanoparticles are incorporated in the substrate layer. This substrate layer has structure which is similar to MMM but on top of the substrate layer, there will be a selective layer which mainly made by crosslinking or aqueous and organic monomer. This TFC membrane with nanocomposite substrate has found to have high porosity support layer and also improving hydrophilicity of TFC membrane as well as lower the internal concentration polarization which often a major issue for FO [12, 13].

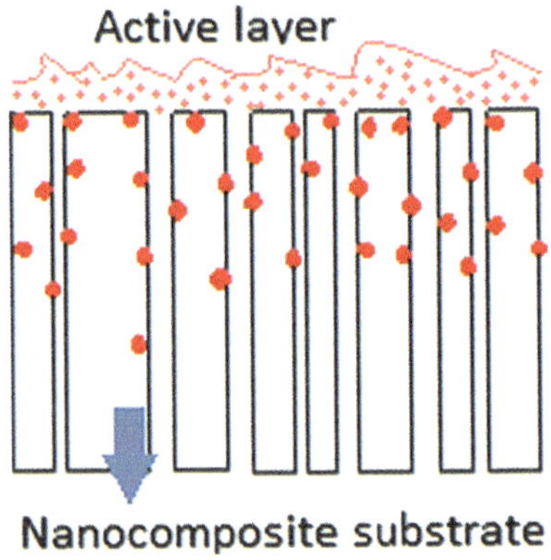

Fig. 6 TFC membrane with nanocomposite substrate

1.5.5 Surface Located Nanocomposite Membrane

Figure 7 shows the surface located nanocomposite membrane. For this type of membrane, membrane surface modification is usually done by depositing nanoparticles on membrane via self-assembly method, chemical grafting, layer-by-layer assembly and coating method. Both self-assembly and coating are simpler methods. In self-assembly method, membrane is soaked completely into solution that contains nanoparticles or by allowing the nanoparticles solution to sit on the membrane surface then follow by drying and solvent evaporation. For coating method, the deposition of nanoparticles is done via filtration or dipping method. However, the nanoparticles are weakly attached to the membrane surface and leaching often occurs during filtration process [45]. Different from coating and self-assembly method, nanoparticles are deposited on the surface of the membrane either via weak Van Der Waals force or electrostatic force in chemical grafting method. Layer-by-layer method on the other hand involved multiple layers of nanoparticles attached to the membrane surface through different interaction, such as chemical bonding, electrostatic force and also hydrogen bonding [30]. Due to these membranes have the minimal effect on membrane intrinsic structure, it has seen to be one of the potential techniques to be implemented in enhancing the membrane surface properties [26].

2 Nanohybrid Polymeric Membranes

In the recent years, there are a number of works have been devoted in designing nanohybrid based polymeric membrane for water purification. One of the approaches is to adopt nanohybrid materials as itcould offer various possibilitiesduring the design of the membrane. Nanohybrid materials is simply defined as a combination of at least two nanomaterials by means of chemical or physical bonding to obtain synergistic effect [46–48]. Up to date, there are a lot of available nanohybrid polymeric membrane for water purification application available in theonline database. This section will discuss two major types of nanohybrid materials that wasused namely graphene oxide (GO) and non-graphene oxide based nanohybrids materials.

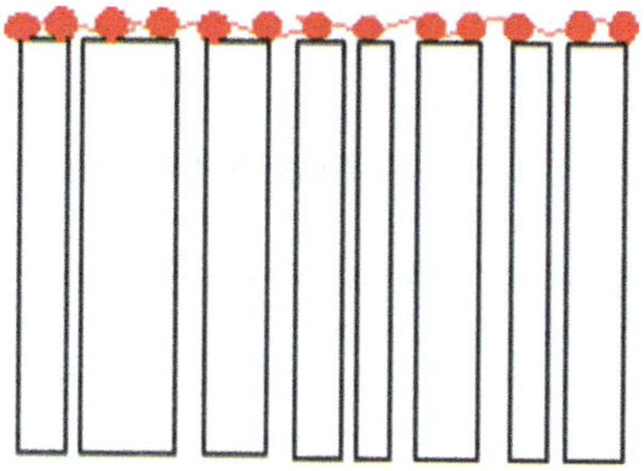

Fig. 7 Surface located nanocomposite membrane

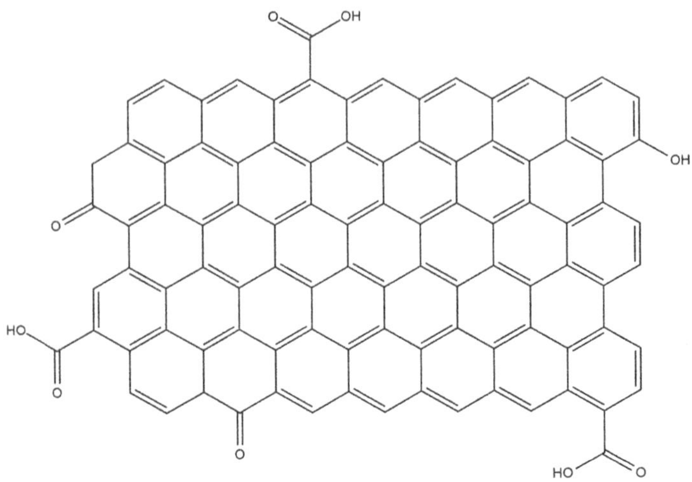

Fig. 8 Graphene oxide structure

2.1 Graphene Oxide

Over the last decades, GO has shown its capability as a unique nanomaterial in designing polymeric based membrane attributed to its good mechanical and physical strength, high flexibility, and abundance in oxygenated functional groups namely carboxyl, hydroxyl and other hydrophilic functional group as shown in Fig. 8. The presence of these functionals group provides GO with good hydrophilicity and tuneable properties that favour the synthesis of polymeric membrane especially in water treatment and purification related process [49]. The positive impact of GO in membrane fabrication process has received tremendous interest from the researchers globally and there are lots of available review papers discussing on the encouraging performance of GO in membrane technology [50–58]. These review papers discussed on the recent development of GO based membrane, morphology and structural properties, modification and performance strategies, applications and preparation of GO based membrane. In view of these, the interest of this paper will outline the recent GO and non-GO nanohybrid polymeric based membrane for water treatment and purification application.

2.1.1 Silicon Dioxide Graphene Oxide Nanohybrid

Silicon dioxide (SiO_2) or known as silica is one of the attractive materials to be embedded into polymer matrices owing to its structural flexibility, excellent physical and thermal properties, and hydrophilicity property that helps to improve permeability. Together with GO, the formed nanohybrid present a stronger hydrophilicity, higher surface area as well as good mechanical-thermo stability. For instance, Wu et al. [46] is one of the earliest researchers attempt of usage of SiO_2-GO nanohybrid in PSF membrane via phase inversion. The findings reveal that by using 0.3 wt% of SiO_2-GO nanohybrid, the pure water flux improved nearly twice as compared to the neat PSF membrane associated with the improved hydrophilicity as seen in the drop of contact angle measurement from approximately 72°–64°. The addition of hydrophilic SiO_2-GO nanohybrid also accelerate the solvent and nonsolvent exchange rate and thus forming a larger pore size that allowed more water molecules to permeate through the membrane barrier. Although the pore size increased, the egg albumin rejections remain high (>98%) and comparable to the neat membrane attributed to the good dispersion and compatibility of SiO_2-GO nanohybrid with the polymer matrices. The author further concludes that the addition of SiO_2-GO nanohybrid help to enhance the antifouling property of the membrane to the highest degree and it was evident through when the flux recovery ratio improved from approximately 61% to 70% [46]. Others reported SiO_2-GO nanohybrid polymeric based membranes for dye wastewater and oily wastewater treatment were also seen in the literatures [59–62].

2.1.2 Titanium Oxide Graphene Oxide Nanohybrid

Among the wide range of materials used for membrane modification process, TiO_2 is one of the nanomaterials that has gained numerous attentions in the field of membrane research attributed to its photocatalytic properties. By coupling it with GO, the TiO_2-GO nanohybrid can produce membrane with synergistic photocatalytic antifouling properties. With this concept, Xu et al. [63], prepared photocatalytic antifouling TiO_2-GO PVDF membrane displayed superior photodegradation efficiency (improved about 50–70%), enhanced photodegradation kinetics (1.0–1.5 times faster) toward the BSA pollutants. During the antifouling test, the subsequent water flux decreased after the BSA filtration due to the occurrence of BSA fouling on the membrane surface. In this case, the presence of the photocatalyst property from the TiO_2-GO nanohybrid equip the modified membrane with self-cleaning property that can effectivelyremove strongly bound BSA. In terms of flux, it recorded more than 2 times improvement from about 150 LMH to 487 LMH while keeping the BSA rejection (>90%). This is as expected as the hydrophilicity of the membrane increased about 20% and this wasdue to the presence of oxygen based functional group in GO/TiO_2 nanohybrid. The supplementation of the nanohybrid caused the formation of larger porous structure due to the acceleration in the exchange rate. This was evidence as the porosity for the modified membrane was seen to expand from 69.6 to 83.1% [63]. To date, there have been numerous efforts seen in the introduction of TiO_2-GO nanohybrid in the membrane research for water treatment and purification as documented in the literatures [64–68].

2.1.3 Silver Graphene Oxide Nanohybrid

Ag has been continuously explored due to its strong antibacterial property that favourable in producing membrane with antibiofouling properties which can effectively prevent formation of microorganisms [27]. GO on the other hand also possessed antimicrobial property and the combination of both these nanomaterials serve as a great auxiliary in forming a desirable membrane [69]. For instance, Mahmoudi et al. [70] prepared an Ag-GO PSF membrane for BSA application and the addition of Ag-GO nanohybrid showed obvious improvement to the membrane antibacterial property. The FESEM images shown that membrane with the presence of Ag-GO nanohybrid prevent the formation of *E-coli* bacteria while PSF neat membrane showed high number of bacteria colony growth. Apart from that, at 0.5 wt% of Ag-GO nanohybrid, the membrane showed twofold fluximprovement as compared to neat membrane this probably associated with the enhanced hydrophilicity property as seen in the significant drop in water contact angle from 83.3° to about 58°. The membrane enjoyed a higher hydrophilicity owing the presence of oxygenated functional groups from GO nanomaterials [70]. Subsequently, the authors continue to apply the Ag-GO nanohybrid in nylon-6,6 membrane and membrane bioreactor applications [71, 72]. Apart from conventional types of MMM, the adoption Ag-GO nanohybrid could also been seen in one interesting work by Ko et al. 2018 whereby the authors coated the Ag-GO nanohybrid on the surface of PVDF membrane via pressurized filtration process. The coating process was performed through filtration of Ag-GO nanohybrid suspensions using batch dead end filtration cell. Through this method the contact angle measurement reduced from 89.45° to 81.55° which indicating the improvement in the membrane hydrophilicity. This improvement facilitates more water uptake across membrane as the permeability increased by 53% and it also showing a comparable bacterial cell removal. The Ag-GO PVDF membrane also demonstrated strong antibiofouling properties as the membrane show low cell viability which is less for 50% for modified membrane and approximate 100% for neat membrane [73]. Other than that, the usage of Ag-GO nanohybrid in hollow fibre and FO membrane for protein, NOM and desalination application was also reported somewhere else [74, 75].

2.1.4 Zinc Oxide Graphene Oxide Nanohybrid

Like Ag nanomaterials, zinc oxide (ZnO) nanomaterials also had been widely adopted in the preparation of polymeric membrane due to its multifunctional properties such as antibacterial property, great hydrophilic property, high surface area, non-toxic and chemically stable [76–78]. To date, works that involved ZnO-GO in polymeric membrane has grown over the years. For instance, Chung et al. 2017 prepared ZnO-GO PSF membrane via PI technique to investigate the antibacterial and the antifouling properties of ZnO-GO nanohybrid. The authors claimed that the usage of ZnO-GO nanohybrid has greatly enhanced the properties of the fabricated membrane in terms of hydrophilicity where the contact angle measurement drops from about

68° (ZnO-GO membrane) to 40° (neat membrane). Associated with the improved hydrophilicity, the permeability showed positive enhancement from 0.89 LMH to 5.11 LMH while the antifouling properties of ZnO-GO membrane recorded more than 90% of FRR as compared to neat membrane at approximately 80%. The better permeability can be correlated with the addition of hydrophilic ZnO-GO nanohybrid and through this addition, it greatly reduces the viscosity of the dope solution and thus enhanced the diffusion rate to produce highly permeable membrane. It is also evident through the morphological study as the addition of ZnO-GO nanohybrid helps to create more finger like channels and macro voids that favour the transportation of water molecules. On the other hand, the ZnO-GO membrane showed outstanding antibacterial properties as compared to the neat membrane as there are no bacterial colonies was seen [79]. Others study that adopted different types of polymers (PVDF and PES), different experiment conditions (concentration of polymers, ZnO-GO nanohybrid concentration, and so forth) and different application such as pharmaceutical wastewater treatment, activated sludge treatment, wastewater reclamation and water desalination were also seen in the literatures [80–83].

2.1.5 Iron Oxide Graphene Oxide Nanohybrid

Iron oxide (Fe_3O_4) nanomaterials is gaining tremendous attention in the wastewater treatment field because of its chemical and thermal stability, low toxicity and possessed hydrophilic property. The presence of these property allowed Fe_3O_4 to be used in membrane fabrication field for water treatment and purification purposes. Combining Fe_3O_4 with GO, this nanohybrid offered a better enhancement towards the membrane performances. For instances, Chai et al. 2019 carried out a performance comparison study of Fe_3O_4 based membrane and Fe_3O_4-GO based membrane with PSF neat membrane and the finding reveals that by adopting Fe_3O_4-GO nanohybrids, it helps to improve the flux from 51.82 LMH to 112.47 LMH while the adoption of single Fe_3O_4 based membrane does not show much positive enhancement on the water flux. The increased of the flux can be related to the improved hydrophilicity as shown in the contact angle measurement for neat membrane drops from 78.80° to 69.97° for the Fe_3O_4-GO based membrane. On the other hand, the lower flux obtained by Fe_3O_4 based membrane was probably due to the fact of aggregation and this provide an insight that with presence of GO, ithelps to minimize the aggregation of the single nanomaterials Fe_3O_4 [84]. On his subsequent works, the authors also continue to study effect of the nanohybrid concentrations and optimal iron to GO molar ratio that provides the optimum flux, rejection and flux recovery ratio in various applications [85–87]. In his recent study, at 0.6 wt% of Fe_3O_4-GO nanohybrid, the fabricated MMM membrane showed a better performance as compared to neat membrane. The flux improved by 67.91%, Congo red rejection improved by 12.64% while the FRR improved 32.11%. Such improvement is related to the improvement of the hydrophilicity properties of the membrane by 11.42% and highly negative charge membrane (from about −3 to −30 mV) that induced by the presence of Fe_3O_4-GO nanohybrid [87]. Others relevant Fe_3O_4-GO nanohybrid based polymeric membrane for various application such as heavy metal removal and different anionic dyes also available in the literatures [88, 89].

Another interesting study was carried out by Huang et al. 2016 wherein the authors utilize magnetic property of Fe_3O_4-GO nanohybrid and altering the movement of the Fe_3O_4-GO nanohybridthrough the presence of external magnetic field. The advantages of this method allowed the nanohybrid to be orderly arrangedand anchored on the porous substrate and the proposed behaviour of the Fe_3O_4-GO nanohybrid during the fabrication process is illustrated in Fig. 9. The migration behaviour of the Fe_3O_4-GO nanohybrid was supported by various analyses in terms of the Fe_3O_4-GO nanohybrid movement recorded by micrograph images, SEM cross sectional images and EDS mapping, and surface roughness. In terms of performance indicator, the membrane prepared in the presence of magnetic field showed enhancement of high-water flux from about 310 LMH to 484 LMH with comparable BSA rejection (>75%). This improvement is strongly related to the migration of the Fe_3O_4-GO nanohybrid to the surface thus induced a more hydrophilic membrane (improved approximately by 20%) and the slight improvement in the porosity that caused by the quick exchange between solvent and non-solvent phase. These two criteria help to facilitate more water transportation across the membrane. Apart from that, the FRR also improved from 43 to 83% which further justify the advantage of Fe_3O_4-GO nanohybrid migration behaviour to the surface [90]. Another similar study that utilizes the external magnetic field at different experimental condition for BSA application could also be seen somewhere else [91].

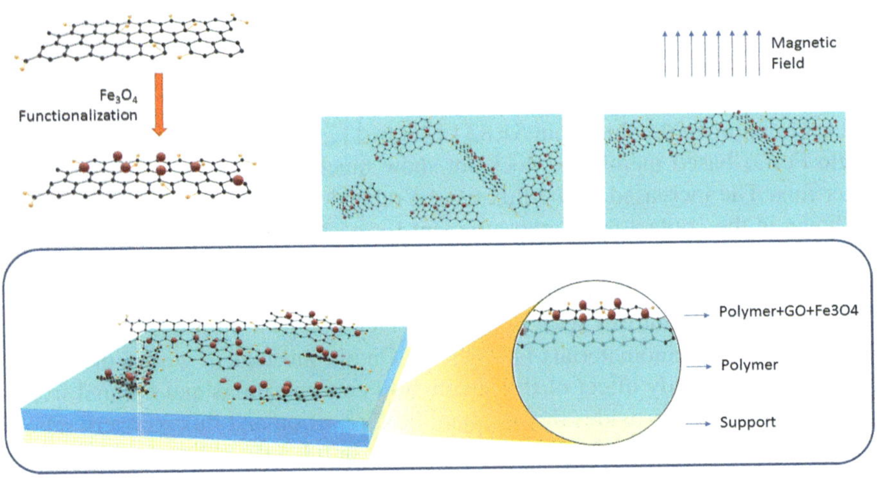

Fig. 9 Fabrication of Fe_3O_4/GO PVDF Membrane under the Magnetic Field

2.1.6 Carbon Nanotubes Graphene Oxide Nanohybrid

Carbon nanotubes (CNTs) is another carbon derivatives family that showed great support while designing a membrane due to its inherent characteristics in terms of resistance, mechanical and chemical properties, high surface area, remarkable water transport property and membrane performance. In particular, CNTs includes single-walled carbon nanotubes (SWCNTs), multi-walled carbon nanotubes (MWCNTs) and oxidized multi-walled carbon nanotubes (OMWCNTs) received a very high attention during the design of membrane for water treatment applications [92–94]. Herein, we report the synergy effect of OWCNTs GO nanohybrid in polymer membrane for water purification application. For example, Yuan et al. 2019, prepared a CNT-GO PVDF membrane for BSA pollutant application. The presence of OWCNTs-GO nanohybrid help to membrane to enjoy high water flux that increased from 75.5 LMH to 125.6 LMH while producing a comparable BSA rejection. This increased of flux was associated with the increased hydrophilicity as shown in the reduction of water contact angle from 79.6° to 62.1°. Also, the synergy effect of the OWCNTs-GO nanohybrid showed better enhancement as compared to single MWCNTs nanomaterials as the wettability measurement reduce from 69.5° to 62.1° which can be explained by the large number of oxidized functionals groups. From the improved hydrophilicity, it further accelerates the PI process thus formed an interconnected finger-like pores that is desirable in promoting water transport [95]. Other relevant CNTs GO based nanohybrid used during preparation of polymeric membrane for BSA, palm oil mill effluent (POME) and desalination are also reported somewhere else [96–98].

2.1.7 Other Graphene Oxide Nanohybrid

The previous section highlights specific types of nanohybrid to be used in membrane research. There are also other interesting studies related with the combination of other GO based nanohybrid in polymeric membrane for water purification. For instance, a study that adopts GO based nanohybrid could also be seen by Du et al. 2020. The authors combine zinc sulfide (ZnS) with GO into PVDF polymer to improve the antifouling property of the membrane. With the incorporation of the nanohybrids, the modified membrane reached 431.9 LMH which is approximately two times higher than neat membrane (227.4 LMH) while maintaining similar BSA rejection rate. On the other hand, the modified membrane also showed enhanced FRR at about 10.2% andboth improvement of flux and antifouling property was largely associated with the enhanced hydrophilicity as observed in the modified membrane. From the water contact angle measurement, it drops from 77.5° to 61.7° and this could be closely related to the presence of the hydrophilic nature of GO. Apart from that, the presence of ZnS in the membrane also helped in adding a self-cleaning effect under UV irradiation and the results under the UV irradiation showed greater FRRcompared to sample without irradiation. Under the UV irradiation, the modified membrane FRR showed notable improvement from 66.7% (without UV) to 84.6% (with UV) while for the neat membrane, the FRR is very close (with and without UV), and this prove the presence of ZnS help to facilitate self-cleaning that able to reduce the tendency of fouling [99].

2.1.8 Non-Graphene Oxide Based Nanohybrid

The growth and demandfor GO nanohybrids polymeric based membrane for water purification and separation has received well and positive response from the researchers. Besides GO based nanohybrids, there are others non-GO based polymeric membrane that worth to be discuss. For instance, Nguyen et al. 2014 deposited Ag and TiO_2 onto the cellulose acetate (CA) forward osmosis membrane. TiO_2 is broadly known as photocatalyst which utilizes the low band gap energy to initiate radicals for organic matter decomposition. On the other hand, negative surface charge, good hydrophilicity and antibacterial properties made Ag nanoparticles a famous component in wastewater treatment research. This surface modification resulted in a greater roughness compared to virgin membrane, reducing the contact angle from 48° to 44°. The membrane with Ag nanoparticle alone did significantly killed bacteria. From adenosine triphosphate (ATP) bioluminescence assay, it is shown to have reduction of bacterial growth from 23.5 μg/L to 6.0 μg/L with Ag nanoparticles coated membrane. However, the antibacterial property of nanocomposite can be even better than Ag nanoparticles alone. Growth was dropped further to 2.0 μg/L possibly due to antifouling property of TiO_2, which increased the interaction between Ag nanoparticles and bacteria. The inhibition of biofouling was further confirmed with the observation of delaying water-flux inhibition by 47% after 3 h of filtration experiment [100].

In another study, Li et al. 2019 alsoimmobilized Ag and TiO_2 coupled with polyethylenimine (PEI) glued with polydopamine (PDA) onto PAN membrane. Comparing with control cases, the decoration of PDA, PEI and TiO_2 decreased water contact angle by ~50% owing to their hydrophilic properties. TiO_2 deposition also reduced surface roughness, promote hydrophilicity. However, the presence of Ag layer increased water contact angle slightly, but the antibacterial advantage outweighs its hydrophobicity. In filtration performance wise, the complex surface modification did not increase dye rejection significantly, but the permeation flux improved about twofold compared to PEI/PAN membrane. It was hypothesized that the incorporation of TiO_2 contributes to looser membrane structure and more voids to increase water penetration. Plate counting method suggested the pristine membrane can kill e. coli at bacteriostasis rate of 20%. This merit arisen from the presence of cationic groups which impede bacteria through intense electrostatic interaction. Significant improvement can be seen with modified membrane case, where incorporation of Ag and TiO_2 achieved bacteriostatic rate of 99.7%, manifesting enhanced antibacterial feature [101]. Another similar work has been developed by Chen et al. [102] to produce TiO_2-AgNP nanocomposite PSF membrane. Interestingly, they worked with rejection test against the pollutant called antibiotic rejection genes (ARG). This is alarming as the uncontrolled usage of antibiotics in livestock has led to rapid development of ARG, suppressing the medication effect. This research team developed the membrane with filtration and antimicrobial to treat ARB contaminated sewage water. The nanoparticles, TiO_2 and Ag were mixed with dopamine and fabricated to obtain the hybrid membrane. Dopamine functioned to reduce Ag ions and immobilize the product, Ag nanoparticles. AgNP has been hypothesized to form schottky barrier with the presence of TiO_2, lowering the possibility for electron-hole recombinant. This hybrid property can increase the availability of reactive oxygen species for rapid photocatalytic effect. The hydrophilicity increased water contact angle analysis, manifested about 20% reduction with the incorporation of Ag and TiO_2. The study observed significant crack on bacterial cell surface in SEM image, proving the antibacterial ability of hybrid membrane. It was proposed that Ag ions released killed bacteria by suppressing RNA polymerase transcriptional ability [102].

Membrane bioreactor often bothered with the extracellular polymeric substances resulted from the microorganisms inhibited inside. These sludge flocs arisen from abundant availability of organic substance and oxygen content in bioreactor, promoted the growth of microorganisms. Therefore, antifouling issue is the hot topic to ensure long term operation of membrane. Ahsani et al. 2020 synthesized Ag-SiO_2PVDF and the presence of Ag-SiO_2 nanoparticles reduce the hydrophobicity of PVDF and thus reduce the cake formation on membrane surface. At the same time, the antibacterial effect of Ag nanoparticles inhibited deposited bacteria growth on membrane surface thereby prevented clogging of membrane pores. There were 87% and 92% less of *E. coli* and *s. aureus* respectively compared to the control sample. To prolong and stabilize the antibacterial operation of Ag nanoparticles, SiO_2 nanoparticles was used as a support to immobilize Ag nanoparticles. This innovative solution helped in preventing Ag nanoparticle agglomeration and this maximizing the effective surface area. In addition to function as the support, silica nanoparticle also acted as physical cross linker thus improving the mechanical strength of the membrane. Under microscopy observation, the foreign nanoparticles changed the surface morphology by increasing surface pore quantity and larger internal void. In this regard, flux has increased, and water diffused at greater rate. At steady filtration state, the modified membrane maintained 59% more permeate flux compared to pristine PVDF membrane. Chemical oxygen demand was reduced to 90% while ammonium removal achieved over 95%. Biofilm formation was controlled in 55% less compared to pristine sample [103].

Besides that, a recent type of nanomaterials namely quantum dots found its interest in membrane fabrication. For instance, Heng et al. [104] modified membrane with-nitrogen doped carbon quantum dots (NCQD) and TiO_2 on PES to reject methylene blue pollutants. Interestingly, the process of making CQD was green method as it was sourced from oil palm frond. It also served as electron reservoir during TiO_2 photocatalysis process. The incorporation of nanoparticles into the membrane is in self-assembly mode onto the PES membrane. Therefore, PES must be surface activated and grafted to enable possible interactions for nanoparticles immobilization. The surface activation method started with irradiation where UV broke the polymer bond and then the cleaved bond promoted the polymerization of acrylic acid. Hence, the surface is grafted and coated with a layer of PAA. PAA contributed carboxyl groups site, serves to bond covalently with titanium cations. The functionalization of nanoparticles reduced the water contact angle from ~70 to ~30. This is attributed to the nature of hydrophilic nature of carboxyl and hydroxyl groups in nanocomposite. The enhanced hydrophilicity hindered the attachment of hydrophobic pollutant, aid in membrane rejection. Pristine membrane rejected at 46.6% of methylene blue while modified one can reject up to 90.9%. The addition of TiO_2 showed a positive effect in antifouling property. Modified membrane reached to steady filtration rate at shorter duration by four-fold, testifying minor attachment of foulants. Most of the pores were safe from clogging owing to surface tension provided by nanocomposite. Moreover, the TiO_2 lengthened the operational efficiency by improving the reusability of membrane. Self-cleaning is anticipated and performed 10% better on flux recovery compared to conventional washing method [104].

Another similar report was also reported by Koe et al. [105] co-doped nitrogen and sulphur to the CQD and mixed with TiO_2 before incorporated to PSF membrane. Doping of N and S is necessary to reduce the band gap energy and inhibit recombination of charges and holes, promoting more photocatalysis. Interestingly, the source of nitrogen was from green material, egg yolks with hydrothermal treatment. While sulphur was from mixture of hydrochloric acid and sulfuric acid. The up converted photoluminescence (UPCL) nature of CQD converts morelight into shorter wavelength useful for the excitation of TiO_2 nanoparticles. In another words, long wavelength light is unable to activate the photocatalysis of TiO_2 and thus addition of CQD can capture more light energy. This was verified by photocatalysis under UV light and visible light. The visible light activated the TIO_2 and performed better as broader light wavelength can be absorbed for pollutant degradation. Although CQD catalyzes reaction, overloaded CQD may covered up the active sites of TIO_2, doing more harm than good. In hydrophilicity test, greater amount of hydroxyl groups presented on membrane test after the modification led to reduction in contact angle. The hydrophilicity played an important role in improving the porosity where the functional groups speeded up the mass transfer in PI process. However, excessive CQD caused aggregation of nanoparticles. In this case, membrane became rougher and hydrophilicity reduced. Moreover, viscous membrane suspension hindered the formation of pores thereby reduced the effective working area and water permeation rate. CQD is hypothesized to function as electron acceptor to hold electrons temporary. This merited photocatalysis through hindering of electron-hole recombination. The hybrid membrane has brought 59% removal efficiency for diclofenac compared to pristine TiO_2 membrane [105].

Apart from TiO_2-CQD based nanohybrid, other CQD based nanohybrid also seen in the literature. For example, extracted pulp-free lemon juice has been used by Gan et al. [106] with hydrothermal treatment to produce CQD. The obtained CQD was mixed with Ag precursor and reduced to form nanohybrid solution. Associated with the involvement of the nanohybrid, the fabricated modifiedmembrane is causing 10% increment in water contact angleto presence of hydrophilic functional group. In pore size analysis, the incorporation of nanoparticles improved the pore diameter, implying the rapid mass transfer between solvent and water bath. However, at very high composition of nanoparticles incorporation, the pore size decreased. High viscosity during the fabrication stage may impede the mass transfer and unable to produce large pore structure. With the effect of hydrogen bonding, the movement of water molecules across the membrane was enhanced and improved the permeability. Tartrazine, a negatively charge pollutant was ejected with electrostatic force by Ag-CQD nanocomposite PSF membrane. The pollutant was rejected at 92% improvement with the use of modified membrane compared to the virgin PSF membrane. Notably, the lowest loading of nanoparticles in PSF membrane has resulted in poorest rejection among all cases. This is due to the formation of big pore diameter and insufficient electronegativity of membrane, led to undesired passage of pollutant molecule through the pore structure. As mentioned, modified membrane has higher electronegativity. This feature builds up repulsive force, prevents the foulant from staying stagnant on membrane surface. In this way, fouling layer formation can be reduced. Thus, the electronegativity improved the reusability of membrane by reducing foulant accumulation [106].

Incorporated Ag doped ZnO/Fe_3O_4/MWCNTS to polyacrylic acid (PAA) modified polyamide disc to remove amoxicillin. ZnO can act as photocatalyst only excitation with light wavelength greater than 400 nm. This reduces the efficiency of light irradiated utilization. Moreover, the photo-generated electron-hole pairs are susceptible for recombination and reduce photocatalyst efficiency. Thus, the plasmonic-metal Ag nanoparticles and MWCNTS can widen the ZnO absorption spectrum range at the same time form Schottky barrier to reduce electron-hole recombinant. These nanoparticles are impregnated over Fe_3O_4 to enable easy separation during the operation. However, the deposition of these nanocomposite onto polyamide disc require grafting with PAA. The PAA reduced the hydrophobicity of polyamide disc, reducing the foulant attachment. Other than that, the carboxyl functional groups of PAA rendered as the site to anchor nanocomposite via covalent bonds. Nitrogen adsorption desorption isotherm test manifested about two-fold increase in specific area and 60% increase in pore volume. The nanocomposite created more active area for membrane. Ionized carboxylic groups presented in both membrane surface and internal structure repel the attachment of pollutant. Incorporation of nanocomposites has inevitable disadvantage in reducing the flux. However, the modified membrane can function better with xenon lamp irradiation. This may be attributed to the photo-wetting effect generated by the functional groups [107].

3 Conclusion and Future Prospects

Over the past years, we have seen a huge amount of effort and research showing positive impacts to the polymeric membrane design in water treatment application. The adoption of nanohybrids in membrane science is definitely a great option to the membrane science. In spite of the advancement of membrane science, more research is still required to address challenges in terms commercialization and sustainability.

In terms of commercialization, to the best of the author's knowledge the nanohybrid incorporated polymer membrane for water treatment are currently not available. Several areas of improvement in terms of nanohybrid stabilities, performance stabilities, life span of the membrane shall be addressed before the point of commercialization can be reached. On the hand, in terms of sustainability, adoption of green materials, biopolymers, green solvent, compatibility of nanohybrid with the polymer should also be considered as during this era of sustainability.

References

1. Saleh TA, Gupta VK (2016) Chapter 1—An overview of membrane science and technology. In: Saleh TA, VKBT-N, Gupta PM (eds). Elsevier, pp 1–23. https://doi.org/10.1016/B978-0-12-804703-3.00001-2
2. Ahmad NNR, Ang WL, Leo CP, Mohammad AW, Hilal N (2021) Current advances in membrane technologies for saline wastewater treatment: a comprehensive review. Desalination 517:115170. https://doi.org/10.1016/j.desal.2021.115170
3. Ang WL, Mohammad AW (2021) 4—Recent development in nanofiltration process applications. In: Hilal N, Ismail AF, Khayet M, Johnson DBT-OE (eds) Elsevier, pp 97–129. https://doi.org/10.1016/B978-0-12-821016-1.00010-3
4. Ang WL, Mohammad AW, Johnson D, Hilal N (2020) Unlocking the application potential of forward osmosis through integrated/hybrid process. Sci Total Environ 706:136047. https://doi.org/10.1016/j.scitotenv.2019.136047
5. Zhu Q, Tang X, Feng S, Zhong Z, Yao J, Yao Z (2019) ZIF-8@SiO2 composite nanofiber membrane with bioinspired spider web-like structure for efficient air pollution control. J Memb Sci 581:252–261. https://doi.org/10.1016/j.memsci.2019.03.075
6. Ahmad NNR, Leo CP, Mohammad AW, Shaari N, Ang WL (2021) Recent progress in the development of ionic liquid-based mixed matrix membrane for CO_2 separation: a review. Int J Energy Res 45:9800–9830. https://doi.org/10.1002/er.6518
7. Rudolph G, Virtanen T, Ferrando M, Güell C, Lipnizki F, Kallioinen M (2019) A review of in situ real-time monitoring techniques for membrane fouling in the biotechnology, biorefinery and food sectors. J Memb Sci 588:117221. https://doi.org/10.1016/j.memsci.2019.117221
8. Lipnizki F(2010) 4.06—Basic aspects and applications of membrane processes in agro-food and bulk biotech industries. In: Drioli E, LBT-CMS, Giorno E (eds) Elsevier, Oxford, pp 165–194. https://doi.org/10.1016/B978-0-08-093250-7.00035-9
9. Van Der Bruggen B, Vandecasteele C, Van Gestel T, Doyen W, Leysen R (2003) A review of pressure-driven membrane processes in wastewater treatment and drinking water production. Environ Prog 22:46–56. https://doi.org/10.1002/ep.670220116
10. Guo W, Ngo H-H, Li J (2012) A mini-review on membrane fouling. Bioresour Technol 122:27–34. https://doi.org/10.1016/j.biortech.2012.04.089
11. Wei X, Sanders KT, Childress AE (2021) Reclaiming wastewater with increasing salinity for potable water reuse: water recovery and energy consumption during reverse osmosis desalination. Desalination 520:115316. https://doi.org/10.1016/j.desal.2021.115316
12. Akyol C, Ozbayram EG, Accoroni S, Radini S, Eusebi AL, Gorbi S, Vignaroli C, Bacchiocchi S, Campacci D, Gigli F, Farina G, Albay M, Fatone F (2021) Monitoring of cyanobacterial blooms and assessing polymer-enhanced microfiltration and ultrafiltration for microcystin removal in an Italian drinking water treatment plant. Environ Pollut 286:117535. https://doi.org/10.1016/j.envpol.2021.117535
13. Ang WL, Mohammad AW (2015) 12 - Mathematical modeling of membrane operations for water treatment. In: Basile A, Cassano A, NKBT-A in M.T. for Rastogi WT (eds) Woodhead

Publ Ser Energy. Woodhead Publishing, Oxford, pp 379–407. https://doi.org/10.1016/B978-1-78242-121-4.00012-5
14. Ismail AF, Khulbe KC, Matsuura T (2019) Chapter 5—RO Membrane module. In: Ismail AF, Khulbe KC, T.B.T.-R.O. Matsuura (eds). Elsevier, pp 117–141. https://doi.org/10.1016/B978-0-12-811468-1.00005-0
15. Yu H, Li X, Chang H, Zhou Z, Zhang T, Yang Y, Li G, Ji H, Cai C, Liang H (2020) Performance of hollow fiber ultrafiltration membrane in a full-scale drinking water treatment plant in China: a systematic evaluation during 7-year operation. J Memb Sci 613:118469. https://doi.org/10.1016/j.memsci.2020.118469
16. Gai W, Zhang Y, Zhao Q, Chung TS (2021) Highly permeable thin film composite hollow fiber membranes for brackish water desalination by incorporating amino functionalized carbon quantum dots and hypochlorite treatment. J Memb Sci 620:118952. https://doi.org/10.1016/j.memsci.2020.118952
17. Eke J, Yusuf A, Giwa A, Sodiq A (2020) The global status of desalination: An assessment of current desalination technologies, plants and capacity. Desalination 495:114633. https://doi.org/10.1016/j.desal.2020.114633
18. Tang Y, Lin Y, Ford DM, Qian X, Cervellere MR, Millett PC, Wang X (2021) A review on models and simulations of membrane formation via phase inversion processes. J Memb Sci 640:119810. https://doi.org/10.1016/j.memsci.2021.119810
19. Ismail N, Venault A, Mikkola J-P, Bouyer D, Drioli E, Tavajohi Hassan Kiadeh N (2020) Investigating the potential of membranes formed by the vapor induced phase separation process. J Memb Sci 597:117601. https://doi.org/10.1016/j.memsci.2019.117601
20. Liu C, Qiao W, Wang C, Wang H, Zhou Y, Gu S, Xu W, Zhuang Y, Shi J, Yang H (2021) Effect of poly (lactic acid) porous membrane prepared via phase inversion induced by water droplets on 3T3 cell behavior. Int J Biol Macromol 183:2205–2214. https://doi.org/10.1016/j.ijbiomac.2021.05.197
21. Tang Y, Lin Y, Ma W, Wang X (2021) A review on microporous polyvinylidene fluoride membranes fabricated via thermally induced phase separation for MF/UF application. J Memb Sci 639:119759. https://doi.org/10.1016/j.memsci.2021.119759
22. Tan JY, Ang WL, Mohammad AW (2021) Hydrophobic polyvinylidene fluoride membrane modified with silica nanoparticles and silane for succinic acid purification using osmotic distillation process. J Kejuruter 33:89–101
23. Gohil JM, Ray P (2017) A review on semi-aromatic polyamide TFC membranes prepared by interfacial polymerization: potential for water treatment and desalination. Sep Purif Technol 181:159–182. https://doi.org/10.1016/j.seppur.2017.03.020
24. Ahmad AL, Otitoju TA, Ooi BS (2019) Hollow fiber (HF) membrane fabrication: a review on the effects of solution spinning conditions on morphology and performance. J Ind Eng Chem 70:35–50. https://doi.org/10.1016/j.jiec.2018.10.005
25. Huang Y, Xiao C, Huang Q, Liu H, Zhao J (2021) Progress on polymeric hollow fiber membrane preparation technique from the perspective of green and sustainable development. Chem Eng J 403:126295. https://doi.org/10.1016/j.cej.2020.126295
26. Yin J, Deng B (2015) Polymer-matrix nanocomposite membranes for water treatment. J Memb Sci 479:256–275. https://doi.org/10.1016/j.memsci.2014.11.019
27. Bassyouni M, Abdel-Aziz MH, Zoromba MS, Abdel-Hamid SMS, Drioli E (2019) A review of polymeric nanocomposite membranes for water purification. J Ind Eng Chem 73:19–46. https://doi.org/10.1016/j.jiec.2019.01.045
28. Goh PS, Ismail AF (2018) A review on inorganic membranes for desalination and wastewater treatment. Desalination 434:60–80. https://doi.org/10.1016/j.desal.2017.07.023
29. Fard AK, Mckay G, Buekenhoudt A, Al Sulaiti H, Motmans F, Khraisheh M, Atieh M (nd) Inorganic membranes: preparation and application for water treatment and desalination. https://doi.org/10.3390/ma11010074
30. Lakhotia SR, Mukhopadhyay M, Kumari P (2018) Surface-modified nanocomposite membranes. Sep Purif Rev 47:288–305. https://doi.org/10.1080/15422119.2017.1386681

31. Dlamini DS, Li J, Mamba BB (2019) Critical review of montmorillonite/polymer mixed-matrix filtration membranes: possibilities and challenges. Appl Clay Sci 168:21–30. https://doi.org/10.1016/j.clay.2018.10.016
32. Ihsanullah (2019) Carbon nanotube membranes for water purification: developments, challenges, and prospects for the future. Sep Purif Technol 209:307–337. https://doi.org/10.1016/j.seppur.2018.07.043
33. Mollahosseini A, Abdelrasoul A (2019) Recent advances in thin film composites membranes for brackish groundwater treatment with critical focus on Saskatchewan water sources. J Environ Sci (China) 81:181–194. https://doi.org/10.1016/j.jes.2019.01.014
34. Li J, Wang H, Yuan X, Zhang J, Chew JW (2020) Metal-organic framework membranes for wastewater treatment and water regeneration. Coord Chem Rev 404:213116. https://doi.org/10.1016/j.ccr.2019.213116
35. Ali S, Rehman SAU, Luan HY, Farid MU, Huang H (2019) Challenges and opportunities in functional carbon nanotubes for membrane-based water treatment and desalination. Sci Total Environ 646:1126–1139. https://doi.org/10.1016/j.scitotenv.2018.07.348
36. Wu W, Shi Y, Liu G, Fan X, Yu Y (2020) Recent development of graphene oxide based forward osmosis membrane for water treatment: a critical review. Desalination 491:114452. https://doi.org/10.1016/j.desal.2020.114452
37. Mamah SC, Goh PS, Ismail AF, Suzaimi ND, Yogarathinam LT, Raji YO, El-badawy TH (2021) Recent development in modification of polysulfone membrane for water treatment application. J Water Process Eng 40:101835. https://doi.org/10.1016/j.jwpe.2020.101835
38. Yan Z, Zhang Y, Yang H, Fan G, Ding A, Liang H, Li G, Ren N, Van der Bruggen B (2020) Mussel-inspired polydopamine modification of polymeric membranes for the application of water and wastewater treatment: a review. Chem Eng Res Des 157:195–214. https://doi.org/10.1016/j.cherd.2020.03.011
39. Sajid M, Sajid Jillani SM, Baig N, Alhooshani K (2021) Layered double hydroxide-modified membranes for water treatment: recent advances and prospects. Chemosphere 287:132140. https://doi.org/10.1016/j.chemosphere.2021.132140
40. Ng LY, Chua HS, Ng CY (2021) Incorporation of graphene oxide-based nanocomposite in the polymeric membrane for water and wastewater treatment: a review on recent development. J Environ Chem Eng 9:105994. https://doi.org/10.1016/j.jece.2021.105994
41. Zare S, Kargari A (2018) Membrane properties in membrane distillation. Elsevier Inc. https://doi.org/10.1016/B978-0-12-815818-0.00004-7
42. Jabbarvand Behrouz S, Khataee A, Safarpour M, Arefi-Oskoui S, Woo Joo S (2021) Carboxymethyl cellulose/polyethersulfone thin-film composite membranes for low-pressure desalination. Sep Purif Technol 269:118720. https://doi.org/10.1016/j.seppur.2021.118720
43. Hamid MF, Abdullah N, Yusof N, Ismail NM, Ismail AF, Salleh WNW, Jaafar J, Aziz F, Lau WJ (2020) Effects of surface charge of thin-film composite membrane on copper (II) ion removal by using nanofiltration and forward osmosis process. J Water Process Eng 33. https://doi.org/10.1016/j.jwpe.2019.101032
44. Chiao YH, Chen ST, Ang MBMY, Patra T, Castilla-Casadiego DA, Fan R, Almodovar J, Hung WS, Wickramasinghe SR (2020) High-performance polyacrylic acid-grafted pvdf nanofiltration membrane with good antifouling property for the textile industry. Polymers (Basel). 12:1–14. https://doi.org/10.3390/polym12112443
45. Shahmirzadi MAAA, Kargari A (2018) Nanocomposite membranes. Elsevier Inc. https://doi.org/10.1016/B978-0-12-815818-0.00009-6
46. Wu H, Tang B, Wu P (2014) Development of novel SiO_2-GO nanohybrid/polysulfone membrane with enhanced performance. J Memb Sci 451:94–102. https://doi.org/10.1016/j.memsci.2013.09.018
47. Wang J, Zhang W, Yue X, Yang Q, Liu F, Wang Y, Zhang D, Li Z, Wang J (2016) One-pot synthesis of multifunctional magnetic ferrite-MoS_2-carbon dot nanohybrid adsorbent for efficient Pb(II) removal. J Mater Chem A. 4:3893–3900. https://doi.org/10.1039/c6ta00269b
48. Alizadeh A, Abdi G, Khodaei MM, Ashokkumar M, Amirian J (2017) Graphene oxide/Fe_3O_4/SO_3H nanohybrid: a new adsorbent for adsorption and reduction of Cr(VI) from aqueous solutions. RSC Adv 7:14876–14887. https://doi.org/10.1039/c7ra01536d

49. Ma J, Ping D, Dong X (2017) Recent developments of graphene oxide-based membranes: a review. Membranes (Basel) 7. https://doi.org/10.3390/membranes7030052
50. Cheng MM, Huang LJ, Wang YX, Tang JG, Wang Y, Zhao YC, Liu GF, Zhang Y, Kipper MJ, Belfiore LA, Ranil WS (2017) Recent developments in graphene-based/nanometal composite filter membranes. RSC Adv 7:47886–47897. https://doi.org/10.1039/c7ra08098k
51. Yang Han Z, Jun Huang L, Jiao Qu H, Xin Wang Y, Jie Zhang Z, Lin Rong Q, Qi Sang Z, Wang Y, Kipper MJ, Guo Tang J (2021) A review of performance improvement strategies for graphene oxide-based and graphene-based membranes in water treatment. J Mater Sci 56:9545–9574. https://doi.org/10.1007/s10853-021-05873-7
52. Wang X, Zhao Y, Tian E, Li J, Ren Y (2018) Graphene oxide-based polymeric membranes for water treatment. Adv Mater Interf 5:1–20. https://doi.org/10.1002/admi.201701427
53. Zhang N, Qi W, Huang L, Jiang E, Bao J, Zhang X, An B, He G (2019) Review on structural control and modification of graphene oxide-based membranes in water treatment: from separation performance to robust operation, Chinese. J Chem Eng 27:1348–1360. https://doi.org/10.1016/j.cjche.2019.01.001
54. Wei Y, Zhang Y, Gao X, Ma Z, Wang X, Gao C (2018) Multilayered graphene oxide membrane for water treatment: a review. Carbon NY 139:964–981. https://doi.org/10.1016/j.carbon.2018.07.040
55. Hegab HM, Zou L (2015) Graphene oxide-assisted membranes: fabrication and potential applications in desalination and water purification. J Memb Sci 484:95–106. https://doi.org/10.1016/j.memsci.2015.03.011
56. Jiang Y, Biswas P, Fortner JD (2016) A review of recent developments in graphene-enabled membranes for water treatment. Environ Sci Water Res Technol 2:915–922. https://doi.org/10.1039/c6ew00187d
57. Song N, Gao X, Ma Z, Wang X, Wei Y, Gao C (2018) A review of graphene-based separation membrane: materials, characteristics, preparation and applications. Desalination 437:59–72. https://doi.org/10.1016/j.desal.2018.02.024
58. Ammar A, Al-Enizi AM, AlMaadeed MAA, Karim A (2016) Influence of graphene oxide on mechanical, morphological, barrier, and electrical properties of polymer membranes. Arab J Chem 9:274–286. https://doi.org/10.1016/j.arabjc.2015.07.006
59. Li ZK, Lang WZ, Miao W, Yan X, Guo YJ (2016) Preparation and properties of PVDF/SiO2@GO nanohybrid membranes via thermally induced phase separation method. J Memb Sci 511:151–161. https://doi.org/10.1016/j.memsci.2016.03.048
60. Chen Du Y, Jun Huang L, Xin Wang Y, Yang K, Jie Zhang Z, Wang Y, Kipper MJ, Belfiore LA, Guo Tang J (2020) Preparation of graphene oxide/silica hybrid composite membranes and performance studies in water treatment. J Mater Sci 55:11188–11202. https://doi.org/10.1007/s10853-020-04774-5
61. Liu Y, Zhang F, Zhu W, Su D, Sang Z, Yan X, Li S, Liang J, Dou SX (2020) A multifunctional hierarchical porous SiO_2/GO membrane for high efficiency oil/water separation and dye removal. Carbon NY 160:88–97. https://doi.org/10.1016/j.carbon.2020.01.002
62. Alkindy MB, Naddeo V, Banat F, Hasan SW (2020) Synthesis of polyethersulfone (PES)/GO-SiO_2 mixed matrix membranes for oily wastewater treatment. Water Sci Technol 81:1354–1364. https://doi.org/10.2166/wst.2019.347
63. Xu Z, Wu T, Shi J, Teng K, Wang W, Ma M, Li J, Qian X, Li C, Fan J (2016) Photocatalytic antifouling PVDF ultrafiltration membranes based on synergy of graphene oxide and TiO_2 for water treatment. J Memb Sci 520:281–293. https://doi.org/10.1016/j.memsci.2016.07.060
64. Gharehbakhsh H, Panahi HA, Toosi MR, Hassani AH, Moniri E (2020) Application of polyamide thin-film composite layered on polysulfone-GO/TiO_2 mixed matrix membranes for removal of nitrotoluene derivatives from petrochemical wastewaters. Environ Sci Pollut Res 27:42481–42494. https://doi.org/10.1007/s11356-020-10210-7
65. Liu Y, Yu Z, Peng Y, Shao L, Li X, Zeng H (2020) A novel photocatalytic self-cleaning TiO_2 nanorods inserted graphene oxide-based nanofiltration membrane. Chem Phys Lett 749:137424. https://doi.org/10.1016/j.cplett.2020.137424

66. Safarpour M, Vatanpour V, Khataee A (2016) Preparation and characterization of graphene oxide/TiO_2 blended PES nanofiltration membrane with improved antifouling and separation performance. Desalination 393:65–78. https://doi.org/10.1016/j.desal.2015.07.003
67. Sirinupong T, Youravong W, Tirawat D, Lau WJ, Lai GS, Ismail AF (2018) Synthesis and characterization of thin film composite membranes made of PSF-TiO_2/GO nanocomposite substrate for forward osmosis applications. Arab J Chem 11:1144–1153. https://doi.org/10.1016/j.arabjc.2017.05.006
68. Guang Wu L, Yang Zhang X, Wang T, Hui Du C, Hong Yang C (2019) Enhanced performance of polyvinylidene fluoride ultrafiltration membranes by incorporating TiO_2/graphene oxide. Chem Eng Res Des 141:492–501. https://doi.org/10.1016/j.cherd.2018.11.025
69. Zhu J, Wang J, Hou J, Zhang Y, Liu J, Van der Bruggen B (2017) Graphene-based antimicrobial polymeric membranes: a review. J Mater Chem A 5:6776–6793. https://doi.org/10.1039/c7ta00009j
70. Mahmoudi E, Ng LY, Ba-Abbad MM, Mohammad AW (2015) Novel nanohybrid polysulfone membrane embedded with silver nanoparticles on graphene oxide nanoplates. Chem Eng J 277:1–10. https://doi.org/10.1016/j.cej.2015.04.107
71. Mahmoudi E, Ng LY, Ang WL, Teow YH, Mohammad AW (2020) Improving membrane bioreactor performance through the synergistic effect of silver-decorated graphene oxide in composite membranes. J Water Process Eng 34:101169. https://doi.org/10.1016/j.jwpe.2020.101169
72. Mahmoudi E, Ng LY, Ang WL, Chung YT, Rohani R, Mohammad AW (2019) Enhancing morphology and separation performance of polyamide 6,6 membranes by minimal incorporation of silver decorated graphene oxide nanoparticles. Sci Rep 9:1–16. https://doi.org/10.1038/s41598-018-38060-x
73. Ko K, Yu YJ, Kim MJ, Kweon J, Chung H (2018) Improvement in fouling resistance of silver-graphene oxide coated polyvinylidene fluoride membrane prepared by pressurized filtration. Sep Purif Technol 194:161–169. https://doi.org/10.1016/j.seppur.2017.11.016
74. Fan Y, Quan X, Zhao H, Chen S, Yu H, Zhang Y, Zhang Q (2017) Poly(vinylidene fluoride) hollow-fiber membranes containing silver/graphene oxide dope with excellent filtration performance. J Appl Polym Sci 134:1–9. https://doi.org/10.1002/app.44713
75. Soroush A, Ma W, Silvino Y, Rahaman MS (2015) Surface modification of thin film composite forward osmosis membrane by silver-decorated graphene-oxide nanosheets. Environ Sci NANO 2:395–405. https://doi.org/10.1039/c5en00086f
76. Sheikh M, Pazirofteh M, Dehghani M, Asghari M, Rezakazemi M, Valderrama C, Cortina JL (2020) Application of ZnO nanostructures in ceramic and polymeric membranes for water and wastewater technologies: a review. Chem Eng J 391:123475. https://doi.org/10.1016/j.cej.2019.123475
77. Mukherjee M, De S (2018) Antibacterial polymeric membranes: a short review. Environ Sci Water Res Technol 4:1078–1104. https://doi.org/10.1039/c8ew00206a
78. Sri Abirami Saraswathi MS, Nagendran A, Rana D (2019) Tailored polymer nanocomposite membranes based on carbon, metal oxide and silicon nanomaterials: a review. J Mater Chem A 7:8723–8745. https://doi.org/10.1039/c8ta11460a
79. Chung YT, Mahmoudi E, Mohammad AW, Benamor A, Johnson D, Hilal N (2017) Development of polysulfone-nanohybrid membranes using ZnO-GO composite for enhanced antifouling and antibacterial control. Desalination 402:123–132. https://doi.org/10.1016/j.desal.2016.09.030
80. Ayyaru S, Dinh TTL, Ahn YH (2020) Enhanced antifouling performance of PVDF ultrafiltration membrane by blending zinc oxide with support of graphene oxide nanoparticle. Chemosphere 241. https://doi.org/10.1016/j.chemosphere.2019.125068
81. Mahlangu OT, Nackaerts R, Mamba BB, Verliefde ARD (2017) Development of hydrophilic GO-ZnO/PES membranes for treatment of pharmaceutical wastewater. Water Sci Technol 76:501–514. https://doi.org/10.2166/wst.2017.194
82. Mahlangu OT, Nackaerts R, Thwala JM, Mamba BB, Verliefde ARD (2017) Hydrophilic fouling-resistant GO-ZnO/PES membranes for wastewater reclamation. J Memb Sci 524:43–55. https://doi.org/10.1016/j.memsci.2016.11.018

83. Rajakumaran R, Boddu V, Kumar M, Shalaby MS, Abdallah H, Chetty R (2019) Effect of ZnO morphology on GO-ZnO modified polyamide reverse osmosis membranes for desalination. Desalination 467:245–256. https://doi.org/10.1016/j.desal.2019.06.018
84. Chai PV, Mahmoudi E, Mohammad AW, Choy PY (2020) Iron oxide decorated graphene oxide embedded polysulfone mixed-matrix membrane: comparison of different types mixed-matrix membranes on antifouling and performance. IOP Conf Ser Earth Environ Sci 463. https://doi.org/10.1088/1755-1315/463/1/012174
85. Chai PV, Mahmoudi E, Teow YH, Mohammad AW (2017) Preparation of novel polysulfone-Fe_3O_4/GO mixed-matrix membrane for humic acid rejection. J Water Process Eng 15:83–88. https://doi.org/10.1016/j.jwpe.2016.06.001
86. Chai PV, Mohammad AW, Teow YH, Ebrahim M (2017) Evaluation of Iron oxide decorated on graphene oxide (Fe_3O_4/GO) nanohybrid incorporated in PSF membrane at different molar ratios for Congo red rejection. J Teknol 79:73–81. https://doi.org/10.11113/jt.v79.10440
87. Chai PV, Law JY, Mahmoudi E, Mohammad AW (2020) Development of iron oxide decorated graphene oxide (Fe_3O_4GO) PSf mixed-matrix membrane for enhanced antifouling behavior. J Water Process Eng 38. https://doi.org/10.1016/j.jwpe.2020.101673
88. Koushkbaghi S, Jafari P, Rabiei J, Irani M, Aliabadi M (2016) Fabrication of PET/PAN/GO/Fe_3O_4 nanofibrous membrane for the removal of Pb(II) and Cr(VI) ions. Chem Eng J 301:42–50. https://doi.org/10.1016/j.cej.2016.04.076
89. Sahraei R, Shahalizade T, Ghaemy M, Mahdavi H (2018) Fabrication of cellulose acetate/Fe_3O_4@GO-APTS-poly(AMPS-co-MA) mixed matrix membrane and its evaluation on anionic dyes removal. Cellulose 25:3519–3532. https://doi.org/10.1007/s10570-018-1800-9
90. Huang Y, Fa Xiao C, Lin Huang Q, Liang Liu Q, Qiang Hao Q, Song L (2018) Magnetic field induced orderly arrangement of Fe_3O_4/GO composite particles for preparation of Fe_3O_4/GO/PVDF membrane. J Memb Sci 548:184–193. https://doi.org/10.1016/j.memsci.2017.11.027
91. Xu Z, Wu T, Shi J, Wang W, Teng K, Qian X, Shan M, Deng H, Tian X, Li C, Li F (2016) Manipulating migration behavior of magnetic graphene oxide via magnetic field induced casting and phase separation toward high-performance hybrid ultrafiltration membranes. ACS Appl Mater Interf 8:18418–18429. https://doi.org/10.1021/acsami.6b04083
92. Salehi E, Heidary F, Daraei P, Keyhani M, Behjomanesh M (2020) Carbon nanostructures for advanced nanocomposite mixed matrix membranes: a comprehensive overview. Rev Chem Eng 36:723–748. https://doi.org/10.1515/revce-2017-0073
93. Ma L, Dong X, Chen M, Zhu L, Wang C, Yang F, Dong Y (2017) Fabrication and water treatment application of carbon nanotubes (CNTs)-based composite membranes: a review. Membranes (Basel) 7. https://doi.org/10.3390/membranes7010016
94. Li C, Yang J, Zhang L, Li S, Yuan Y, Xiao X, Fan X, Song C (2021) Carbon-based membrane materials and applications in water and wastewater treatment: a review. Environ Chem Lett 19:1457–1475. https://doi.org/10.1007/s10311-020-01112-8
95. Yuan XT, Xu CX, Geng HZ, Ji Q, Wang L, He B, Jiang Y, Kong J, Li J (2020) Multifunctional PVDF/CNT/GO mixed matrix membranes for ultrafiltration and fouling detection. J Hazard Mater 384. https://doi.org/10.1016/j.jhazmat.2019.120978
96. Wang Y, Liu Y, Yu Y, Huang H (2018) Influence of CNT-rGO composite structures on their permeability and selectivity for membrane water treatment. J Memb Sci 551:326–332. https://doi.org/10.1016/j.memsci.2018.01.031
97. Ho KC, Teow YH, Mohammad AW, Ang WL, Lee PH (2018) Development of graphene oxide (GO)/multi-walled carbon nanotubes (MWCNTs) nanocomposite conductive membranes for electrically enhanced fouling mitigation. J Memb Sci 552:189–201. https://doi.org/10.1016/j.memsci.2018.02.001
98. Kim HJ, Lim MY, Jung KH, Kim DG, Lee JC (2015) High-performance reverse osmosis nanocomposite membranes containing the mixture of carbon nanotubes and graphene oxides. J Mater Chem A 3:6798–6809. https://doi.org/10.1039/c4ta06080f

99. Du J, Li N, Tian Y, Zhang J, Zuo W (2020) Preparation of PVDF membrane blended with graphene oxide-zinc sulfide (GO-ZnS) nanocomposite for improving the anti-fouling property. J Photochem Photobiol A Chem 400:112694. https://doi.org/10.1016/j.jphotochem.2020.112694
100. Nguyen A, Zou L, Priest C (2014) Evaluating the antifouling effects of silver nanoparticles regenerated by TiO_2 on forward osmosis membrane. J Memb Sci 454:264–271. https://doi.org/10.1016/j.memsci.2013.12.024
101. Li J, Yuan S, Zhu J, Van der Bruggen B (2019) High-flux, antibacterial composite membranes via polydopamine-assisted PEI-TiO_2/Ag modification for dye removal. Chem Eng J 373:275–284. https://doi.org/10.1016/j.cej.2019.05.048
102. Chen H, Huang M, Wang Z, Gao P, Cai T, Song J, Zhang Y, Meng L (2020) Enhancing rejection performance of tetracycline resistance genes by a TiO_2/AgNPs-modified nanofiber forward osmosis membrane. Chem Eng J 382:123052. https://doi.org/10.1016/j.cej.2019.123052
103. Ahsani M, Hazrati H, Javadi M, Ulbricht M, Yegani R (2020) Preparation of antibiofouling nanocomposite PVDF/Ag-SiO_2 membrane and long-term performance evaluation in the MBR system fed by real pharmaceutical wastewater. Sep Purif Technol 249:116938. https://doi.org/10.1016/j.seppur.2020.116938
104. Heng ZW, Chong WC, Pang YL, Sim LC, Koo CH (2021) Novel visible-light responsive NCQDs-TiO_2/PAA/PES photocatalytic membrane with enhanced antifouling properties and self-cleaning performance. J Environ Chem Eng 9:105388. https://doi.org/10.1016/j.jece.2021.105388
105. Koe WS, Chong WC, Pang YL, Koo CH, Ebrahim M, Mohammad AW (2020) Novel nitrogen and sulphur co-doped carbon quantum dots/titanium oxide photocatalytic membrane for in-situ degradation and removal of pharmaceutical compound. J Water Process Eng 33:101068. https://doi.org/10.1016/j.jwpe.2019.101068
106. Gan JY, Chong WC, Sim LC, Koo CH, Pang YL, Mahmoudi E, Mohammad AW (2020) Novel carbon quantum dots/silver blended polysulfone membrane with improved properties and enhanced performance in tartrazine dye removal. Membranes (Basel) 10:1–21. https://doi.org/10.3390/membranes10080175
107. Irani E, Amoli-Diva M (2020) Hybrid adsorption–photocatalysis properties of quaternary magneto-plasmonic ZnO/MWCNTs nanocomposite for applying synergistic photocatalytic removal and membrane filtration in industrial wastewater treatment. J Photochem Photobiol A Chem 391:112359. https://doi.org/10.1016/j.jphotochem.2020.112359

Magnetic Nanohybrid Materials for Water Purification

Anindita Bhuyan and Md. Ahmaruzzaman

Abstract In recent years, the constant population growth worldwide has put pressure on the need for clean and safe water for human consumption. Magnetic nanomaterials (MNMs) have emerged as a promising material in the past few decades due to their unique physiochemical properties. MNMs possess the desired characteristics for their application in water purification. However, it is important to develop well defined magnetic nanomaterials for efficient removal of water pollutants. This requires effective synthetic methods for the synthesis of shape, size and morphology-controlled nanomaterials. Various physical, chemical and biological methods including ball milling, gas phase deposition, thermal decomposition, hydrothermal, solvothermal, co-precipitation, sol–gel, etc. have been explored over the years for their synthesis. The presence of organic and inorganic pollutants in water even in trace concentration has extreme adverse effects on human and environmental health. Hence, it is the need of the hour to develop effective and economical methods for application in water remediation. In this chapter, the problem of water pollutants, their threatening effects and the use of MNMs in water purification have been addressed. These MNMs are characterized using methods such as UV, IR, XPS, XRD, SEM, TEM, VSM, AFM, RS, ^{1}HNMR, etc. Finally, the application of these materials for water purification have been discussed in detail—highlighting the removal of pesticides, dyes, pharmaceutical drugs, inorganic anions, heavy metals and oil spill from water.

Keywords Magnetic nanomaterials · Water remediation · Morphology

1 Introduction

Nanotechnology is one of the most emerging and widely researched areas in modern times. It aims at developing new materials and devices of structures having nanoscale dimension. Nanoscience explores materials in the nanoscale range (1–100 nm) in atleast one dimension. Confined structure and dimensionality of nanomaterials

A. Bhuyan · Md. Ahmaruzzaman (✉)
Department of Chemistry, National Institute of Technology Silchar - 788010, Assam, India
e-mail: mda2002@gmail.com

S. K. Swain (ed.), *Nanohybrid Materials for Water Purification*, Composites Science and Technology, https://doi.org/10.1007/978-981-19-2332-6_8

makes them an attractive as well as a promising material for various applications in industries, water remediation and health care sector among others.

Today water pollution is one of the major concerning environmental issues. Water covers a vast majority of the earth surface, yet only a small amount is fit for human consumption. Modernization and industrialization have led to the widespread influx of pollutant into the water bodies leading to serious environmental consequences including adverse effect on human health. Some of the major contributors for water contaminants are industrial effluent, agricultural runoff, domestic sewage and hospital effluent. Contaminants are present as inorganic, organic and biological form and it is extremely crucial to implement methods to purify wastewater [63].

The removal of these pollutants is a quite challenging field and possesses serious issues. One of the major concerns is the removal of contaminants present in low concentrations which are often missed out in conventional water treatment methodologies. Nanotechnology is able to solve this problem to quite an appreciable level owing to their unique properties which include size selective adsorbent phenomenon and quantum confinement. Nanomaterials can be utilized for efficient uptake of organic (dyes pharmaceuticals, pesticides and insecticides), inorganic (heavy metals) and biological (microbes) pollutants. Another issue in the removal of toxic pollutants is to ensure that no degradation by-product and nanomaterial residue are left behind in the water source. Henceforth, it is necessary for the nanomaterials to not only have high affinity toward pollutants for their efficient removal but it is also important to develop materials that are easy to recover from the purified water. Magnetic nanomaterials can be a promising candidate for the water purification purpose [41].

2 Magnetic Nanomaterials

One class of nanomaterials that have attracted intensive research in recent times is magnetic nanomaterials (MNMs). These nanomaterials are utilised in variety of applications such as catalysis, chemical sensors, magnetic hyperthermia, drug delivery, magnetic resonance imagimg, etc. (Fig. 1). MNMs are generally composed of one magnetic element which can either be a monocomponent such as (i) metallic nanoparticles like Fe, Co or Ni (ii) metal alloys such as FePd or FePt (iii) metal oxides such as Fe_3O_4, Fe_2O_3 (iv) metal carbides like Fe_2C or Fe_3C_2 or they can be bicomponent materials like $CoFe_2O_4$ and $NiFe_2O_4$ They can also be heterostructure materials such as Fe_3O_4@Ag, Fe_3O_4@SiO_2, Fe_3O_4@biochar, etc. [136].

Magnetic particles in nanoscale range display unique physical and chemical properties. Owing to their small size, large surface area and high porosity, these materials have gained interest in their widespread application in adsorption and catalysis. These materials can be easily synthesized and modified into various nanocomposites for targeted applications in different fields. Further, these nanomaterials can be easily controlled using an external magnetic field. Conventional magnetic materials generally lose their permanent magnetic properties when they are reduced to nanoscale range. Superparamagnetic nanoparticles (NPs) show magnetic properties only in

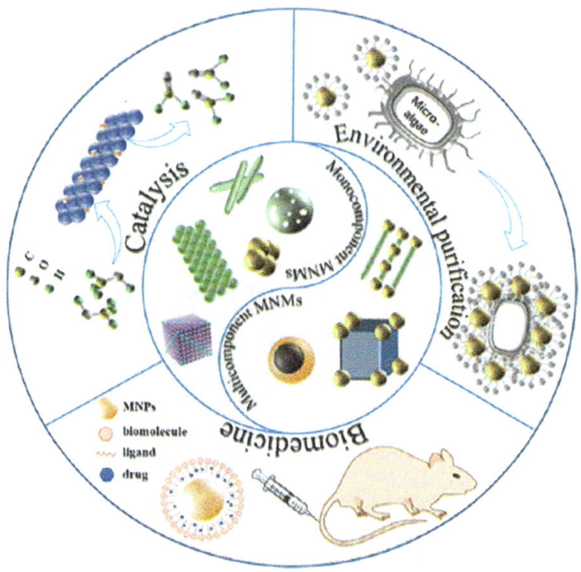

Fig. 1 Application of magnetic nanomaterial in environmental purification, catalysis and biomedicine (Reprinted with permission from [136])

the presence of external magnetic field (EMF). Hence, MNMs can also be used in biological mediums for biomedical application.

Iron based nanomaterials constitute the most widely studied magnetic particles. They show excellent magnetization and biocompatibility. Further they show unique properties including superparamagnetism, high crystallinity, coercivity, high specific surface area and dispersity [4]. However, these materials are prone to agglomeration and corrosion. The MNMs are often capped with polymers like polyethylene glycol or incorporated with inorganic components like activated carbon, silica, alumina, Au, or Pt. This increases the stability of the MNMs and prevents the agglomeration, oxidation and provides protection against corrosion. The properties of these materials are highly dependent on the size, shape and morphology of the NPs. The synthesis methods and conditions thereby influence their properties and application.

3 Synthesis Techniques

It is quite necessary to obtain well-defined structure and morphology of MNMs to achieve its unique properties. Over the years, synthesis of MNM has been well developed. A Variety of methods such as hydrothermal, solvothermal, co-precipitation, sol–gel process, thermal decomposition, etc. have been utilized in the preparation of MNMs. The size, shape and morphology can be controlled by varying the reaction parameters. MNMs can be synthesized either by using physical methods (milling), chemical methods (hydrothermal, co-precipitation, sol–gel) [55, 72] or biological methods. Some of the commonly used chemical synthesis method includes.

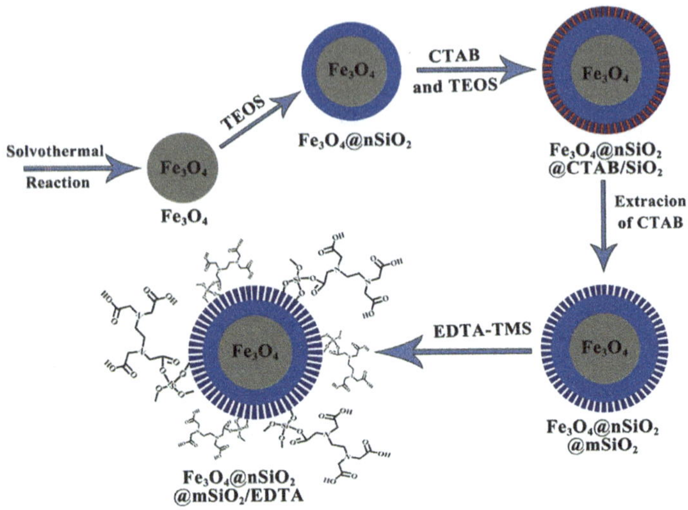

Fig. 2 Schematic diagram of the preparation of $Fe_3O_4@nSiO_2@mSiO_2$/EDTA (Reprinted with permission from [119]

3.1 Hydrothermal/Solvothermal

Hydrothermal method is one of the most commonly used synthetic approaches used in the synthesis of shape-controlled and stable nanoparticles [92]. Generally, hydrothermal method is carried out in a reaction chamber at high temperature and pressure. The solvothermal method [119] follows basically the same principle as hydrothermal with the only difference in the solvent used (Fig. 2). Elemental metals, metal oxides and alloys have all been synthesized using this route. A novel hydrothermal synthesis of γ-Fe_2O_3, Fe_3O_4 and α-Fe_2O_3 [54] was reported by Jayanthi et al.

3.2 Sol–Gel Method

Sol-gel method is based on hydroxylation and condensation of precursors. Gao et al. reported the synthesis of Fe_3O_4 using this method [42]. Small sized and monodispersed nanoparticlesof nearly 8 nm were obtained. This method has also been used in the synthesis of Co-Cr-Ferrite with excellent properties [71]. Some other synthesis of magnetic nanoparticles such as Fe_3C/C [120], $NiFe_2O_4$ [112] $Mg_{0.5}Zn_{0.5}FeMnO_4$ [86] etc. have been reported using sol-gel method.

3.3 Co-Precipitation

It is a very common method used in the synthesis of Fe_3O_4 [25]. The process generally involves preparation of iron oxide nanoparticles from aqueous solution of ferrous and ferric salt by addition of alkaline substances [79]. Although this method is easy to use, it has poor control over size and shape of the nanoparticles. However, the addition of surfactant or polymer coating helps control the size of the nanoparticles [38, 81].

3.4 Thermal Decomposition

This method generally involves the decomposition of organometallic precursors using capping agents (surfactants) in the presence of organic solvents. The synthesis processes are carried out under anaerobic conditions. Manipulation of reaction conditions like reaction temperature, reaction time and annealing temperatures helps to control the morphology and size of the nanoparticles [31]. The advantage of this method is that it is useful to synthesize morphology controlled, high crystalline NPs in high yields [114]. However, the use of surfactants during synthesis process makes it difficult for further surface modification for improved properties [78].

3.5 Polyol Method

It is a liquid phase synthesis method, where polyols (example: ethylene glycol or N-methyldiethanolamine) simultaneously acts as solvent, surfactants and reducing agent. Here precursors are dispersed in polyol and reflexed at high temperature [23]. This method is especially useful to synthesize uniform magnetic nanoparticles as well as elemental NPs such as Fe, Co and Ni from their oxides [103]. The morphology and size of the NPs can also be modified by changing the reaction time or the solvent used. MNMs possessing interesting properties have been synthesized using this method.

3.6 Biological Induced Methods

Magnetic nanomaterials are also synthesized via biological systems. Simple, eco-friendly and cost-effective techniques make it an emerging area of research. Nanomaterials are either synthesized using plant extracts as precursors or the biological entities are used as stabilizing agents and modification for the magnetic nanomaterials. Siddiqui and Chaudhry used a seed mediated coprecipitation technique to grow $MnFe_2O_4$ on *Nigella sativa* plant seeds (Black cumin) [50]. The composite was applied for adsorption of organic pollutants (methylene blue) and also inhibits bacterial growth. Huang et al. carried out an enzyme immobilization on magnetic chitin nanofibre composite (MCNC) [124]. Enzymes being excellent biocatalyst, these nanomaterials can be applied onto various applications. Mixed fungal biomass coated with magnetic nanoparticles (MNP-FB) [100] were used for the removal of toxic Cr(VI) ions from aqueous solution. They isolated two fungal biomass, namely *Aspergillus fumigatus* and *Aspergillus niger* and used them as precursor for synthesis of an adsorbent. A maximum adsorption of 249.9 mg/g was achieved indicating it to be a proficient and environment friendly adsorbent material.

Several other physical, chemical and biological methods are also explored for the synthesis of magnetic nanomaterials especially super-paramagnetic iron oxide nanoparticles (SPIONS), as depicted in Fig. 3.

4 Characterization Techniques

The properties of MNMs are directly influenced by size, shape and structures of the nanoparticles. Several methods are adopted for analysis and characterization of the MNMs.

Fourier Transform Infrared Spectroscopy (FTIR) is a spectroscopic method used to determine the quality of material as well as used to confirm the identity and presence of functional groups present in the modified MNMs. The stretching vibration of Fe–O bond of iron oxide NPs changes on the formation of nanocomposites with other materials such as SiO_2 or ZnO [133], thereby confirming the formation of hybrid composites. Fe_3O_4 synthesized by Vojoudi *et al.* showed adsorption band around 457 and 573 cm^{-1} which correspond to Fe–O stretching frequency [117]. However, after coating the iron oxide with mesoporous silica, new adsorption peaks around 805 and 1095 cm^{-1} were obtained. Further, by analyzing the formation and nature of bonds formed between the adsorbent material and the adsorbate (substrate), the mechanism of adsorption can also be determined.

The size of the magnetic NPs can be determined by several methods such as X-ray diffraction (XRD), Transmission Electron Microscopy (TEM) and dynamic light scattering (DLS). DLS method determines the size of the MNMs where the nanoparticle suspension is exposed to an electromagnetic wave and scattering is recorded. Apart from size, the crystalline structure of the MNMs can be analyzed using XRD as the diffraction patterns correspond to the plane of the NPs [73].

TEM is one of the best methods to analyze the size of nanomaterials. TEM is based on the technique where an electron beam is transmitted through the specimen, and these electrons interact with the specimen as they pass through it. It is used for the determination of internal structure of nanomaterials. The core dimension and coating dimension directly influence the magnetic nature of the magnetic NPs. The major disadvantage of TEM is the tedious sample preparation process and selection of suitable volatile and inert solution for sample preparation. For core shell MNMs, TEM can be used to clearly observe and confirm the core–shell structures [57]. TEM images also successfully show the presence of any agglomeration that may occur in

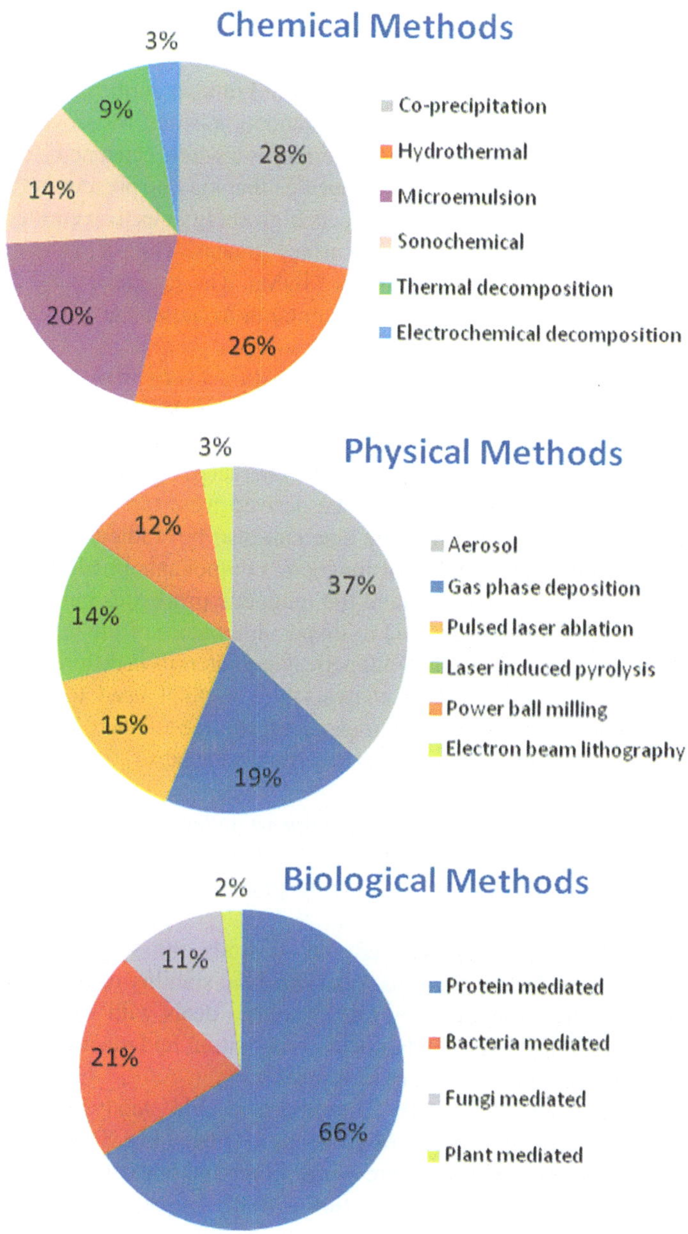

Fig. 3 Synthesis of SPIONs by three different routes viz., Physical, chemical and biological (Preprinted with permission from [5])

the MNM dispersion. Moreover, it is also helpful to study the effect of core shell on the agglomeration of the MNMs.

Energy dispersive X-ray (EDX) analysis of TEM images confirms the presence of the various elements present in the magnetic nanocomposites. Moreover, the weight percentage of the components and nanomaterials as a whole can also be determined. It is also useful in the detection of any impurities that may be present in the MNMs. Scanning electron Microscopy (SEM) detects high energy electron emitted from the surface of the nanoparticles and creates surface images. The SEM image analysis determines the surface morphology of the MNMs. The layers of the NPs and the surface, whether planar, smooth or rough; regular or irregular can all be observed by studying the images.

Vibrating sample magnetometry (VSM) analysis is a very important technique to study the magnetic properties of magnetic nanomaterials. VSM plots the variation of magnetization (M) against the applied magnetic field. Here, the magnetic hysteresis loop is studied at room temperature. Magnetic characteristic of the MNM, whether ferromagnetic or superpaarmagnetic can be derived from the readings. Moreover, magnetic properties variation with core size can also be studied using VSM. It is also helpful to quantify the number of magnetic components in MNMs. This technique thus helps to not only understand the magnetic properties but also considers magnetic accountability, recovery and re-dispersible nature of the nanoparticles for their potential application as suitable magnetic nanomaterials in different fields [126].

Brunauer–Emmett–Teller (BET) analysis is a method used to determine the specific surface area (SSA), pore diameter and volume of the MNMs. The pore size and SSA is a major parameter which determines the adsorption capacities of the MNMs.

Photoluminescence (PL) spectroscopy helps to understand the transport, migration and recombination of electron–hole pairs. Understanding the recombination process and electron transfer helps to study the photocatalytic activity of the MNMs [46].

Zeta potential study is also carried out to study the surface charge of the nanocomposite which further helps to understand the stability of the NCs [48].

X-ray photoelectron spectroscopy (XPS) studies deals with surface characterization and help to determine the chemical composition and electronic state of the individual elements in the hybrid nanocomposites.

Raman spectroscopy (RS) [24], Thermogravimetric Analysis (TGA) [113], Elemental mapping, Atomic Force Microscopy (AFM) [43], Scanning Tunneling Microscopy (STM), ^{1}HNMR [99] are some other methods used for analysis of magnetic nanomaterials and their interaction with the targeted substrate materials.

5 Application in Water Purification

Water contamination is one of the major environmental issues in recent era. Millions of deaths in human are caused every year due to inadequate or unsafe access to pure water. Consumption of polluted water leads to a variety of diseases and even deaths of living organisms. It also leads to deterioration of natural environment in various aspects. Hence, it is very important to seek removal of these toxic contaminants as a topic of critical concern.

Recently, magnetic nanomaterials have gained interest for their unique physiochemical properties. These nanomaterials show high adsorption and removal efficiency towards toxic contaminants. In addition, easy recovery and reusability of the MNMs makes them an efficient and cost-effective nanomaterial. Magnetic nanomaterials have been widely explored for their adsorption and catalytic action toward various types of pollutants (Fig. 4). Organic pollutants including pesticides, fertilizers, dyes pharmaceutical and personal care products; inorganic metals and anions, biological pathogens impose major threat to the environment as well as human health. MNMs through their adsorption and catalytic properties form a suitable candidate for waste water remediation.

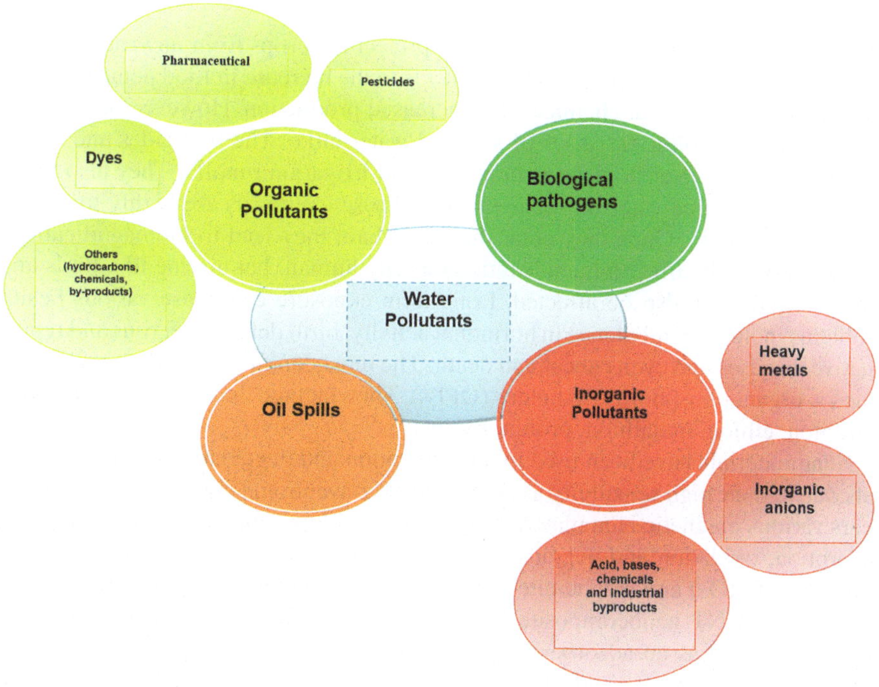

Fig. 4 Overview of magnetic nanomaterial application in water purification

5.1 Organic Pollutants

Organic pollutants can be considered a critical environmental threat due to its persistent nature. Organic pollutants are hence often referred to as persistent organic pollutants (POPs). These POPs can bioaccumulate and undergo biomagnifications as it enters the food chain, thereby harming not only the aquatic life but also causing adverse effect on human. Unlike inorganic pollutants which mainly include heavy metals, cations and anions, organic pollutants comprise of a wide range of compounds and functional groups having different chemical and physical properties. Organic pollutants include hydrocarbons, dyes, pesticides, insecticides, herbicides, pharmaceuticals, health and personal care products, detergents and other toxic chemicals used in our everyday life.

5.1.1 Pesticides/Insecticides

These chemicals are used in agriculture to protect the crops from unwanted weeds, insects and mites. Increasing population leads to the increase in food demand which leads to extensive use of chemicals for increased production. However, this has led to some serious pollution issues of soil and water alike. The pesticides and insecticides enter the water bodies via rainwater and irrigation runoffs. They also make way into the groundwater through leaching. These chemicals are highly toxic and non-biodegradable. Once they enter the food chain they lead to biomagnifications and can adversely affect human health. Not only human, but marine life, birds and other animals can also be affected. Long term exposure can cause various health problems including imbalance in hormonal activity, birth defects, nervous and reproductive diseases and even cancer and death. The most commonly used class of pesticides is organophosphorus pesticides (OPPs). They include diazinon, chloropyrifos, parathion, ethion, malathion, profenofos, etc.

Nanomaterials have been used in the adsorption and degradation of these chemicals due to their high sensitivity as compared to conventional materials [9]. Over the years various magnetic nanomaterials are being developed and used in the detection, adsorption, oxidation and degradation of pesticides. Among various MNMs used, magnetite (Fe_3O_4) and maghemite (Fe_3O_2) have gained the most popularity.

A Fe_3O_4/CNT nanocomposite [76] was synthesized by hydrothermal method and was utilized as an adsorbent for the removal of organophosphorus pesticides, fenithrothion, profenofos and ethion. The adsorption process was carried out by solid phase extraction and liquid chromatography. Another Fe_3O_4/Red mud nanoparticle (NP) [13] was researched for the removal of organophosphorus pesticides (diazonin, parathion, malathion). During the same time a Fe_3O_4/C nanocomposite [97] was synthesized using a novel technique. They obtained the magnetic Fe_3O_4 using a typical wet chemical synthesis method and the carbon NPs were synthesized using walnut shells. The Fe_3O_4 was then added to the carbon solution under continuous stirring to form the nanocomposite (NCs). They applied the as-synthesized NCs for the removal of diazonin, a pesticide commonly used against sucking and chewing mites and insects. By using 0.155 g of the Fe_3O_4/C NCs at optimum pH 6 and 35 min contact time, 100% efficiency was obtained for solution containing 0.5 ppm diazonin.

An amino-functionalized Fe_3O_4-WO_3 NP [82] was synthesized by using two methods simultaneously, namely co-precipitation and hydrothermal. The tested the NPs for the photolytic degradation of diazonin from both synthetic and real water sample in the presence of UV-radiation. Although in the absence of any other organic compounds removal efficiency of 96.4% was obtained, however in the presence of citric acid, folic acid, EDTA, phenol and humic acid, it decreased to 70.3, 69.39, 67.37, 60.12 and 3.22%. Very recently an iron oxide decorated carbon fiber composite (Fe-ACF/CNF) [107] was synthesized by chemical vapour deposition (CVD) and was utilized for the extraction of 29 pesticides using gas chromatography. The adsorbent was reusable upto 4 cycles and had high accuracy of 70–120%.

Several other MNMs like silica supported Fe_3O_4 [85], Fe_3O_4/grapheme [17], Fe_3O_4/SiO_2 core shell [36], Fe_3O_4/biochar NCs [113], Fe_3O_4 decorated β-cyclodextrin polymer [99], g-C_3N_4/Fe_3O_4/Ag [46], Fe_3O_4@SiO_2-MWCNT [123] Fe_3O_4@ZIF-8@polymer core–shell shell [118], and Fe_3O_4@N-ZnO [6] have been utilized in the removal and degradation of pesticides.

5.1.2 Dyes

Dyes are coloured organic compounds that bonds chemically to the substances to which it is applied and imparts colour to it. They are different from pigments in the sense that pigments do not bind to the substances chemically. Dyes can be broadly classified into ionic dyes which include cationic and anionic dyes and non-ionic dyes comprising of vat dyes and disperse dyes.

Dyes have been used by mankind over several decades for various applications. Earlier natural dyes derived from plants and insects were used. However, this tends to fade away with time when exposed to sunlight and water. Synthetic dyes were first discovered in 1856 and since then it has been produced and utilized on large scales. Artificial dyes have complicated structure and are persistent in nature. When released into the waterbodies, they consume oxygen and increase the biological oxygen demand. Also due to their lower density they form a layer on the water surface cutting off sunlight and air from water. This imposes major threat to the aquatic life. Moreover, aromatic structures of dyes show mutagenic and carcinogenic effect towards human. Dye traces may also cause allergies and skin irritation to living organisms. Therefore, removal of dyes is a major environmental concern. Some of the major industries that contribute to the discharge of dye effluent into the waterbodies include textile, paper and pulp, food industry, tanning industry, cosmetic, plastic, paints and dye manufacturer and many others. The textile industry forms the highest contributor among all [62].

Several methods have been researched for the removal of dyes from water, which include chemical oxidation, degradation, adsorption, biological degradation, etc.

Adsorption has an advantage compared to other methods owing to its simple technique, cost effectiveness and most importantly the fact that it does not leave behind any by-product behind provided it is efficiently removed [35]. Most commonly used adsorbents include silica, activated carbon, zeolites, clays and other polymeric materials.

The importance towards the concept of easy and fast recovery and reusability of adsorbent materials, MNMs has gained interest towards the removal of dyes. Magnetite (Fe_3O_4) is one of the most widely explored magnetic materials for dye removal. These materials have affinity towards external magnetic field which allows their easy separation. A novel Fe_3O_4@GPTM@Gly NP [133] was synthesized and utilized for the adsorption of cationic and anionic dyes (MB, OR, AR, MEB, AZ). Another hybrid shell adsorbent, κ-carrageenan-Silica core shell encapsulated Fe_3O_4nanoparticles (Fe_3O_4@SiO_2/SiCRG) [110] were researched for the adsorption of MB. High adsorption capacity of 530 mg/g was observed at 25 °C and pH 9, which was higher than many other magnetic materials. Magnetite-bioadsorbent nanocomposite (MLMC) [51] of Moringa oleifera leaves and cotton shell powder was explored for removal of brilliant green dye from water. Recently a novel chitosan-vanadium-titanium-magnetite (CS-VTM) nanocomposite [131] was synthesized through a one pot synthesis method to study its adsorption towards congo red. For CR dose of 100 mg/L, an adsorption capacity of 62.2 mg/g was achieved at optimum pH 6 and 65 °C. Overall, the CS-VTM was found to be efficient and environment friendly nanoadsorbent. A porous magnetite (PMNs) [32] was also synthesized and used as a highly efficient adsorbent for removal of AR57 and RR dyes. The adsorption of the dyes was found to be endothermic in nature. The q_m obtained was 888.8 mg/g for AR57 and 808.43 mg/g for RR dye. Magnetite-bioadsorbent NPs are also explore for removal of dyes from wastewater.

Panda et al. reviewed various other uses of magnetite nanomaterial composites as adsorbent for dye removal [91]. Several other works of magnetite incorporated composite used as dye adsorbents include Fe_3O_4/rGO [111], Co_3O_4@SiO_2 NC [45],Fe_3O_4/GPTMS@P-Lys [127], Fe_3O_4/chitosan [49], $NiFe_2O_4$/ZnO [135], Fe_3O_4/Pectin; Fe_3O_4/silica pectin/hybrid [11], Mn doped Fe_3O_4 [64], Au-Fe_3O_4-NCs-AC [29], rGO/Fe_3O_4 [59], Ggh-g-PAcM/ Fe_3O_4 [66], Humic Acid(HA)- Fe_3O_4 NP [1], etc.

5.1.3 Pharmaceuticals

Over the last few decades, urbanization and modernization has taken over human life. With increased human activity, pharmaceutical products have become a vital part in improving the quality of human life. The various classes of pharmaceutical compounds (PhCs) include antibiotics (tetracycline, amoxicillin, and sulfonamides), antihistamines, analgesics and antipyretics, antiseptics, non-steroidal anti-inflammatory drugs (NSAIDs) (diclofenac, ibuprofen), beta-blockers (propranolol, metoprolol), hormones, steroids, etc. Pharmaceuticals not only form a major part of human life, but are also widely used in several other sectors such as industries, veterinary, agriculture [69] livestock and animal husbandry, etc. Pharmaceutical products and byproducts, as emergent pollutants, have gained interest in recent times due to its persistent nature, low degradability and its potential effect on human health even when present in trace amounts.

These contaminants make way into the waterbodies mainly through drug manufacture industries, hospitals and municipal effluents. High amounts of pharmaceuticals have been detected in surface water, groundwater as well as drinking water. The adverse effects they may cause to humans include allergies, chronic diseases, antibiotic resistance, metabolic perturbations, and endocrine disruption among other. Prolonged presence of antibiotics in water is observed to produce harmful bacteria. Moreover, they result in bioaccumulation and eutrofication thereby adversely affecting marine life which includes disruption in fish spawning and reduced fertility. Pharmaceutical wastewater release is often unregulated. Moreover, due to its versatility and trace concentration makes it go undetected in conventional waste water treatment plants. Therefore, it is very important to develop high effluent and adequate removal techniques for pharmaceutical waste treatment. Several methods used to remove these contaminants from water include ultrafiltration, membranes, biodegradation, oxidation, adsorption, etc. Activated carbon, Grapheme oxides, zeolites, nanoclay, biochar, chitosan, etc. and their hybrid nanocomposites are some commonly used materials for the adsorption and degradation of this target molecules.

Magnetic nanomaterials and their composite counterparts are a major breakthrough for their removal process. Various MNMs like Fe_3O_4 [19], Fe_3O_4@C [60], Fe_3O_4-red mud NCs [14], Fe_3O_4-MoO_3-AC [77], $MnFe_2O_4$@TiO_2-rGO [22] have been used in the removal of common antibiotics such as ciprofloxacin, rimapicin, cephalexin, levofloxacin and amoxicillin. Raha and Ahmarruzaman synthesized g-C_3N_4@ Fe_3O_4/ZnO nanorods [95] and g-C_3N_4/NiO/ZnO/Fe_3O_4 [94] for the photolytic degradation of pantaprazole and esomeprazole. These are a class of drugs known as proton pump inhibitors and are commonly used against acidity in stomach. NSAIDs such as Ibuprofen, naproxen, diclofenac, etc., form a very important and commonly used class of drugs. Hence it forms a major part of the pharmaceutical contaminants in water [89]. Magnetic nanoparticles like γ-Fe_3O_4-Zeolite [12], $CoFe_2O_4$ [27], Fe_3O_4@decanoic acid [10], Bi_2O_4/ Fe_3O_4 [126], $NiFe_2O_4$/AC [39], Fe_3O_4@Ag [116], etc. have been explored for their removal.

5.2 *Inorganic Pollutants*

Inorganic pollutants in wastewater mainly comprises of heavy metals, non-metallic salts, acids and bases, fluorides and plant nutrients such as phosphates, nitrates, ammonium ion, etc.

5.2.1 Heavy Metals

Metallic chemical elements having high density which are generally toxic and present low concentrations in the environment can be referred to as heavy metals. Although heavy metals such as Cd, Hg, As, Pd, Pb, Cr, Al, Co, Ni, Cu, Fe, Mn are mostly present in trace amounts, they are one of the most widely spread and toxic water pollutants. Heavy metals enter the waterbodies through industrial discharge (tanning, paints, pesticides, etc.), mining and through agricultural runoffs (fertilizers and pesticides). These pollutants accumulate in the environment and cause severe threat to human life. These are non-biodegradable components and some metals like Cd have infinite lifetime. They also accumulate in the food chain and result in biomagnifications and are highly toxic and carcinogenic. Heavy metals such as Ni, As and Cr can cause skin diseases, pulmonary fibrosis, nausea, dizziness, etc. Other metals such as Pb, Cd and Hg can affect lung and kidney functions, cause mental retardation, vision loss, anorexia, thyroid dysfunction and birth defects. Moreover, accumulation in the water bodies adversely affects the aquatic life damaging gills of fish and even deaths. Removal of these toxins, therefore, is a topic of critical concern.

Magnetic nanomaterials are being used for heavy metal removal from wastewater. Magnetic properties of the nanomaterials aid in the easy extraction of pollutants as well as the recovery of the adsorbent. Fe_3O_4 silica core shell MNM and modifications of the MNM with cetylmethylammonium bromide (CTAB) and bis(3-triethoxysilylpropyl)tetrasulphide (MSCMNP-S_4) was researched as a novel and efficient adsorbent for removal of heavy metals [117]. They could successfully remove Hg(II), Pd(II) and Pb(II) with 303.03, 256.41 and 270.27 mg/g maximum adsorption capacity (Q_m) and efficiency of 98.8, 96.4 and 95.7%. A novel magnetic nanoparticle (NP), iron oxide magnetic NP grafted hyperbranced polyglycerol (HPG-MNP) was synthesized and used for removal of Cu, Ni and Al ions from industrial wastewater. [7].

Surface modification of iron oxide NPs with materials like silica result in increased stability of the NP and also prevents aggregation. Magnetic iron oxide nanoparticles are considered a suitable material for removal of As due to the strong interaction between Fe and As. Table 1 summarizes the various MNMs used for heavy metal removal from water and wastewater.

5.2.2 Inorganic Anions (Phosphates/Nitrates/Fluorides)

Phosphates and nitrates are vital nutrients for plants. Phosphorus is commonly present as phosphates in nature. Phosphates are commonly used in fertilizers. It is also found in toothpaste, detergents, water softener, cured meat, pharmaceuticals and processed cheese. Nitrates are common nitrogenous compounds that are used as plant nutrients. Nowadays, fertilizers are used on large scale in agricultural farms to increase production. Excessive use of fertilizers and pesticides has resulted in release of nitrates into the waterbodies. These nutrients are detected in rivers, lakes, reservoirs, groundwater and even drinking water. High nitrate and phosphate concentration in water lead to

Table 1 Heavy metal removal from water and wastewater using various MNMs

MNM	Target Metal	Q_m (mg/g)	References
Fe_3O_4@rGO	As (III) As (V)	13.10 5.83	[21]
Thiol@magnetic $mSiO_2$	Pb(II) Hg(II)	91.5 260	[70]
Fe_3O_4@APS@AA-Co-CA MNPs	Pb(II) Cd(II) Cu(II)	166.1 29.6 126.9	[44]
Ascorbic acid@Fe_3O_4	As(III)	46.06	[37]
Fe_3O_4	As (III) As(V)	188.69 153.8	[75]
Fe_3O_4@graphene	Cr(IV) Pb(II) Hg(II) Cd(II) Ni(II)	17.29 27.95 23.03 27.83 22.07	[47]
MDA-magMCM-48	Pb(II) Cu(II) Cr(VI) Cd(II)	127.24 125.80 115.60 114.08	[8]
Fe_3O_4@SiO_2-SH	Hg((II)	132	[121]
MSCMNP-S_4	Hg(II) Pd(II) Pb(II)	303.0 256.41 270.27	[117]
Fe_3O_4 Si- Fe_3O_4	As(III)	88.19 67.92	[52]
Fe_3O_4@$mSiO_2$ core shell	Cu(II)	84.4	[57]
Fe_3O_4	As(III) As(V)	–	[18]
FSP FSBP	Pb(II)	202.8 143.7	[130]
Fe_3O_4@biochar	As(III)	5.49	[87]
MNPs35@SiO_2	Pb(II)	14.9	[88]
Fe_3O_4@$nSiO_2$@$mSiO_2$/EDTA	Cr(III)	30.59	[119]

eutrophication. The nutrients enter the waterbodies through irrigation and rainwater runoff. They also enter the ground water through leaching. They accumulate in the water and result in algal bloom excessive growth of aquatic plants. This results in the depletion of dissolved oxygen in water. This further leads to the death of aquatic organisms. In human, increase in amount of phosphorus in the body can lead to phosphorus retention. This led to chronic kidney diseases, cardiovascular morbidity and even deaths in people undergoing hemodialysis. Exposure to nitrates can cause blue baby syndrome or methemoglobinemia, a condition common in infants that occurs due to lack of enough oxygen in blood.

Various magnetic NPs such as Fe_3O_4/ZrO_2/chitosan [56], Fe_3O_4@SO_2@MPS@poly(4-vinylpyridine) core shell-shell [53], rectorite/Fe_3O_4-CTAB [121], AFMCS [68] Fe_3O_4@chitosan core–shell [40], Fe_3O_4@GO-Sr [102] and Fe_3O_4@GO-CMC [61] have been researched for the removal of phosphates and nitrates from water.

Fluorides are another important water pollutant which affects human in a major way. All water generally contains some fluoride. Natural causes of high fluoride concentration in water include weathering of minerals like fluorospars (CaF_2), flurapatite ($Ca_5(PO_4)_3F$) and cryolite (Na_3AlF_6) as well as through volcanic eruption. Fluorides are used in industries for production of glass, semiconductors, ceramics, rubber and fertilizers. Fluoride upto a concentration of 1.0 mg/L is an essential component of human health. Instances have shown that fluoride is even added to municipal water because it is believed to prevent tooth decay in local population. It is also present in toothpastes and mouthwashes. However, it become hazardous if the concentration exceeds 1.5 mg/L [125]. Consumption of fluoride contaminated water can cause ache and other skin problems, cardiovascular diseases, muscle spasm, abdominal pain, nausea and high blood pressure. Exposure to fluoride for longer time can also result in skeletal and dental fluorosis, early puberty and low fertility in women, thyroid dysfunction, neurological problems and bone cancer.

Some of the defluoridation technique used for drinking water is ion-exchange, reverse osmosis, adsorption and precipitation. Fluorides are hard base having small size and high electronegativity and has high affinity towards multivalent metallic ions like Fe(III) and Al(III). Hence magnetic nanoadsorbents are a good candidate for its removal. Aluminium when is combined with iron, very efficient adsorbent having advantages of both components is obtained. Chai *et al.* synthesized Fe_3O_4embedded Al_2O_3 for fluoride adsorption and obtained maximum adsorption of 88.48 mg/g at pH 6.5. Following similar route Fe_3O_4/Al_2O_3 NPs were developed for removal of fluoride from drinking water. Following similar route, a novel sulfate doped Fe_3O_4/Al_2O_3 was developed for fluoride removal from drinking water [20]. Similarly, batch adsorption of fluoride was carried out using magnetic corn stover biochar [84]. Magnetic Ce-Ti@Fe_3O_4 nano core shell [80] was also used for fluoride treatment of drinking water.

A biochar Fe_2O_3/Fe_2O_4nanoadsorbent [28] was also developed and used for the adsorption of nitrates and fluorides. Mohammadi et al. also synthesized carboxylated chitosan-iron complex [83] and studied its action for simultaneous removal of fluorides, phosphates and nitrates.

5.3 Oil Spills

Oil spills are generally accidental discharge for petroleum products into the water bodies. It has become a concerning environmental issue. Oil and petroleum forms one of the most important and primary source of energy and provides raw material for a number of chemical industries. With the depletion oil reserves in the land, men have moved their eyes to exploit the ocean. Majority of the oil spills occur during extraction and transport, through leakage in drilling rigs, pipelines and oil vessels, transfer of oil, oil spills from ships and tanker accidents, etc. The common components of oil spills include crude oil, gasoline, fuel oil diesel, petrol, kerosene, etc.

The oil spilled into the water bodies cut the oxygen supply from the water which adversely affects the biological oxygen demand (BOD). This hampers the respiration of aquatic organisms and photosynthesis of aquatic plants. Ultimately, this may lead to the devastating loss of aquatic life cycle. The oil spills also undergo weathering [98] which leads to the evaporation, emulsification, sedimentation and biodegradation and may result in the spread of oil components to other part of the environment, harming not only the marine life but also adversely affecting human. Hence it is an absolute necessity for the removal of oil spills to conserve the environment.

Nowadays, a variety of methods are employed for cleanup of oil spills. These can be categorized into chemical, mechanical and bioremediation methods [105]. The various techniques include use of booms, shimmers, surfactants, dispersants and adsorbents like silica, zeolite, polyethylene, grapheme, cellulose, etc. An efficient adsorbent material must be cost effective, must have high adsorption capacity and must be reusable. A number of researches have been conducted to develop novel and efficient adsorbents for oil spills. The major issue faced by the currently used adsorbents is the removal and recovery of the sorbent material along with the oil. Secondary pollution, low efficiency and long process time are some other problems faced by other removal methods.

This disadvantage may be overcome by the utilization of magnetic nanoparticles. Magnetic nanomaterials are considered as a promising candidate for oil spill remediation. These materials can either be employed alone or functionalized into nanocomposites. The nanocomposites are generally composed of an inorganic magnetic component dispersed/coupled with an organic polymer, surfactant, biomolecule, carbon, silica, ionic liquid among others. The composites do not lose their magnetic properties. Additionally, the addition of hydrophobic component further increases the potential and efficiency of the magnetic nanoparticle. Superparamagnetism, high magnetic susceptibility with high stability, non-sinking properties, high adsorption capacity and recyclability makes them a suitable material for oil remediation. The advantage of using magnetic nanomaterial is that the resultant aggregation of the oil with the MNMs exhibit magnetic properties which can then be easily extracted by external magnetic field. However, these materials must be biocompatible so that it does not lead to any secondary pollution. The nanoadsorbent can then be reused after washing them with solvents such as hexane and ethanol. Lu et al. demonstrated an excellent recyclability of upto 7 cycles was exhibited by a chitosan capped magnetic nanomaterial [74].

The most common materials researched are the maghemite (Fe_2O_3) and magnetite (Fe_3O_4) based nanomaterials. Acetylated curauafibres/Fe_2O_3 incorporated magnetic

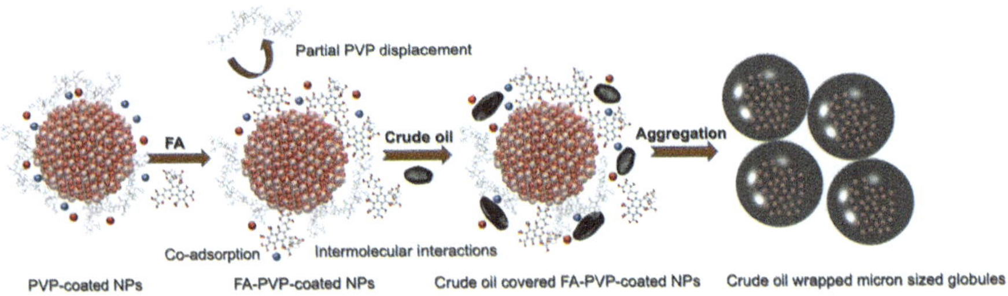

Fig. 5 Schematic diagram of adsorbtion of crude oil on PVP coated Fe_3O_4 NPs (Reprinted with permission from [15])

polymer nanocomposite [33] was developed for removal of oil. Another MNM developed from coconut shell-based AC-iron oxide NC [96] was developed for magnetic separation of oil. A novel sulfonated asphalteen coated magnetic NP [3] was used for removal of crude oil. The asphaltene capping agent forms colloidal particles with the crude oil through strong hydrogen bond, electrostatic and π-π^*stacking forces. The as—synthesized nanomaterials were successful in removal of crude oil with adsorption capacity of 22.5 g/g. Flaked polyolefin-based adsorbents (PA) shows excellent properties for oil removal. However, these materials, due to their structure are not fully recoverable. To overcome this problem, very recently Kim et al. synthesized polyolefin based nanoadsorbent (PMA) [65]. This facilitated easy removal of the nanoadsorbent. Further, the oil-PMA composite can be fully paralyzed and converted to refined oil. Hence it forms a promising material for oil spill removal. Other magnetic nanocomposites used for these purpose includes SPION/β-cyclodextrin core shell [67], Fe_3O_4@OA (Saber et al., 2015), Fe_3O_4@SiO_2 [128], Fe_3O_4@PS [128], CS-grafted Fe_3O_4 [74], yeast-magnetite bionanocomposite (YB-MNP) [26], SPIONS-chitosan NC [106], magnetic ferrogels [101] and PVP- Fe_3O_4 nanoparticles [15] (Fig. 5).

5.4 *Biological Pathogens*

Pathogenic microorganisms present in water for human consumption can cause adverse effect on human. Thus, elimination and disinfection of these water pollutants are a necessary step to ensure safety to human health. Bacteria, cyanobacteria, viruses, protozoa and fungi are various pathogens that can cause severe harm to human being as well as pose threat to aquatic life. Health issues including cholera, typhoid, jaundice, abdominal and intestinal infections, respiratory disorders, gastroenteritis, respiratory and digestive system inflammation, polio, hepatitis, encephalitis, giardiasis, etc. are all caused by microbes present in water.

Every year, a million people face death caused by these waterborne diseases (WBD). Hence WBD can be considered as a major global challenge. Various techniques such as filtration, flocculation, chlorination, disinfection, adsorption, etc.

have been used for elimination of biological contaminants. The inefficiency of these methods calls for the development of more efficient systems for pathogen treatment. Magnetic nanomaterials and their biologically compatible functionalized nanocomposites are a newly emerging material that shows high efficiency for the removal of a wide range of pathogens.

Abdolmaleki et al. synthesized β-cyclodextrin-poly(isophthalamide)-magnetic nanoparticles (CDPA-MNP) [2] that exhibited effective removal of both gram positive (*Baccilus cereus*) and gram negative (*Escherichia coli*) bacteria with high efficiency. The $MnFe_2O_4$/BC composite synthesized by Siddiqui and Chaudhry (Siddiqui and Chaudhry, 2018) showed bacterial growth inhibition against Straphylococcus aureus and Escherichia coli cells. This may be due to the phytogenic content on the black cumin (BC) seeds which interact with the lipid layer of bacteria cell membrane and inhibit their growth. NiO nanoparticles was synthesized by a sol-gel method using Salvia macrosiphon Boiss extract as limiting agent. The nanoadsorbent shows excellent photocatalytic properties and was used for degradation of methylene blue with 80% efficiency. Similarly, Pinto et al., synthesized a series of magnetic nanoparticles (FeO, FeO/AC, MnFeO, CuFeO, CoFeO) [93] and used them as low cost and simple materials for the elimination of both gram positive and gram-negative bacterial strains from the water. Mesoporous $Co_2O_3/Cu_2O_3:Al_2O_3:SiO_2$ magnetic NC was used as a novel material for water disinfection from 7 bacterial strains (*Escherichia coli, Salmonella enterica, Pseudomonas aeruginosa, Listeria monocytogenes, Staphylococcus aureus, Enterococcus faecalis, Bacillus subtilis*).

PQA-MNP [30], PEG-Fe_3O_4 [108], Fe_3O_4@Arg, Fe_3O_4@Lys, and Fe_3O_4@PLL [58], Fe_3O_4-SiO_2-NH_2 [129], chitosan-olligosaccharide/Fe_3O_4 [104], Graphene/Fe_3O_4 [137], AF-$CoFe_2O_4$ [16] and SWCNT-iron oxide [34] are other magnetic nanomaterials that have been used for the removal of pathogens from waste water.

The magnetic nanomaterials can be considered as a promising material with potent disinfection properties and their viable application in waste water treatment. These nanomaterials pose additional advantage such that they can be easily separated and reused.

6 Perspective and Conclusion

Over the last few decades, nanotechnology has emerged as a promising tool for their application in various fields. Magnetic nanomaterials have attracted much attraction recently. MNMs are highly stable and efficient nanoparticles owing to their high specific area, high selectivity and fast separation. Moreover, these materials can be synthesized and modified using simple techniques. These materials are widely explored for their interesting and unique physiochemical properties. The size, shape and morphology of these nanoparticles directly influence the properties of the MNMs. Hence, synthesis methods are primary requisite for development for materials with desired properties. Monodispersed magnetic nanomaterials can be successfully synthesized using different methods like hydrothermal, sol–gel, co-precipitation, etc. Further, surface modification of the MNMs with inorganic materials and organic coatings play a crucial role for the development of nanomaterials with improved fuctionalization. The properties of the nanomaterials also influence the interaction of the MNMs with the substrates. The application of these nanomaterials for adsorption, degradation and removal of different water pollutants is a topic of research. MNMs are being used for water decontamination and removal of pollutants like pesticides, pharmaceuticals, dyes, chemicals, heavy metals, inorganic anions, etc. However, the eco-friendly synthesis and environmental fate of these nanomaterials must also be considered for their efficient and widespread application for water purification. The MNMs reduces the use of chemicals to a certain level by reducing the number of extraction steps since they can be extracted using external magnetic field.

Despite the fact that magnetic nanomaterials are a promising material for their application in water purification, certain aspects need to be addressed. The excessive use of hazardous chemicals for the synthesis and fabrication of MNMs caused stress for the proper disposal of the harmful chemicals and solvents. If fabrication is done taking safety measures and the hazardous toxins are discharged without treatment, they may lead to secondary pollution and also transform into more toxic substances. Biological approaches are being explored to overcome this issue. Biological synthetic methods are often cheap, safe, non-toxic and eco-friendly in nature.

Although there is a rise in the interest and research of these nanomaterials, most work is based on batch experiments carried out in small scale. Therefore, sufficient effort must be taken to explore their potential in large scale industrial conditions. Efficient adsorption, regeneration, performance stability and economical availability must also be addressed during their large-scale application.

In conclusion, MNMs opened up new development in environment application and can be considered as a promising candidate for their application in water purification. Finally, the cost effectiveness, enhanced adsorption, regeneration and reusability of the MNMs are important criteria that require intense study and research for development of efficient materials for water remediation.

References

1. Abate GY, Alene AN, Habte AT, Addis YA (2021) Adsorptive removal of basic green dye from aqueous solution using humic acid modified magnetite nanoparticles: Kinetics, equilibrium and thermodynamic studies. J Polym Environ 29:967–984
2. Abdolmaleki A, Mallakpour S, Mahmoudian M, Sabzalian MR (2017) A new polyamide adjusted triazinyl-β-cyclodextrin side group embedded magnetic nanoparticles for bacterial capture. Chem Eng J 309:321–329
3. Abdullah MMS, Al-Lohedan HA, Atta AM (2016) Novel magnetic iron oxide nanoparticles coated with sulfonated asphaltene as crude oil spill collectors. RSC Adv 6:59242–59249
4. Akbarzadeh A, Samiei M, Davaran S (2012) Magnetic nanoparticles: preparation, physical properties, and applications in biomedicine. Nanoscale Res Lett 7:144

5. Ali A, Zafar H, Zia M, Ul Haq I, Phull AR, Ali JS, Hussain A (2016) Synthesis, characterization, applications, and challenges of iron oxide nanoparticles. Nanotechnol Sci Appl 9:49–67
6. Allam EA, Ali ASM, Elsharkawy RM, Mahmoud ME (2021) Framework of nano metal oxides N-NiO@N-Fe_3O_4@N-ZnO for adsorptive removal of atrazine and bisphenol-A from wastewater: Kinetic and adsorption studies. Environ NanotechnolMonitManag 16:100481
7. Almomani F, Bhosale R, Khraisheh M, Kumar A, Almomani T (2019) Heavy metal ions removal from industrial wastewater using magnetic nanoparticles (MNP) Appl. Surf Sci 506:144924
8. Anbia, Kargosha K, Khoshbooei K (2014) Heavy metal ions removal from aqueous media by modified magnetic mesoporous silica MCM-48. Chem Eng Res Des 93:779–788
9. Aragay G, Pino F, Merkoci A (2012) Nanomaterials for sensing and destroying pesticides. Chem Rev 112:5317–5338
10. Asgharinezhad AA, Mollazadeh N, Ebrahimzadeh H, Mirbabaei F, Shekari N (2014) Magnetic nanoparticles based dispersive micro-solid-phaseextraction as a novel technique for coextraction of acidic and basic drugs from biological fluids and waste water. J Chromatogr A 1338:1–8
11. Attallah OA, Al-Ghobashy MA, Nebsen M, Salemb MY (2016) Removal of cationic and anionic dyes from aqueous solution with magnetite/pectin and magnetite/silica/pectin hybrid nanocomposites: kinetic, isotherm and mechanism analysis. RSC Adv 6:11461
12. Attia TMS, Hu XL, Qiang YD (2013) Synthesized magnetic nanoparticles coated zeolite for the adsorption of pharmaceutical compounds from aqueous solution using batch and column studies. Chemosphere 93:2076–2085
13. Aydin S (2016) Removal of organophosphorus pesticides from aqueous solution by magnetic Fe_3O_4/red mud-nanoparticles. Water Environ Res 88(12):2275–2284
14. Aydin S, Aydin ME, Beduk F, Ulvi A (2019) Removal of antibiotics from aqueous solution by using magnetic Fe_3O_4/red mud-nanoparticles. Sci Total Environ 670:539–546
15. Boateng K, Mirshahghassemi S, Wu H, Flora JRV, Grassian VH, Lead JR (2020) Mechanistic study of oil adsorption onto PVP-coated magnetic nanoparticles: an integrated experimental and molecular dynamics study to inform remediation. Environ Sci Nano 8:485–492
16. Bohara RA, Throat ND, Mulla NA, Pawar SH (2017) Surface-modified cobalt ferrite nanoparticles for rapid capture, detection, and removal of pathogens: a potential material for water purification. Appl Biochem Biotechnol 182(2):598–608
17. Boruah PK, Sharma B, Hussain N, Das MR (2017) Magnetically recoverable Fe_3O_4/graphene nanocomposite towards efficient removal of triazine pesticides from aqueous solution: Investigation of the adsorption phenomenon and specific ion effect. Chemosphere 168:1058–1067
18. Cai G, Zhu X, Li K, Qi X, Wei Y, Wang H, Hao F (2018) Self-enhanced and efficient removal of arsenic from waste acid using magnetite as an in situiron donator. Water Res 15(157):269–280
19. Cai W, Weng X, Chen Z (2019) Highly efficient removal of antibiotic rifampicin from aqueous solution using green synthesis of recyclable nano-Fe_3O_4. Environ Pollut 247:839–846
20. Chai L, Wanga Y, Zhao N, Yang W, You X (2013) Sulfate-doped Fe3O4/Al2O3 nanoparticles as a novel adsorbent for fluoride removal from drinking water Water Res 47(12):4040–4049
21. Chandra V, Park J, Chun Y, Woo Lee J, In-Chul Hwang I-C, Kim KS (2010) Water-dispersible magnetite-reduced graphene oxide composites for arsenic removal. ACS Nano 4(7):3979–3986
22. Chang L, Pu Y, Jing P, Cui Y, Zhang G, Xu S, Cao B, Guo J, Chen F, Qiao C (2021) Magnetic core-shell $MnFe_2O_4$@TiO_2 nanoparticles decorated on reduced graphene oxide as a novel adsorbent for the removal of ciprofloxacin and Cu(II) from water. Appl Surf Sci 541:148400
23. Cheng C, Xu F, Gu H (2011) Facile synthesis and morphology evolution of magnetic iron oxide nanoparticles in different polyol processes New J Chem 35:1072–1079
24. Cvejic Z, Rakic S, Kremenovic A, Antic B, Jovalekic C, Colomban P (2006) Nanosize ferrites obtained by ball milling: Crystal structure, cation distribution, size-strain analysis and Raman investigations. Solid State Sci 8:908–915

25. Darwish MSA, Kim H, Lee H, Ryu C, Lee JY, Yoon J (2019) Synthesis of magnetic ferrite nanoparticles with high hyperthermia performance via a controlled co-precipitation method. Nanomater 9(8):1176
26. Debs KB, Cardonaa DS, Silvaa HDT, Nassar NN, Carrilho ENVM, Haddada PS, Labuto G (2019) Oil spill cleanup employing magnetite nanoparticles and yeast-based magnetic bionanocomposite. J Environ Manag 230:405–412
27. Deng J, Shao Y, Gao N, Tan C, Zhou S, Hu X (2013) $CoFe_2O_4$ magnetic nanoparticles as a highly active heterogeneous catalyst of oxone for the degradation of diclofenac in water. J Hazard Mater 262:836–844
28. Dewagea NB, Liyanagea AS, Pittman Jr CU, Mohan D, Mlsnaa T (2018) Fast nitrate and fluoride adsorption and magnetic separation from water on α-Fe2O3 and Fe3O4 dispersed on Douglas fir biochar. Bioresour Technol 263:258–265
29. Dil EA, Ghaedi M, Asfaram A, Bazrafshan AA (2018) Ultrasound wave assisted adsorption of congo red using gold-magnetic nano-composite loaded on activated carbon: optimization of process parameters. Ultrason Sonochem 46:99–105
30. Dong H, Huang J, Koepsel RR, Ye P, Russell AJ, Matyjaszewski K (2001) Recyclable antibacterial magnetic nanoparticles grafted with quaternized poly (2- (dimethylamino) ethyl methacrylate) brushes. Biomacromol 12:1305–1311
31. Effenberger FB, Couto RA, Kiyohara PK, Machado G, Masunaga SH, Jardim RF, Rossi LM (2017) Economically attractive route for the preparation of high quality magnetic nanoparticles by the thermal decomposition of iron (III) acetylacetonate. Nanotechnol 28:115603
32. El-Desouky MG, Hassan N, Shahat A, El-Didamony A, El-Bindary AA (2021) Synthesis and characterization of porous magnetite nanosphere iron oxide as a novel adsorbent of anionic dyes removal from aqueous solution. Biointerface Res Appl Chem 11(5):13377–13401
33. Elias E, Costa R, Marques F, Oliveira G, Guo Q, Thomas S, Souza FG Jr (2014) Oil-spill cleanup: The influence of acetylated curaua fibers on the oil-removal capability of magnetic composites. J Appl Polymer Sci 132(13):41732
34. Engel M, Hadar Y, Belkin S, Lu X, Elimelech M, Chefetz B (2018) Bacterial inactivation by a carbon nanotube-iron oxide nanocomposite: a mechanistic study using *E. coli* mutants. Environ Sci 5:372–380
35. Fan L, Luo C, Sun M, Qiu H, Li X (2013) Synthesis of magnetic β-cyclodextrin–chitosan/graphene oxide as nanoadsorbent and its application in dye adsorption and removal. Colloids Surf B: Biointerfaces 103:601–607
36. Farmany A, Mortazavi SS, Mahdavi H (2016) Ultrasound-assisted synthesis of Fe_3O_4/SiO_2 core/shell with enhanced adsorption capacity for diazinon removal. J Magn Magn Mater 416:75–80
37. Feng L, Cao M, Ma X, Zhu Y, Hu C (2012) Superparamagnetic high-surface-area Fe3O4 nanoparticles as adsorbents for arsenic removal. J Hazard Mater 217:439–446
38. Freitas JCD, Branco RM, Lisboa IGO, Costa TPD, Campos MGN, Jafelicci Júnior M, Marques RFC (2015) Magnetic nanoparticles obtained by homogeneous coprecipitation sonochemically assisted. Mater Res 18:220–224
39. Fröhlich AC, Foletto EL, Dotto GL (2019) Preparation and characterization of $NiFe_2O_4$/activated carbon composite as potential magnetic adsorbent for removal of ibuprofen and ketoprofen pharmaceuticals from aqueous solutions. J Clean Prod 229:828–837
40. Fu C-C, Tran HN, Chen XH, Juang RS (2019) Preparation of polyaminated Fe3O4@Chitosan core-shell magnetic nanoparticles for efficient adsorption of phosphate in aqueous solutions. J Ind Eng Chem 83:235–246
41. Gallo-Cordova A, Lemus J, Palomares FJ, Morales MP, Mazarío E (2020) Superparamagnetic nanosorbent for water purification: Assessment of the adsorptive removal of lead and methyl orange from aqueous solutions. Sci Total Environ 711:134644
42. Gao L, Tang Y, Wang C, Yao L, Zhang J, Gao R, Tang X, Chong T, Zhang H (2019) Highly-efficient amphiphilic magnetic nanocomposites based on a simple sol-gel modification for adsorption of phthalate esters. J Colloid Interface Sci 552:142–152

43. Gautam RK, Tiwari I (2020) Humic acid functionalized magnetic nanomaterials for remediation of dye wastewater under ultrasonication: Application in real water samples, recycling and reuse of nanosorbents. Chemosphere 245:125553
44. Ge F, Li M-M, Ye H, Zhao B-X (2012) Effective removal of heavy metal ions Cd^{2+}, Zn^{2+} Pb^{2+}, C^{u2+} from aqueous solution by polymer-modified magnetic nanoparticles. J Hazard Mater 211:366–372
45. Ghafar HHA, Ali GAM, Fouad OA, Makhlouf SA (2015) Enhancement of adsorption efficiency of methylene blue on Co_3O_4/SiO_2 nanocomposite. Desalination Water Treat 53(11):2980–2989
46. Ghodsi S, Esrafili A, Kalantary RR, Gholami M, Sobhi HR (2020) Synthesis and evaluation of the performance of g-C_3N_4/Fe_3O_4/Ag photocatalyst for the efficient removal of diazinon: Kinetic studies. J Photochem Photobiol A 389:112279
47. Guo X, Du B, Wei Q, Yang J, Hu LY, L, Xu W, (2014) Synthesis of amino functionalized magnetic graphenes composite material and its application to remove Cr(VI), Pb(II), Hg(II), Cd(II) and Ni(II) from contaminated water. Hazard Mater 278:211–220
48. Hajdú A, Tombácz E, Illés E, Bica D, Vékás L (2008) Magnetite nanoparticles stabilized under physiological conditions for biomedical application. Prog Colloid Polym Sci 135:29–37
49. Hosseini F, Sadighia S, Hosseini-MonfaredH MNM (2016) Dye removal and kinetics of adsorption by magnetic chitosan nanoparticles. Desalination Water Treat 57(51):24378–24386
50. Ilahi S, Saif S, Chaudhry A (2018) *Nigella sativa* plant based nanocomposite-$MnFe_2O_4$/BC: An antibacterial material for water purification. J Clean Prod 200:996–1008
51. Imran M, Islam AU, Tariq MA, Siddique MH, Shah NS, Khan ZH, Amjad M, Din SU, Shah GM, Naeem MA, Nawaz M, Rizwan M (2019) Synthesis of magnetite-based nanocomposites for effective removal of brilliant green dye from wastewater. Environ Sci Pollut Res 26:24489–24502
52. Jamali-Behnam F, Najafpoor AA, Davoudi M, Rohani-Bastami T, Alidadi,H, Esmaily H, Dolatabadi M (2017) Adsorptive removal of arsenic from aqueous solutions using magnetite nanoparticles and silica-coated magnetite nanoparticles. Environ Prog Sustain Energy 37(3):951–960
53. Javaheri F, Hassanajili S (2016) Synthesis of Fe3O4@SiO2@MPS@P4VP nanoparticles for nitrate removal from aqueous solutions. Appl Polym Sci 133:44330
54. Jayanthi SA, Nathan DMGT, Jayashainy J, Sagayaraj P (2015) A novel hydrothermal approach for synthesizing α-Fe_2O_3, γ-Fe_2O_3 and Fe_3O_4 mesoporous magnetic nanoparticles. Mater Chem Phys 162:316–325
55. Jesus ACB, Jesus JR, Lima RJS, Moura KO, Almeida JMA, Duque JGS, Meneses CT (2020) Synthesis and magnetic interaction on concentrated Fe_3O_4 nanoparticles obtained by the co-precipitation and hydrothermal chemical methods. Ceram Int 46(8)A:11149–11153
56. Jianga H, Chenc P, Luoa S, Tua X, Caoc Q, Shuc M (2013) Synthesis of novel nanocomposite Fe3O4/ZrO2/chitosan and its application for removal of nitrate and phosphate. Appl Surf Sci 284:942–949
57. Jin S, Park BC, Ham WS, Pan L, Kim YK (2017) Effect of the magnetic core size of amino-functionalized Fe_3O_4 mesoporous SiO_2 core-shell nanoparticles on the removal of heavy metal ions Colloids Surf. A Physicochem Eng Asp 531:133–140
58. Jin Y, Liu F, Shan C, Tong M, Hou Y (2014) Efficient bacterial capture with amino acid modified magnetic nanoparticles. Water Res 50:124–134
59. Kahsay MH, Belachew N, Tadesse A, Basavaiah K (2020) Magnetite nanoparticle decorated reduced graphene oxide for adsorptive removal of crystal violet and antifungal activities. RSC Adv 10:34916
60. Kakavandi B, Esrafili A, Mohseni-Bandpi A, Jafari AJ, Kalantary RR (2014) Magnetic Fe_3O_4@C nanoparticles as adsorbents for removal of amoxicillin from aqueous solution. Water Sci Technol 69(1):147–155
61. Karthikeyan P, Meenakshi S (2019) In situ fabrication of magnetic particles decorated biopolymeric composite beads for the selective remediation of phosphate and nitrate from aqueous medium. J Environ Chem Eng 8(2):103530

62. Katheresan V, Kansedo J, Lau Y (2018) Efficiency of various recent wastewater dye removal methods: A review. J Environ Chem Eng 6(4):4676–4697
63. Kaur R, Hasan A, Iqbal N, Alam S, Saini MK, Raza SK (2014) Synthesis and surface engineering of magnetic nanoparticles for environmental cleanup and pesticide residue analysis: a review. J Sep Sci 37(14):1805–1825
64. Kiani M, Bagheri S, Karachi N, Dil EA (2019) Adsorption of purpurin dye from industrial wastewater using Mn-doped Fe_2O_4 nanoparticles loaded on activated carbon. Desalination Water Treat 152:366–373
65. Kim H, Zhang G, Wu M, Guo J, Nam C (2021) Highly efficient and recyclable polyolefin-based magnetic sorbent for oils and organic solvents spill cleanup. J Hazard Mater 419:126485
66. Kulal P, Badalamoole V (2020) Efficient removal of dyes and heavy metal ions from waste water using Gum ghatti–graft–poly(4-acryloylmorpholine) hydrogel incorporated with magnetite nanoparticles. J Environ Chem Eng 8(5):104207
67. Kumar A, Sharma G, Naushad M, Thakur S (2015) SPION/β-cyclodextrin core–shell nanostructures for oil spill remediation and organic pollutant removal from waste water. Chem Eng J 280(15):175–187
68. Kumar IA, Viswanathan N (2018) Development and reuse of amine-grafted chitosan hybrid beads in the retention of nitrate and phosphate. J Chem Eng Data 63(1):147–158
69. Kumar K, Gupta SC, Chander Y, Singh AK (2005) Antibiotic use in agriculture and its impact on the terrestrial environment. Adv Agron 87:1–54
70. Li G, Zhao Z, Liu J, Jiang G (2011) Effective heavy metal removal from aqueous systems by thiol functionalized magnetic mesoporous silica. J Hazard Mater 192:277–283
71. Lin J, Zhang J, Sun H, Lin Q, Guo Z, Yang H, He Y (2018) Structural and magnetic property of Cr^{3+} substituted cobalt ferrite nanomaterials prepared by the sol-gel method. Materials 11(11):2095
72. Liu Y, Cui T, Wu T, Li Y, Tong G (2016) Excellent microwave-absorbing properties of elliptical Fe_3O_4nanorings made by a rapid microwave-assisted hydrothermal approach. Nanotechnol 27(16):165707
73. Lobato NCC, Mansur MB, Ferreira AM (2017) Characterization and chemical stability of hydrophilic and hydrophobic magnetic nanoparticles. Mater Res 20:736–746
74. Lü T, Chen Y, Qi DM, Cao ZH, Zhang D, Zhao HT (2017) Treatment of emulsified oil wastewaters by using chitosan grafted magnetic nanoparticles. J Alloy Compd 696:1205–1212
75. Lunge S, Singh S, Sinha A (2013) Magnetic ironoxide(Fe_3O_4) nanoparticles from tea waste for arsenic removal. J Magn Magn Mater 356:21–31
76. Maddah B, Hasanzadeh M (2017) Fe_3O_4/CNT Magnetic nanocomposites as adsorbents to remove organophosphorus pesticides from environmental water. Int J Nanosci Nanotechnol 13(2):139–149
77. Mahmoud ME, Saad SR, El-Ghanam AM, Mohamed RHA (2021) Developed magnetic Fe_3O_4–MoO_3-AC nanocomposite for effective removal of ciprofloxacin from water. Mater Chem Phys 257:123454
78. Maity D, Ding J, Xue J (2008) Synthesis of magnetite nanoparticles by thermal decomposition: time, temperature, surfactant and solvent effects. Funct Mater Lett 1(3):189–193
79. Majidi S, Sehrig FZ, Farkhani SM, Goloujeh MS, Akbarzadeh A (2016) Current methods for synthesis of magnetic nanoparticles. Artif Cells Nanomed Biotechnol 44(2):722–734
80. Markeb AA, Alonso A, Sánchez A, Font X (2017) Adsorption process of fluoride from drinking water with magnetic core-shell Ce-Ti@Fe3O4 and Ce-Ti oxide nanoparticles. Sci Total Environ 598:949–958
81. Mascolo M, Pei Y, Ring T (2013) Room temperature co-precipitation synthesis of magnetite nanoparticles in a large pH window with different bases. Materials 6:5549–5567
82. Mohagheghian A, Ayagh K, Godini K, Shirzad-SiboniM (2017) Using amino-functionalized Fe_3O_4-WO_3 nanoparticles for diazinon removal from synthetic and real water samples in presence of UV irradiation. J Adv Oxid Technol 20(2):20160153
83. Mohammadi E, Daraei H, Ghanbari R, Athar SD, Zandsalimi Y, Ziaee A, Maleki A, Yetilmezsoy K (2019) Synthesis of carboxylated chitosan modified with ferromagnetic

nanoparticles for adsorptive removal of fluoride, nitrate, and phosphate anions from aqueous solutions. J Mol Liq 273:116–124
84. Mohan D, Kumar S, Srivastava A (2014) Fluoride removal from ground water using magnetic and nonmagnetic corn stover biochars. EcolEng 73:798–808
85. Moliner-Martinez Y, Vitta Y, Prima-Garcia H, González-Fuenzalida RA, Ribera A, Campíns-Falcó P, Coronado E (2014) Silica supported Fe_3O_4 magnetic nanoparticles for magnetic solid-phase extraction and magnetic in-tube solid-phase microextraction: Application to organophosphorous compounds. Anal Bioanal Chem 406(8):2211–2215
86. Moradnia F, Fardood ST, Ramazani A, Min B, Joo SW, Varma RS (2021) Magnetic $Mg_{0.5}Zn_{0.5}FeMnO_4$ nanoparticles: Green sol-gel synthesis, characterization, and photocatalytic applications. J Clean Prod 288:125632
87. Navarathnaa CM, Karunanayakea AG, Gunatilakec SR, Pittman CU Jr, Perezd F, Mohane D, Mlsnaa T (2019) Removal of Arsenic(III) from water using magnetite precipitated onto Douglas fir biochar. J Environ Manag 250:109429
88. Nicola R, Costi O, Ciopec M, Negrea A, Lazau R, Iana C, Picioru E, Len A, Almásy L, Szerb EI, Putz A (2020) Silica-coated magnetic nanocomposites for Pb^{2+} removal from aqueous solution. Appl Sci 10:2726
89. Oba SN, Ighalo JO, Aniagor CO, Igwegbe CA (2021) Removal of ibuprofen from aqueous media by adsorption: A comprehensive review. Sci Total Environ 780:146608
90. Osama S, Nermen HM, Aljaafari A (2015) Synthesis of magnetic nanoparticles and nanosheets for oil spill removal. NanosciNanotechnol Asia 5:32–43
91. Panda SK, Aggarwal I, Kumar H, Prasad L, Kumar A, Sharma A, Vo DN, Thuan DV, Mishra V (2021) Magnetite nanoparticles as sorbents for dye removal: a review. Environ Chem Lett 19:2487–2525
92. Pandi K, Viswanathan N, Meenakshi S (2019) Hydrothermal synthesis of magnetic iron oxide encrusted hydrocalumite-chitosan composite for defluoridation studies. Int J Biol Macromol 729(132):600–605
93. Pinto M, Ramalho PSF, Moreira NFF, Goncalves AG, Nunes OC, Pereira MFR, Soares OSGP (2020) Application of magnetic nanoparticles for water purification. Environmental Advances 2:100010
94. Raha S, Ahmaruzzaman M (2020) Enhanced performance of a novel superparamagnetic g-C_3N_4/NiO/ZnO/Fe_3O_4 nanohybrid photocatalyst for removal of esomeprazole: Effects of reaction parameters, co-existing substances and water matrices. Chem Eng Technol 395:124969
95. Raha S, AhmaruzzamanM, (2020) Facile fabrication of g-C_3N_4 supported Fe_3O_4 nanoparticles/ZnO nanorods: A superlative visible light responsive architecture for express degradation of pantoprazole. Chem Eng J 387:123766
96. Raj KG, Joy PA (2015) Coconut shell based activated carbon–iron oxide magnetic nanocomposite for fast and efficient removal of oil spills. J Environ Chem Eng 3(3): 2068–2075
97. Razaghi Z, Aghajani Z, Faal AY (2016) Novel synthesis and characterization of Fe_3O_4/C nanocomposits and investigation of its application in removal of pesticide diazinon. J Mater Sci: Mater Electron 27:13119–13126
98. Reed M, Johansen E, BrandvikPJ DP, Lewisà A, Fioccoà R, Mackay D, Prentki R (1999) Oil spill modeling towards the close of the 20th century: overview of the state-of-the-art spill. Sci Technol Bull 5(1):3–16
99. Salazar S, Guerra D, Yutronic N, Jara P (2018) Removal of aromatic chlorinated pesticides from aqueous solution using β-cyclodextrin polymers decorated with Fe_3O_4 nanoparticles. Polymers 10:1038
100. Saravanan A, Kumar PS, Govarthanan M, George CS, Vaishnavi S, Moulishwaran B, Kumar SP, Jeevanantham S, Yaashikaa P (2020) Adsorption characteristics of magnetic nanoparticles coated mixed fungal biomass for toxic Cr(VI) ions in aquatic environment. Chemosphere 267:129226
101. Scheverín N, Fossati A, Horst F, Lassalle F, Jacob S (2020) Magnetic hybrid gels for emulsified oil adsorption: an overview of their potential to solve environmental problems associated to petroleum spills. Environ Sci Pollut Res Int 27(1):861–872

102. Sereshti H, Afsharian EZ, Bidhendi ME, Nodeh HR, Kamboh MA, Yilmaz M (2019) Removal of phosphate and nitrate ions aqueous using strontium magnetic graphene oxide nanocomposite: Isotherms, kinetics, and thermodynamics studies. Environ Prog Sustain Energy 39(2e):13332
103. Shanmugam S, Subramanian B (2020) Evolution of phase pure magnetic cobalt ferrite nanoparticles by varying the synthesis conditions of polyol method. Mater Sci Eng B 252:114451
104. Shukla S, Arora V, Jadaun A, Kumar J, Singh N, Jain, (2015) Magnetic removal of Entamoeba cysts from water using chitosan oligosaccharide-coated iron oxide nanoparticles. Int J Nanomed 10:4901–4917
105. Singh H, Bhardwaj N, Arya SK, KhatrI M (2020) Environmental impacts of oil spills and their remediation by magnetic nanomaterials. Environ Nanotechnol Monit Manag 14:100305
106. Singh H, Jain A, Kaur J, Arya SK, Khatri M (2019) Adsorptive removal of oil from water using SPIONs–chitosan nanocomposite: kinetics and process optimization. Appl Nanosci 10:1281–1295
107. Singh M, Pandey A, Singh S, Singh SP (2021) Iron nanoparticles decorated hierarchical carbon fiber forest for the magnetic solid-phase extraction of multi-pesticide residues from water samples. Chemosphere 282:131058
108. Singh S, Barick KC, Bahadur D (2001) Surface engineered magnetic nanoparticles for removal of toxic metal ions and bacterial pathogens. J Hazard Mater 192:1539–1547
109. Sivakami M, Renuka Devi K, Renuka R, Thilagavathi T (2020) Green synthesis of magnetic nanoparticles via *Cinnamomum verum* bark extract for biological application. J Environ Chem Eng 8(5):104420
110. Soares SF, Simões, TR, Trindade T, Daniel-da-Silva AL (2017) Highly efficient removal of dye from water using magnetic carrageenan/silica hybrid nano-adsorbents. Water Air Soil Pollut 228(87)
111. Sun H, Cao L, Lu L (2011) Magnetite/reduced graphene oxide nanocomposites: One step solvothermal synthesis and use as a novel platform for removal of dye pollutants. Nano Res 4:550–562
112. Taha TA, Azab AA, Sebak MA (2019) Glycerol-assisted sol-gel synthesis, optical, and magnetic properties of $NiFe_2O_4$ nanoparticles. J Mol Struct 1181:14–18
113. Tassya TS, Matos JS, Khan MY, Zanoelo EF, Mangrich AS, Araújo BR, Navickienea S, Romão LPC (2017) Using magnetized (Fe_3O_4/Biochar nanocomposites) and activated biochar as adsorbents to remove two neuro-active pesticides from waters. J Braz Chem Soc 28(10):1975–1987
114. Tomar D, Jeevanandam P (2020) Synthesis of cobalt ferrite nanoparticles with different morphologies via thermal decomposition approach and studies on their magnetic properties. J Alloys Compd 843:155815
115. Vedernikova IA (2015) Magnetic nanoparticles: Advantages of using, methods for preparation, characterization, application in pharmacy. Ref J Chem 5:256–280
116. Vicente-Martínez Y, Caravaca M, Soto-Meca A, Solana-González R (2020) Magnetic core-modified silver nanoparticles for ibuprofen removal:an emerging pollutant in waters. Sci Rep 10:18288
117. Vojoudi H, Badiei A, Bahar S, Ziarani GM, Faridbod F, Ganjali MR (2017) A new nano-sorbent for fast and efficient removal of heavy metals from aqueous solutions based on modification of magnetic mesoporous silica nanospheres. J Magn Magn Mater 441:193–203
118. Wan M, Xiang F, Liu Z, Guan D, Shao Y, Zheng Y, Jin M, She Y, Cao L, Jin F, Chen R, Wang S, Wu Y, Abd El-Aty AM, Wang J (2021) Novel Fe_3O_4@metal-organic framework@polymer core-shell-shell nanospheres for fast extraction and specific preconcentration of nine organophosphorus pesticides from complex matrices. Food Chem 365:130485
119. Wang J, Tong X, Chen Y, Sun T, Liang L, Wang C (2020) Enhanced removal of Cr(III) in high salt organic wastewater by EDTA modified magnetic mesoporous silica. Microporous Mesoporous Mater 303:110262.

120. Wang X, Zhang P, Gao J, Chen X, Yang H (2015) Facile synthesis and magnetic properties of Fe_3C/C nanoparticles via a sol–gel process. Dyes Pigm 112:305–310
121. Wang Z, Xu J, Hu Y, Zhao H, Zhou J, Liu Y, Lou Z, Xu X, (2016) Functional nanomaterials: Study on aqueous Hg(II) adsorption by magnetic Fe_3O_4@SiO_2-SH nanoparticles. J Taiwan Inst Chem Eng 60:394-402
122. Wanga F, Liu D, Zheng P, Ma X (2016) Synthesis of rectorite/Fe3O4-CTAB composite for the removal of nitrate and phosphate from water. J Indus and Eng Chem 41:165-174
123. Wanjeri VWO, Gbashi S, Ngila JC, Njobeh P, Mamo MA, Ndungu PG (2019) Chemical vapour deposition of mwcnt on silica coated fe_3o_4 and use of response surface methodology for optimizing the extraction of organophosphorus pesticides from water. Int J Anal Chem 2019:4564709
124. Wen-Can Huang W-C, Wang W, Xue C, Mao X (2018) Effective enzyme immobilization onto a magnetic chitin nanofiber composite. ACS Sustain. Chem. Eng 6(7):8118–8124
125. WHO 2006. Chemical fact sheets: fluoride. Guidelines for drinking-water quality (electronic resource): Incorporation first addendum, third ed. 375–377 Geneva
126. Xia D, Lo IMC (2016) Synthesis of magnetically separable Bi_2O_4/Fe_3O_4 hybrid nanocomposites with enhanced photocatalytic removal of ibuprofen under visible light irradiation. Water Res 100:393–404
127. Yu LH, Hao GZ, Gu JJ, Zhou S, Zhang N, Jiang W (2015) Fe_3O_4 /PS magnetic nanoparticles: synthesis, characterization and their application as sorbents of oil from waste water. J MagnMagn Mater 394:14–22
128. Yu LH, Hao GZ, Liang QQ, Jiang W (2015) Fabrication of magnetic porous silica submicroparticles for oil removal from water. Ind Eng Chem Res 54:9440–9449
129. Zhan S, Yang Y, Shen S, Shan J, Li Y, Yang S, Zhu D (2014) Efficient removal of pathogenic bacteria and viruses by multifunctional amine-modified magnetic nanoparticles. J Hazard Mater 274:115–123
130. Zhang F, Tang X, Huang Y, Keller AA, Lan J (2018) Competitive removal of Pb^{2+} and malachite green from water by magnetic phosphate nanocomposites Water Research 150:442–451
131. Zhang W, Lana Y, Maa M, Chaia S, Zuob Q, Kimg K, Gaoa Y (2020) A novel chitosan–vanadium-titanium-magnetite composite as a superior adsorbent for organic dyes in wastewater. Environ Int 142:105798
132. Zhang X, Qian J, Pan B (2015) Fabrication of novel magnetic nanoparticles for efficient and simultaneous removal of multiple pollutants from water fabrication of novel magnetic nanoparticles for adsorption of heavy metals. Environ Sci Technol 50(2):881–889
133. Zhang Y, Wang S, Shen S, Zhao B (2013) A novel water treatment magnetic nanomaterial for removal of anionic and cationic dyes under severe condition. Chem Eng J 233:258–264
134. Zhang YR, Su P, Huang J, Wang QR, Zhao, (2015) A Magnetic Nanomaterial Modified with Poly-Lysine for efficient removal of anionic dyes from water. Chem Eng J 262:313–318
135. Zhu H, Jiang R, Fu Y, Li R, Yao J, Jiang S (2016) Novel multifunctional $NiFe_2O_4$/ZnO hybrids for dye removal by adsorption, photocatalysis and magnetic separation. Appl Surf Sci 369:1–10
136. Zhu K, Ju Y, Xu J, Yang Z, Gao S, Hou Y (2018) Magnetic nanomaterials: Chemical design, synthesis, and potential applications. Acc Chem Res 51(2):404–413
137. Zhan S, Zhu D, Ma S, Yu W, Jia Y, Li Y, Yu H, Shen Z (2015) Highly efficient removal of pathogenic bacteria with magnetic graphene composite. ACS Appl Mater Interfaces 7(7):4290–4298

Carbon Nanotubes Reinforced Polymeric Hybrid Materials for Water Purification

Anny Leudjo Taka, Michael Klink, Xavier Yangkou Mbianda, Fanyana Mtunzi, and Eliazer Bobby Naidoo

Abstract The accessibility of safe drinking water is currently a critical concern; the challenge is to develop an effective method to remove pollutants to ultra-low levels and recycle wastewater through purification. Among the water purification methods that have already been employed, nanotechnology has been considered the most promising method. This is because the nanomaterials used as adsorbents have been proven to remove different classes of pollutants, including emerging water pollutants, and these nanomaterials can be regenerated and reused. Particularly carbon nanotubes-based polymeric nanocomposite materials have attracted significant research attention because they possess multifunctional properties helpful in removing different types of pollutants from wastewater. Therefore, this chapter reviews recent studies reported on carbon nanotubes modified with natural polymers (biopolymers) such as chitosan, cellulose, and cyclodextrin used in water treatment. The cost and economic value of carbon nanotubes modified with polymeric hybrid materials used as nano-sorbents for water purification are also discussed.

Keywords Nanotechnology · Polymer hybrid · Nanocomposite · Nano-sorbents · Water purification

---

A. Leudjo Taka (✉) · F. Mtunzi · E. Bobby Naidoo
Institute of Chemical & Biotechnology, Southern Gauteng Science and Technology Park, Vaal University of Technology, Sebokeng Campus, Vanderbijlpark 1983, South Africa
e-mail: lytany04@yahoo.fr

A. Leudjo Taka · M. Klink (✉) · F. Mtunzi · E. Bobby Naidoo
Department of Chemistry, Vaal University of Technology, Vanderbijlpark Campus, Vanderbijlpark 1900, South Africa
e-mail: michaelk1@vut.ac.za; lytany04@yahoo.fr

M. Klink
Department of Biotechnology, Vaal University of Technology, Vanderbijlpark Campus, Vanderbijlpark 1900, South Africa

X. Yangkou Mbianda
Department of Chemical Sciences, University of Johannesburg, Doornfontein, Johannesburg 2028, South Africa

S. K. Swain (ed.), *Nanohybrid Materials for Water Purification*, Composites Science and Technology, https://doi.org/10.1007/978-981-19-2332-6_9

1 Introduction

Nanotechnology has been demonstrated as a promising approach to provide effective nano-sorbent materials for water purification. Among nanomaterials, carbon nanotubes are well known for their excellent chemical, thermal, mechanical, chelating, and many other properties valuable for their use as nano-sorbent for the removal of pollutants from the contaminated water [1–9]. Carbon nanotubes (CNTs) were first described by Iijima [10]; they can be defined as large molecules with cylindrical shapes. They are made up of a hexagonal arrangement of hybridized carbon atoms and can be formed by rolling up a single sheet or multiple sheets of graphene [3, 11]. Therefore, there exist different types of CNTs, such as single-walled CNTs (SWCNTs), double-walled CNTs (DWCNTs), and multiwalled CNTs (MWCNTs). In addition, there are three well-established methods used to produce a wide variety of CNTs. These methods are electric arc discharge, laser vaporization, and chemical vapor deposition techniques. Among these methods, chemical vapor deposition (CVD) is one of the most popular used methods [2, 12].

CNTs have attracted tremendous attention in various fields such as electronic, medical, energy, and water purification. CNTs are characterized by their very high surface area to volume ratios, having diameters in the order of a few nanometers and lengths in the range of hundreds of nanometers [3, 13–15]. Besides that, CNTs are versatile materials in the sense that they can be modified with different types of chemical moieties in order to improve their properties and usefulness in various applications, particularly in water purification [3, 13, 14]. For instance, many studies have demonstrated the use of functionalized CNTs as adsorbents for the removal of heavy metal ions and organic pollutants [16–18]. However, using only Functionalized CNTs for water purification is expensive; additionally, there are currently emerging water pollutants that are complex and require efficient nano-sorbent materials with multifunctional properties. In this perspective, special attention has been giving to polymer nanocomposites research, particularly on carbon nanotubes reinforced with polymer hybrid materials [2, 19–21].

Polymer nanocomposites can be defined as multiphase materials with excellent multifunctional properties resulting from the combination of each component in the polymer composite [21]. These multifunctional properties, such as thermal stability, physical, mechanical, organic, inorganic, and antimicrobial properties, are vital in water purification. Furthermore, natural polymer-based nanocomposites are mostly preferred among polymer nanocomposites because natural polymers (or biopolymers) are biocompatible, biodegradable, non-toxic, and possess oxygen-rich-functional groups useful for the removal of various classes of water pollutants [21, 22]. Research on CNTs modified with natural biodegradable polymer nanocomposites has attracted a lot of interest. In general, literature has demonstrated that the modification of CNTs with polymer materials usually resulted in nanocomposites and hybrid materials. Therefore, the hybrid material can be defined as a new characteristic different from the original and results from combining different materials (organics and inorganics) at the atomic and molecular levels. In addition, the nanocomposite refers to a material in which the properties of the original material (e.g., CNTs) are reinforced by combining the original material with a dispersion material (e.g., polymer matrix) [19].

However, one of the key challenges during the synthesis of CNT–polymer nanocomposites is to achieve a proper dispersion of CNTs within the polymer matrix. Many approaches have been reported to achieve the dispersion of CNTs in polymer matrix or composite. Some of these approaches include polymer wrapping, chemical functionalization of CNTs, or the addition of surfactant (in a mixture solution of CNTs and polymer material) [23–27]. Another method could be the interlaying/interleaving method involving a thin layer of CNT/polymer in the laminate's midplane [28, 29]. For instance, previous literature has demonstrated that the dispersion of pristine CNTs was practically impossible because the surface of pristine CNTs is hydrophobic and inert [14, 19, 30–32]. In order to resolve the issue, pristine CNTs must first be chemically functionalized to allow the introduction of oxygen-containing groups on their surface. This will ease their dispersion and chemical reactivity with the polymer matrix [3, 12, 33].

Moreover, the modification of polymer (or biopolymer) with CNTs is advantageous because of the ability of CNTs to enhance the polymer properties to a higher order of magnitude than the non-modified polymer. The properties of polymers which can be improved are their thermal stability, surface area to volume ratio, strength, crystallization, chemical, mechanical, physical properties, and many other properties [14, 19, 30–32]. In this regard, various methods can be employed to achieve the synthesis of CNTs/polymer nanocomposites. These methods include the in-situ approach (e.g., cross-linked polymerization), direct mixing (solution mixing or melt compounding), and film casting-dip coating-physical mixing. However, the melt compounding method is mostly used for the thermoplastic polymer; in this method, CNTs are sheared mechanically, compounded in the molten polymer matrix. On the other hand, the solution mixing method required that the polymer be soluble in the solvent used to disperse the CNTs [34].

Hence, the objectives of this chapter are to deliver a review evaluation on the different methods to modify CNTs with a natural polymer such as chitosan, cyclodextrin, and cellulose and their application for the removal of pollutants from contaminated water. The Cost and economic value of these CNTs reinforced with natural polymeric hybrid materials applied in water purification are also examined.

2 Carbon Nanotubes/Cyclodextrin Polymer Nanocomposites for Water Treatment

Cyclodextrin (CD) are cyclic oligosaccharides derived from enzymatic reactions of starch. They are also known as molecular chelating agents, and they are multipurpose materials because they can be modified to enhance their properties and usefulness in various applications (e.g., drug delivery, food, water purification, cosmetics, textiles).

β-Cyclodextrin is most widely used among the three types of CD (α-CD, β-CD, and γ-CD) because of its good reactivity and it is less costly [35–42]. Additionally, there are many reports on the studies of CDs; this section focuses mainly on the insoluble nanosponge CD polymers modified with CNTs for application as adsorbent material in water purification. The synthesis of insoluble nanosponge CD polymers can be achieved using dehydration, condensation, and deprotonation methods. However, many studies on the modification of CDs with CNTs were conducted by the condensation method [2, 37, 43–45].

The condensation method involves the reaction of CD or CD and CNTs with an excess of the bifunctional linker (cross-linker agent) such as diisocyanates (e.g., diphenylmethane diisocyanate (DPMDI), toluene diisocyanate (TDI), hexamethylene diisocyanate (HMDI)) through cross-linking polymerization. The successful reaction can be confirmed by FTIR, proving the disappearance of the cross-linker (isocyanate) peak and the incorporation of the C=O, NH(CO), and other functional groups into the insoluble nanosponge CD polymer or CD/CNT nanocomposite insoluble polymer. The cross-linker agent such as diisocyanate was used to alter the solubility of CD in water resulting in insoluble CD polyurethanes [2, 35, 46, 47].

Insoluble nanosponge CD polymers have demonstrated the capacity to absorb various groups of organic pollutants by forming a host–guest inclusion complex [35, 48–50]. However, these insoluble nanosponge CD polymers have been demonstrated not to be efficient for uptake of complex organic molecules and large inorganic pollutants. Besides that, the literature has shown that the structural integrity of insoluble nanosponge CD polymers was affected after long cycles of regeneration and reuse. Therefore, the development of insoluble nanosponge CD modified with CNTs has been reported to be a valuable way to improve CD polymers' properties and resolve their limitations as adsorbents for water purification [46, 47, 51, 52].

Research studies on the modifications of CD polymers with CNTs to obtain nanocomposite polymers have been reported. These modifications involve mainly the chemical modification of ß-CD with MWCNTs by cross-linking polymerization using HMDI or TDI linker agent. For instance, Salipira et al. have reported on the functionalization of MWCNTs by acid oxidation method and the copolymerization of oxidized-MWCNTs with β-CD. They have demonstrated that oxidized-MWCNTs/β-CD nanocomposite insoluble polyurethane obtained could remove organic pollutants (trichloroethylene and p-nitrophenol) by column studies at 99% [46, 47, 51]. However, in their column studies, Salipera et al. have not conducted detailed column adsorption studies to investigate the effect of pH, contact time, adsorbent dosage, initial concentration, temperature. Additionally, they did not evaluate the performance of their oxidized-MWCNTs/β-CD nanocomposite through column adsorption models. Hence, one can say that this study reported by Salipera and co-workers still presents some limitations.

Lukhele et al. [53] have oxidized MWCNTs, then impregnated the oxidized MWCNTs with Ag nanoparticles; the composite Ag-MWCNTs that they have obtained was copolymerized with β-CD to produce Ag-MWCNTs/β-CD polymer nanocomposite. They have tested the ability of Ag-MWCNTs/β-CD polymer nanocomposite to remove *Escherichia coli* (*E-coli*) in model water solution by a cell viability counting technique [54]. From the results they have obtained, oxidized MWCNTs impregnated with Ag and copolymerized with CD was able to remove 95% of *E-coli* [53, 54].

Krause and co-workers have prepared Fe–Ni bimetallic nanoparticles embedded on MWCNTs, then through cross-linking polymerization using HMDI as a linker, they have reinforced Fe–Ni–MWCNTs composite with β-CD to obtained a nanocomposite polymer. They have used the synthesized nanocomposite polymer Fe–Ni–MWCNTs/β-CD as an adsorbent to remove trichloroethylene (TCE) from a model pollutant solution by column adsorption. They were able to achieve a 98% reduction of TCE using Fe–Ni–MWCNTs/β-CD polymer nanosponge composite. However, the column adsorption study was not thoroughly conducted; that is, the effect of pH, contact time, adsorbent dose, and other critical parameters affecting the column adsorption study were not investigated. Furthermore, they have not used column adsorption models to establish the mechanism of adsorption of TCE unto their Fe–Ni–MWCNTs/β-CD polymer nanosponge composite.

Moreover, Mamba et al. have reported on the modifications of oxidized MWCNTs (oxi-MWCNTs) and phosphonate-functionalized-MWCNTs (pMWCNT), respectively, by cross-linking polymerization withβ-CDusing HMDI as a linker. Then, they have characterized the oxidized-MWCNTs/β-CD (oxiMWCNT/β-CD) [55] and phosphonate-functionalized-MWCNTs/β-CD (pMWCNT/β-CD) [52] polymer nanocomposites obtained by FTIR, SEM, and Brunauer Teller method (BET) for surface area analysis. FTIR analysis has confirmed the successful polymerization of oxidized-MWCNT and pMWCNT, respectively, with β-CD. SEM analysis revealed that polymers' surface morphologies were non-uniform; they have noted a combination of porous surface and granular appearance. They have also demonstrated the ability of the oxidized-MWCNTs/β-CD polymer nanocomposite to adsorb metal ions pollutants (lead and cobalt) model polluted solutions by batch adsorption study thorough the investigation of the effect of initial concentration, contact time, and competing ions. Oxidized-MWCNTs/β-CD polymer nanocomposite was able to remove both lead and cobalt ions from model polluted solutions with a greater adsorption capacity for lead [55].

Mamba et al. have also demonstrated by batch adsorption study that pMWCNT/β-CD polymer was also efficient for removing of 4-chlorophenol and cobalt ions [52]. They have also shown that isotherms models such as Langmuir best explained the mechanism of adsorption of metal ions, while Freundlich was the good fit model for the adsorption of 4-chlorophenol by pMWCNT/β-CD polymer nanocomposite. However, Mamba et al. reported that the low surface area of oxidized-MWCNTs/β-CD and pMWCNT/β-CD polymer nanocomposites negatively affect the adsorption efficiency at higher concentrations of pollutants in contaminated water. In addition, they have not evaluated the effect of adsorbent dosage, pH effect, desorption, thermodynamics, kinetics, and regeneration studies helping to establish the mechanism of adsorption. They have also not tested the ability of their polymer nanocomposites to remove organics or inorganics (heavy metal ions) from real industrial wastewater samples.

In order to overcome the above limitations on the functionalized carbon nanotubes reinforced with β-CD nanosponge polymer, recent studies have been conducted and published by Leudjo taka et al. [56–58]. They have conducted detailed studies on CNTs reinforced with nanosponge cyclodextrin and metal nanoparticles. They have investigated the synthesis of phosphorylated MWCNTs cross-linked with β-CD and doped with titanium and silver nanoparticles using combined methods of phosphorylation (through amidation reaction), cross-linking polymerization, and sol–gel process. In brief, the pristine MWCNTs were first functionalized by acid oxidation and phosphorylation (Fig. 1). Then the phosphorylated MWCNTs obtained were co-polymerized with β-CD using HMDI as a linker (Fig. 2), and the resulting nanocomposite polymer pMWCNTs-CD was decorated with metal nanoparticles by the sol–gel process to produce pMWCNT-CD/TiO_2-Ag (Fig. 3).

Moreover, Leudjo Taka and co-workers have also thoroughly characterized the obtained nanosponge composite polymer, pMWCNT-CD/TiO_2-Ag, using thermal gravimetric analysis (TGA), differential scanning calorimetry (DSC), BET surface

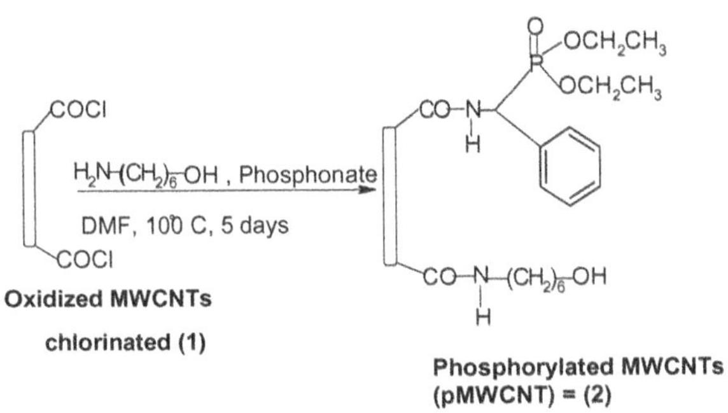

Phosphonate=
Diethyl(α-aminobenzyl)-phosphonate

Fig. 1 Functionalization of MWCNTs by phosphorylation method (reprinted with permission from [14])

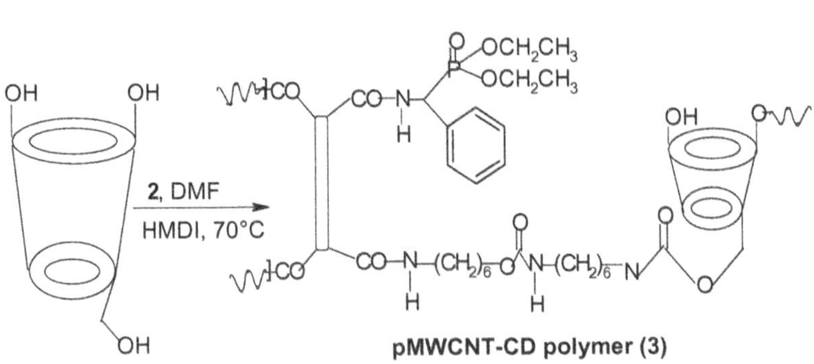

Fig. 2 Cross-linking polymerization of phosphorylated MWCNTs with β-CD (reprinted with permission from [14])

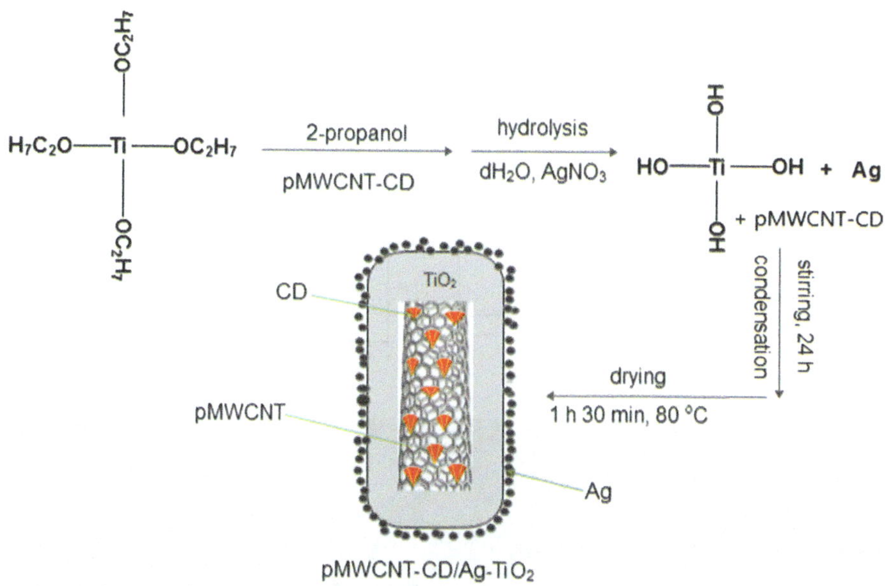

Fig. 3 Modification of pMWCNT-CD with metal nanoparticles using sol–gel process (reprinted with permission from [57])

area analysis, electron microscopy techniques, X-ray diffraction (XRD), and complementary spectroscopy techniques. FTIR spectroscopy confirmed the presence of functional groups (C–H, OH, N–H, O=P–OH, P–O, P=O, Ti–O–C, and NH(CO) of the carbamate linkage and characteristic of the nanosponge polyurethane) on the surface of pMWCNT-CD/TiO_2-Ag nanosponge composite polymer. TGA and DSC analyses were used to assess the thermal stability of the nanosponge composite polymer, which was proved to be stable at a temperature below 100 °C. SEM results revealed that pMWCNT-CD/TiO_2-Ag surface morphology is rough with an aggregation of granular particles, while the native CD and pMWCNT-CD polymers have a porous and sponge-like structure [33]. X-ray photoelectron spectroscopy (XPS) was employed to determine the element composition, binding energy, and respective percentages for each functional group depicted on the surface of the nanosponge composite polymer pMWCNT-CD/TiO_2-Ag. XRD analysis was conducted to confirm the crystallinity in the polymeric chain of developed MWCNT-CD/TiO_2-Ag (due to reinforcement of CD polymer by both pMWCNT and $TiO_2$and Ag nanoparticles) and to identify the dominant crystalline phase (TiO_2 anatase) (Fig. 4II). In addition, from the Raman spectroscopy analysis results, the developed pMWCNT-CD/TiO_2-Ag was a clear combination of CD polymer, MWCNTs, rutile, and anatase TiO_2 (Fig. 4I) [57].

Additionally, Leudjo Taka et al. have also proven that the development of pMWCNT-CD/TiO_2-Ag nanosponge composite polymer had could act as a filter

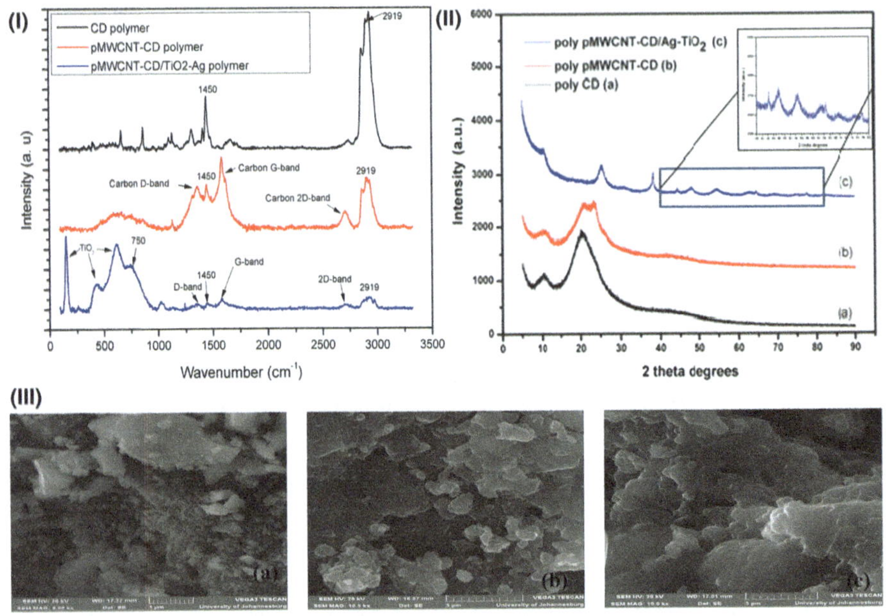

Fig. 4 **I** Raman spectra, **II** XRD patterns and **III** SEM micrographs (**a** pMWCNT-CD/TiO_2-Ag, **b** pMWCNT-CD, **c** CD) of the synthesized nanosponge polymers (reprinted with permission from [14])

to eliminate the organic pollutants (e.g., dyes, chlorinated compounds) [56], inorganic pollutants (e.g., heavy metal ions) [58] and pathogenic microorganism contaminants (e.g., bacteria and fungi) [57]. The batch adsorption studies were conducted to remove organic and metal ions pollutants through the investigation of the factors such as pH, adsorbent dosage, initial concentration, contact time, temperature, and desorption and regeneration studies. The adsorption mechanisms for these pollutants were established through thermodynamic studies and evaluation of the data obtained using various isotherm and kinetic models. Besides that, to prove the efficiency of pMWCNT-CD/TiO_2-Ag to act as a disinfectant material, the antimicrobial studies were conducted by microdilution methods and agar platting techniques. Leudjo Taka et al. have also reported that the highest adsorption capacity and potency of pMWCNT-CD/TiO_2-Ag against pathogens and other pollutants resulted from the reinforcement of CD nanosponge with both pMWCNTs and metal nanoparticles which help to improve the surface area (352.546 m^2/g), adsorption capacity and antimicrobial efficiency [57].

3 Carbon Nanotubes/Chitosan Polymer Nanocomposites for Water Purification

Chitosan is a biopolymer originating from the hydrolysis of chitin (Fig. 5), and it is the N-deacetylated form of chitin with two repeating units bonded together through β-(1,4)-glycosidic bond [34, 59–62]. Chitin is usually obtained from the waste of crustacean shells [59–61, 63]. Chitosan is a biodegradable, environmentally green, and biocompatible polymer. In addition, chitosan is a polycationic and semi-crystalline polymer that is hydrophobic. It has demonstrated excellent chemical and biological properties due to the presence of hydroxyl and amino functional groups, which play an essential role in the reactivity of chitosan [61, 64]. These hydroxyls and amino groups also give the flexibility to chemically alter the structure of chitosan with specific functionalities favoring the introduction of new or improved properties [63, 64]. Besides that, chitosan is hydrophobic in its original form (pristine form), i.e., it is insoluble at neutral pH in water and organic solvents [60, 65]. Hence, it is crucial to improve its solubility in various solvents over a wide pH range through physical and chemical functionalizations useful in further chemical reactions and applications. These chitosan modifications also help introduce the required physical, mechanical, and chemical properties necessary to enhance the reactivity of chitosan

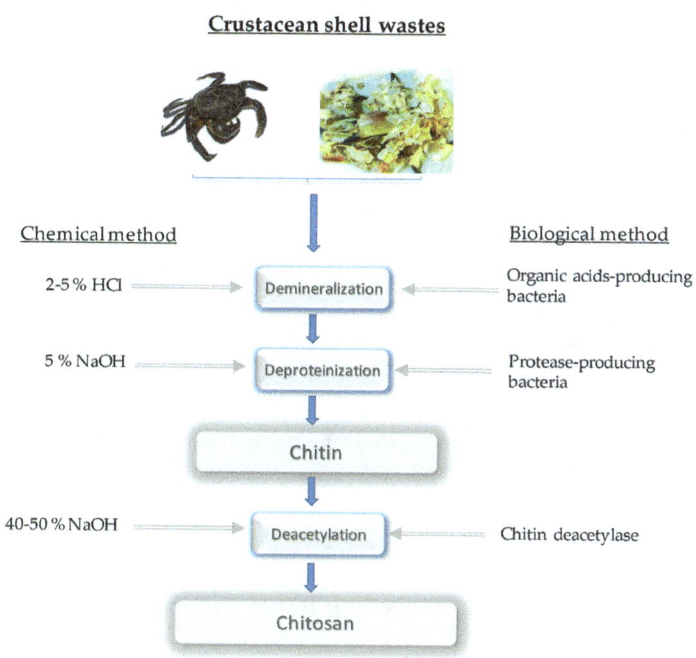

Fig. 5 Extraction methods of chitin and chitosan from crustacean shell wastes (reprinted with permission from [59])

with nanomaterials to produce a chitosan-based nanomaterials composite polymer with high adsorption efficiency and affinity for water contaminants [59–61, 66].

Additionally, chitosan could be chemically functionalized by cross-polymerization, surface grafting, carboxymethylation, internal hydrogen bonding with the aid of solvents, chemical reagents, vapor, coupling agents, and surface-active agents. And, the physical approaches previously used to modify chitosan include sputtering, plasma irradiation, blending, physical mixing, electron beam irradiation, Y-ray irradiation, ultrasounds, and ultraviolet irradiation [60, 61, 66–69]. Figure 6 shows the different methods which can be used for the chemical modifications of chitosan. Chitosan also has different deacetylation degrees, which influences its quality grade, physicochemical modifications, and properties. The degree of deacetylation is the ratio of 2-acetamino-2-deoxy-D-glucopyranose to

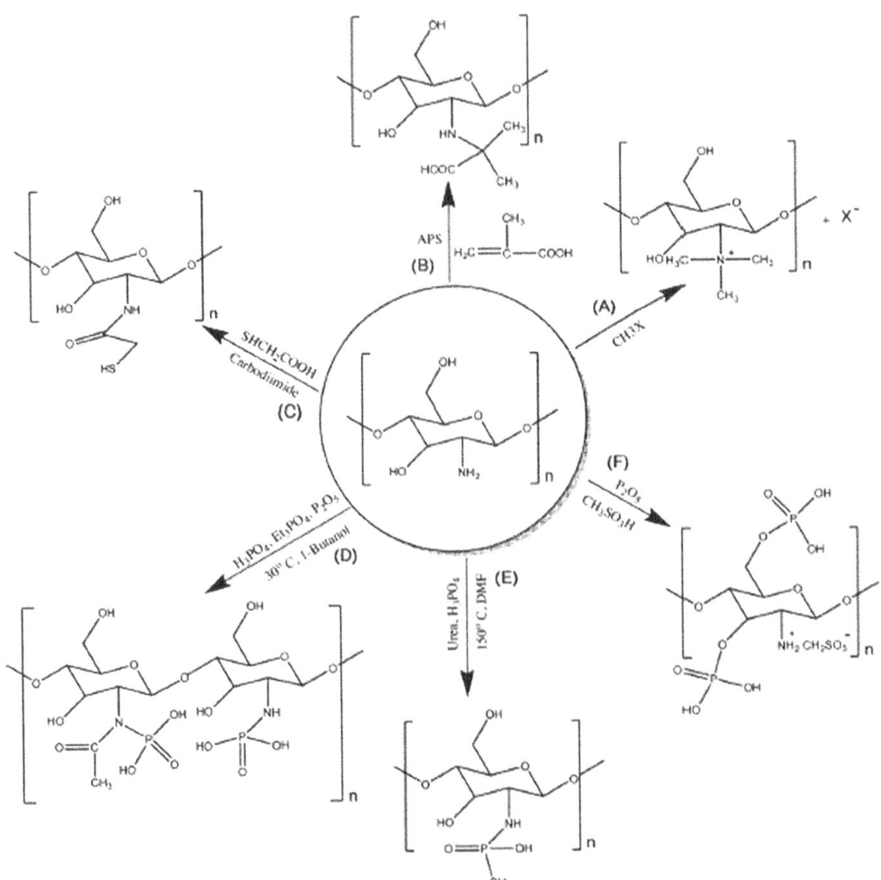

Fig. 6 Schematic representation of the different methods to chemically modify pristine chitosan: methylation (**A**), co-polymerization (**B**), thiolation (**C**), phosphorylation (**D**), (**E**) and (**F**) (reprinted with permission from [20])

2-amino-2-deoxy-D-glucopyranose. Thus, chitosan can be easily soluble in dilute acidic solutions at higher deacetylation's degree (>55%) [59–61, 64].

The research on the synthesis of chitosan-based nanocomposite polymers has drawn a lot of interest because their excellent properties offer a wide range of opportunities for their application [22]. These chitosan-based nanocomposite polymers can be synthesized using various techniques based on *in-situ* or *ex-situ* methods. In the *ex-situ* method, the nanoparticles are first prepared, then added into the polymer matrix solution used as a dispersion medium. On the other hand, the *in-situ* technique includes approaches such as oxidation, cross-polymerization [33, 58], reduction [70–73], precipitation [74], electrospinning methods [75–77], and sol–gel [78]. This *in-situ* technique means that inorganic nanoparticles are directly produced in the polymer matrix solution used as a reaction medium. The *in-situ* technique has been most widely used, and in this technique, the polymer matrix (e.g., chitosan) can act both as a capping agent or stabilizer to prevent the agglomeration of nanoparticles, control the nanoparticles' shape and size during the synthesis [2, 79–81].

In particular, chitosan nanocomposites containing carbon nanotubes and/or nanoparticles fixed into the biopolymer matrix have been developed for their use as nano-sorbents and nanofiltration membranes for the removal of contaminants from wastewater and desalination of water. According to literature, chitosan-based nanocomposite polymers have been demonstrated to enhance the adsorption and degradation of pollutants as well as antibacterial activity [34, 82, 83]. This can be explained by the exceptional mechanical and physicochemical properties resulting from the reinforcement of chitosan with these nanostructured carbons and/or inorganic nanoparticles, which help improve chitosan properties [84–90]. For example, the use of chitosan-based carbon nanotube nanocomposite polymer has been reported by Salam et al. [87] and other researchers to remove toxic metal ions. Salam et al. [87] prepared homogeneous nanocomposites MWCNTs/chitosan through cross-linking polymerization of MWCNTs using glutaraldehyde. The prepared MWCNTs/chitosan was tested for the removal of metal ions pollutants (Cd, Cu, Zn, and Ni) from aqueous solutions by column adsorption. The study proved that MWCNTs/chitosan nanocomposite could efficiently remove metals ions from aqueous solutions [87].

Beheshti et al. [91] have demonstrated the removal of chromium (Cr(VI))using MWCNTs reinforced with both biopolymer chitosan and nanoparticles (Fe_3O_4). The polymer nanocomposite chitosan/MWCNT/Fe_3O_4 was achieved using combined methods of acid oxidation of pristine MWCNTs, then their impregnation with Fe_3O_4 nanoparticles finally dispersion of MWCNTs/Fe_3O_4 composite in chitosan polymer matrix followed by electrospinning process to produce chitosan/MWCNT/Fe_3O_4. The surface morphology and structural analysis were assessed by SEM, TEM, FTIR, and XRD. The performance of chitosan/MWCNT/$Fe_3O_4$was evaluated by batch and fixed-bed column adsorption to eliminate of Cr (VI) from contaminated water solutions. The factors such as pH, initial pollutant concentration, contact time, temperature, and flow rate affecting the adsorption efficiency, were investigated [91].

From the results obtained, the structural analysis and surface morphology analysis had confirmed that chitosan/MWCNT/Fe_3O_4 was a smooth, uniform nanofibrous (Fig. 7a) and crystalline material with various functional groups (amino, amide, hydroxyl, Fe–O, and C–O) on their surface. Beheshti et al. have shown that in the column system, as the flow rate increase, the adsorption efficiency of chitosan/MWCNT/Fe_3O_4 also increased (Fig. 7e). The adsorption of Cr(VI) was favorable at high-temperature 45 °C. Langmuir isotherm and pseudo-second-order

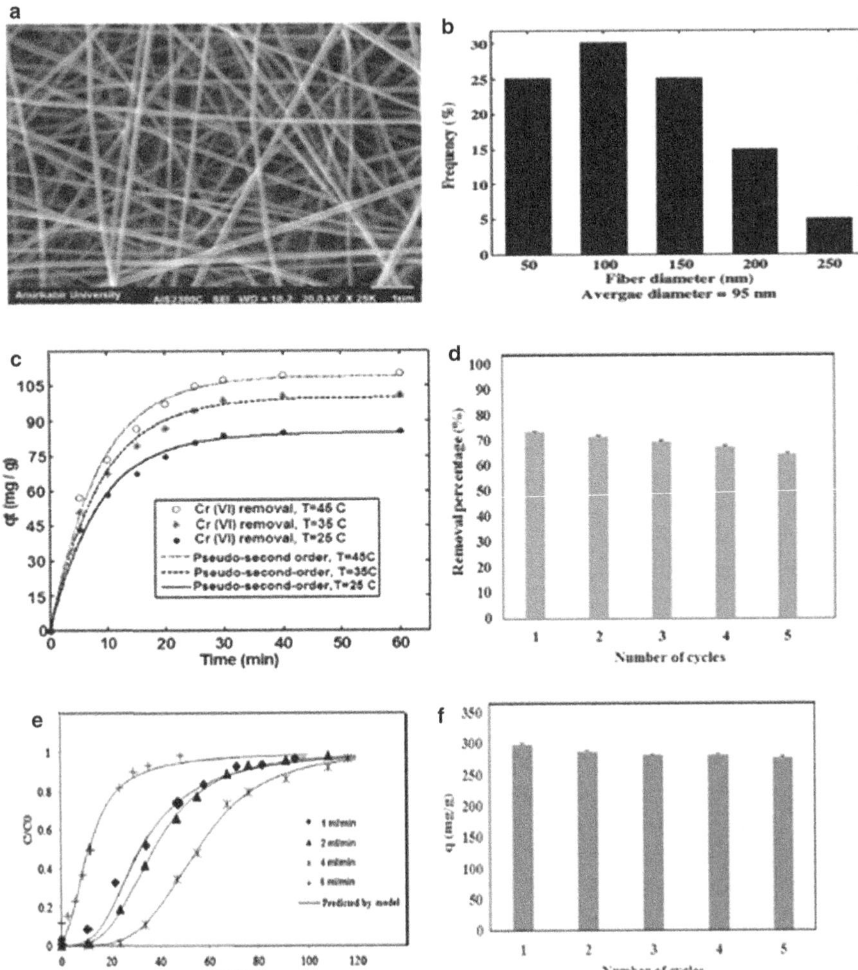

Fig. 7 **a** SEM image and **b** fiber size distribution of chitosan/MWCNTs/Fe_3O_4; **c** kinetic and **d** regeneration studies results for the removal of Cr(VI) by chitosan/MWCNTs/Fe_3O_4 in batch adsorption; **e** column adsorption experiment and predicted breakthrough curves for the removal of Cr(VI) by Chitosan/MWCNTs/Fe_3O_4; **f** regeneration study by column adsorption (reprinted with permission from [91])

kinetic models (Fig. 7c) best explained the mechanism of Cr(VI) removal in batch adsorption system while Thomas model was the best fit model to predict the sorption of Cr(VI) by chitosan/MWCNT/Fe_3O_4 in column system. After five cycles of desorption and regeneration, the adsorption efficiency was not affected in both column and batch adsorption systems (Fig. 7d and f) [91].

The published work of Alsuhybani et al. [92] revealed that nanocomposite membrane MWCNT/chitosan could be utilized as a nanofiltration membrane for the desalination of water by nanofiltration/reverse osmosis system. The nanocomposite membrane was prepared by evaporative casting method, and the characteristics of the developed membrane were determined through analysis of the surface charge, surface morphology, mechanical properties. From the results obtained, Alsuhybani et al. have shown that the combination of the biopolymer chitosan with MWCNT help to enhance the crystallinity, mechanical properties and increase the flux rate and salt rejection capability during the desalination process. These positive changes were due to incorporating MWCNTs bringing about the packing of hydrophilic chains in the biopolymer matrix. It is also essential to know that nanofiltration is one of the membrane processes which required low pressure and possess the properties such as salt rejection, water permeability, pore size between ultrafiltration and reverse osmosis [92].

Furthermore, Shawky et al. [93] have demonstrated the use of polymer chitosan/CNTs nanocomposite beads for the removal of mercury (Hg(II)) by batch adsorption study. They have synthesized different chitosan/CNTs nanocomposite beads using impregnation and protected cross-linking method, respectively. SEM, FTIR, and TGA techniques were employed to characterize the prepared composite chitosan/CNTs beads nanocomposites. Shawky et al. have shown that the nanocomposite beads prepared by the protected cross-linking technique have good thermal stability and were very effective for removing Hg(II) than the other composite beads prepared using impregnation and normal cross-linking method. The removal of Hg(II) was dependent on solution pH, temperature, and contact time. Langmuir isotherm model was the best fit model to explain the mechanism of Hg(II) sorption onto Chito/CNTs nanocomposite beads. In addition, chitosan/CNTs nanocomposite beads could be considered as cost-effective sorbents for water treatment because they can be recycled several times without any significant loss in adsorption capacity [93].

Zhao et al. [31] have recently developed a new polymer nanocomposite made of chitosan, CNTs, and octa-aminopolyhedral oligomeric silsesquioxanes (POSS) using a self-assembled method. The developed polymer nanocomposite chitosan/POSS/CNTs was characterized using FTIR, TGA, XRD, and TEM techniques. The performance of the developed chitosan-based polymer nanocomposite was tested for the degradation of organic pollutants (Congo red and methyl orange dyes) from model water solutions. Zhao et al. have proven that the intermolecular forces and the modifications of chitosan with both CNTs and POSS enhanced the thermal stability and adsorption capacity of chitosan/POSS/CNTs. The maximum sorption capacities obtained were 314.97 mg/g and 63.23 mg/g for Congo red and methyl orange respectively [31].

Morsi et al. [94] have developed polymer nanocomposites (chitosan/AgNPs, Chitosan/CuNPs, and chitosan/CNTs) and a multifunctional polymer nanocomposite made of chitosan, silver nanoparticles, copper nanoparticles, and CNTs (chitosan/AgNPs/CuNPs/CNTs) (Fig. 8a) which were used as antimicrobial agents to eliminate pathogenic microorganism pollutants from wastewater. The results obtained have shown that the combination of chitosan, nanoparticles, and CNTs produced different dimensional shapes with spherical NPs and CNTs dispersed uniformly into the polymer matrix (Fig. 8b). Additionally, they have depicted that at low concentration of pollutants (microbes) and short contact time (10 min), the antimicrobial activity of the multifunctional polymer nanocomposite was the highest (Fig. 8c) against gram-positive and gram-negative bacteria as well as fungal strain (isolated from wastewater) [94].

4 Carbon Nanotubes/Cellulose Polymer Nanocomposites for Water Decontamination

On earth, cellulose is well known as the most abundant biopolymer (or natural polymer) because it is readily available, and its cost is lower than other natural polymers. Cellulose's surface is rich in functional groups containing oxygen which favor its modifications with various chemical moieties or materials. Cellulose is made of D-anhydroglucopyranose units linked by glycosidic bonds, and it is very biocompatible and hydrophilic [95]. Cellulosecan be found into four types of crystals which include: cellulose I derived from plant, cotton, straw, and wood; cellulose II prepared by alkali treatment from natural cellulose and recrystallization, thus it is regenerated; cellulose III (III_1 and III_2) is obtained via liquid ammonia treatment of cellulose I and II; cellulose IV is derived from heating of cellulose III with glycerol. Nanocellulose is the smallest structural unit of cellulose, and it is constituted of bacterial nanocellulose, cellulose nanofibers, and cellulose nanocrystals [19, 95, 96]. In addition, literature has demonstrated that nanocellulose (or cellulose) easily aggregate in a reaction medium, to facilitate its dispersibility in water and other solvents, it is important to chemically modify its surface through processes such as sulphonation, oxidation, phosphorylation, and carboxymethylation just to site a few (Fig. 9). These surface chemical modifications of the cellulose significantly improve the properties of the cellulose for their further modifications with other materials to produce polymeric hybrid composite materials [19, 97, 98].

In this regard, the reinforcement of functionalized CNTs (e.g., oxidized CNTs) with cellulose biopolymer has resulted in novel polymeric hybrid nanocomposite materials with many benefits. From previous studies, it has been proved that CNT/cellulose polymeric hybrid nanocomposite materials possess better properties (water wettability, thermal performance or stability, electrical conductivity, mechanical, adsorption efficiency, salt discharge, and permeation flux properties) than each original component present in the polymer nanocomposites [95]. For instance, Chen

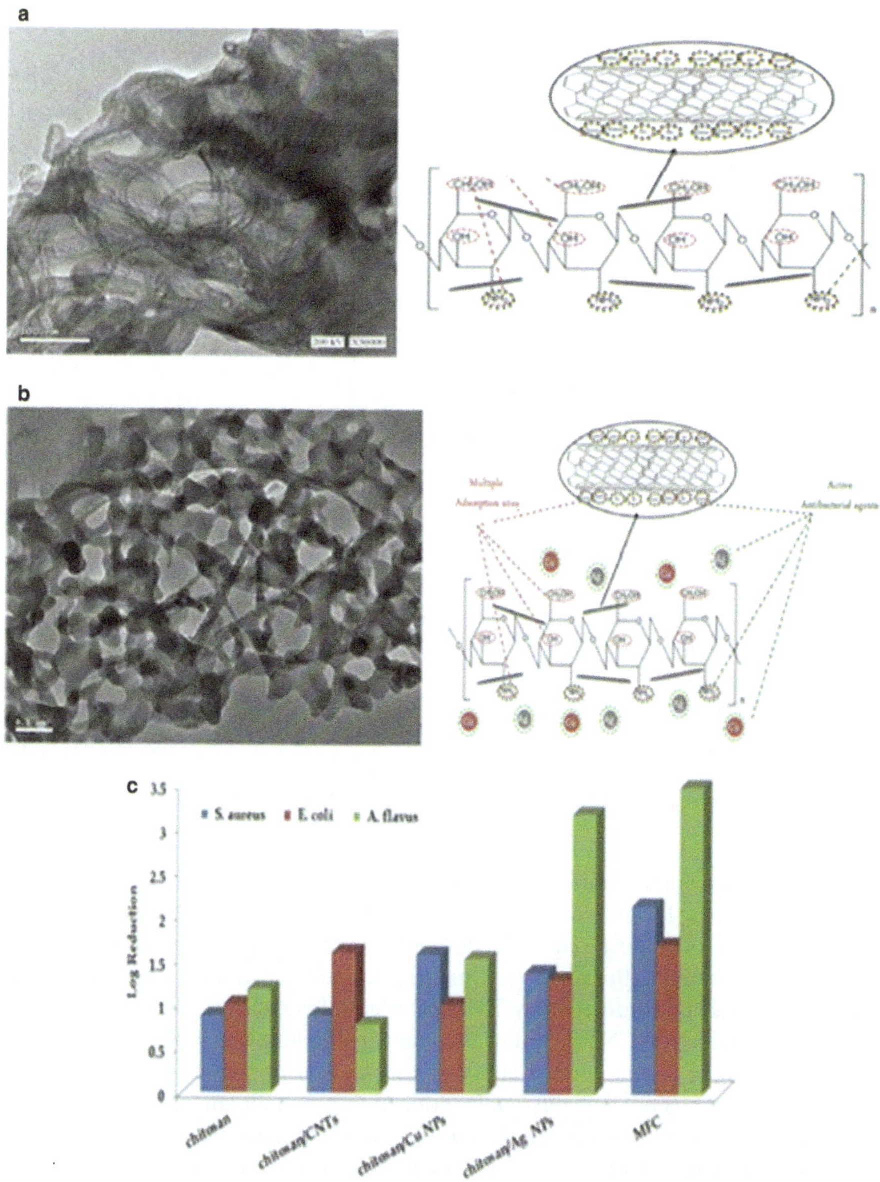

Fig. 8 Illustration of the graphical structural images and TEM micrographs of **a** chitosan/CNTs binanocomposites, **b** Chitosan/AgNPs/CuNPs/CNTs multifunctional nanocomposite and **c** the bar chart summarizing their antimicrobial performance (reprinted with permission from [94])

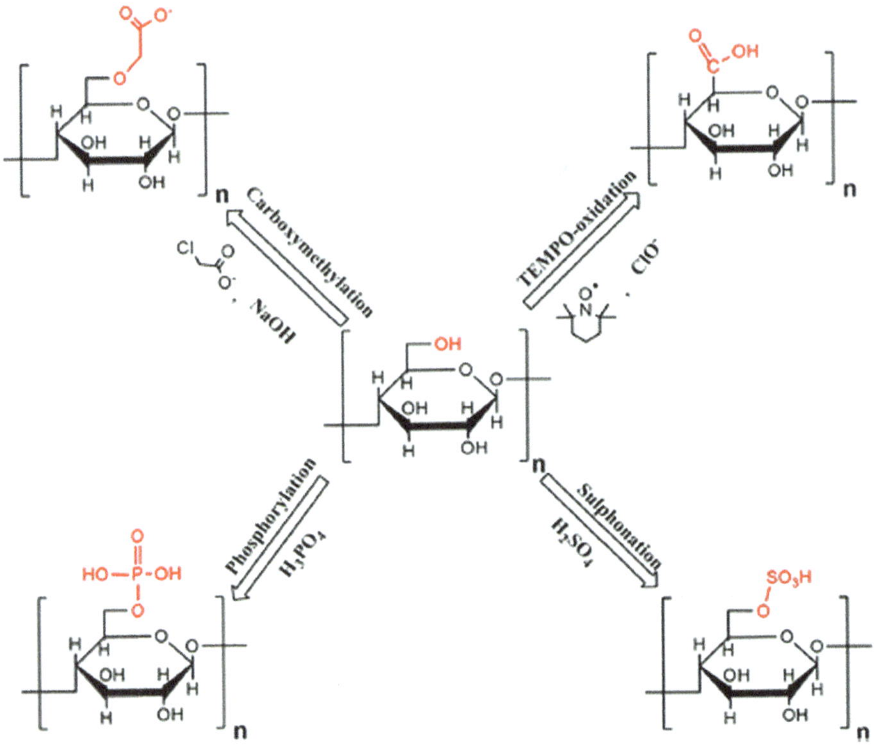

Fig. 9 Different ways to chemically modify the cellulose surface [19]

et al. [99] have prepared a polymer nanocomposite fiber made of regenerated bacterial cellulose (BC) and 1.0 wt% MWCNTs (BC/MWCNTs) by the dry jet-wet spinning method. The nanocomposite fiber has demonstrated much-enhanced properties than the original bacterial cellulose [99]. Similarly, Lu and Hsieh have used the electrospinning method to modified MWCNTs with cellulose acetate. They have shown that the obtained MWCNTs/cellulose acetate nanocomposite fiber has improved properties such as higher surface area and mechanical properties than the original cellulose acetate [100]. Hence, it can be noticed that a small amount MWCNTs can be uniformly be loading into cellulose to produce a polymer nanocomposite fiber which has a positive effect on the water wettability, surface structure, and mechanical properties of the composite fiber [95].

Besides, CNTs can be modified with cellulose to obtain polymer nanocomposite membranes which have potential in water purification. In this perspective, Silva et al. [101] have prepared a cellulose-based nanocomposite membrane by mixing pristine MWCNTs with cellulose acetate following the non-solvent induced phase separation method. They have characterized the product obtained using scanning electron microscopy, TGA, and various spectroscopy techniques. SEM analysis confirmed that their developed pristine-MWCNT/cellulose acetate nanocomposite membrane consisted of a dense layer on top and a porous support layer with a sponge structure. From the TGA result, they have shown that the nanocomposite membrane is very stable up to 320 °C; and the FTIR result confirmed the presence of functional groups resulting from the interaction between the acetate groups of cellulose and CNTs. They have proven that these functional groups were responsible for the effective removal of methylene blue from model wastewater solutions. They have validated that the combination of CNTs with biopolymer cellulose help to adjust the properties of each component in the composite and enhance the properties of polymer nanocomposite membrane, favoring an increase in adsorption capacity [101].

Sabir et al. [102] have also reported on the synthesis of functionalized MWCNTs/Cellulose acetate-polyethylene glycol nanocomposite membrane (Fig. 10) and its application in the desalination of drinking water. To achieve their synthesis, the surface of pristine MWCNTs was firstly engineered (functionalized) by acid oxidation method to introduce oxygen-containing groups helping to facilitated MWCNTs dispersion and reaction with the polymer material. Secondly, the cellulose acetate (CA) was copolymerized with polyethylene glycol 400 (PEG400), then a small amount (0.1–0.5 wt%) of surface engineered MWCNTs (SE-MWCNTs) was

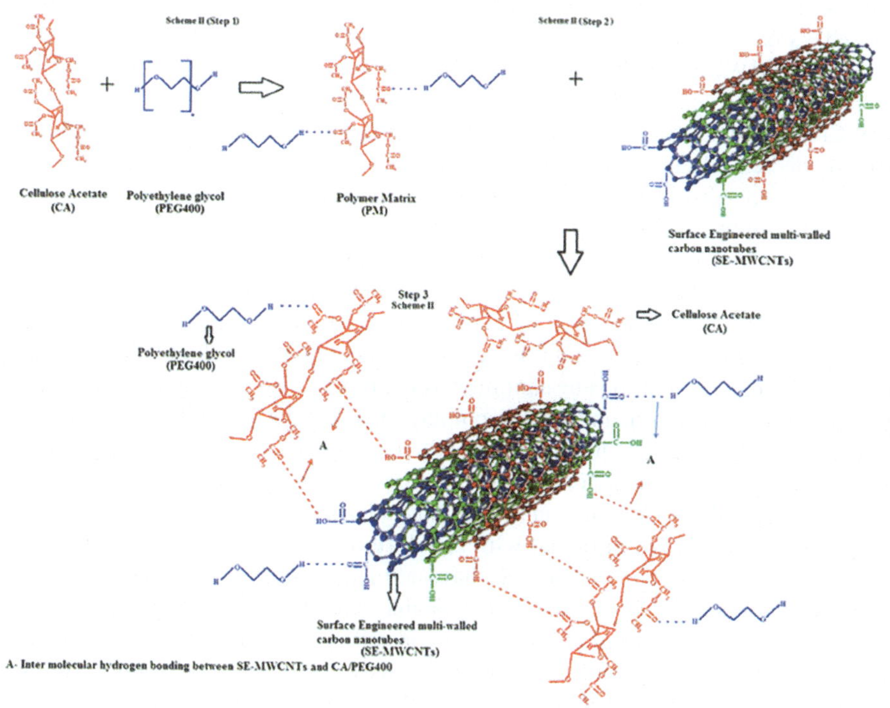

Fig. 10 Schematic illustration of the synthesis of functionalized MWCNTs/cellulose acetate-polyethylene glycol polymer nanocomposite membrane by combined methods of copolymerization and solution casting (reprinted with permission from [102])

dispersed uniformly in CA/PEG400 polymer matrix solution. Finally, the nanocomposite membrane (SE-MWCNTs/CA-PEG400) was obtained by the dissolution casting method. The thermal stability, surface morphology, and structural analysis were investigated by TGA, SEM, and FTIR spectroscopy, respectively. SEM results confirmed the formation of uniform, dense, structured SE-MWCNTs/CA-PEG400 nanocomposite membrane. The performance of the developed nanocomposite membrane was evaluated for removing salt from drinking water by reverse osmosis. From the structural analysis results, Sabir et al. have further reported that SE-MWCNTs/CA-PEG400 polymer nanocomposite membrane had better properties than CA-PEG400 polymer membrane due to the strong interaction between the SE-MWCNTs and CA-PEG400 polymer. This reinforcement of CA-PEG400 polymer with SE-MWCNTs had improved the thermal stability, reduced the permeation flux, increased the desalination and demineralization rates of the SE-MWCNTs/CA-PEG400 nanocomposite membrane [102].

Badawi et al. [103] have also reported on preparing MWCNTs/cellulose acetate, polymer nanocomposite membrane by phase inversion method using deionized water as a non-solvent and acetone as solvent. In their preparation, MWCNTs were first purified and oxidized by acid treatment to introduce functional groups on CNTs' surface, favoring their dispersion into the polymer matrix. They have successfully loaded a small amount of CNTs uniformly into the membranes. They have characterized the MWCNTs/cellulose acetate membrane and demonstrated its capability in desalinating a sodium chloride water solution. They have reported that the lowest MWCNT content in the nanocomposite membrane enhanced the permeation rate and slightly decreased salt retention percentage [103].

CNTs were also modified with biopolymer materials (chitosan, cellulose) and metal nano-photocatalyst to produce polymer nanocomposites, which can be used as photocatalysts to degrade wastewater pollutants. For instance, the synthesis of electrospun cellulose acetate/chitosan/SWCNT/Ferrite/Titanium dioxide nanofiber polymer composite was achieved by Zabihisahebi et al. for the elimination of organic (azo dyes) and inorganic (metal ions) contaminants via photoreduction and adsorption processes. The composite nanofiber was prepared by a combined method of impregnation, sonication, and electrospinning method. The use of the photocatalytic reduction method was adequate to efficiently degrade the azo dyes (e.g., Congo red, methylene blue) at both higher and lower concentrations as well as to remove metal ions (e.g., Cr (VI), As(V)) only at higher concentrations [86].

Furthermore, CNTs reinforced with cellulose or nanocellulose to form polymer nanocomposite hydrogel or aerogels have attracted tremendous attention due to their excellent properties, making them valuable candidates as adsorbent material to resolve the problem of water pollution from oil spillage, chemical leakage, and other water pollutants. For example, Li et al. [104] have fabricated a magnetic cellulose nanofiber (CNF)/polyvinyl alcohol (PVA)/MWCNT carbon aerogel (m-CPMCA) by combined methods of freeze-drying process, carbonization process, and direct immobilization of Fe_3O_4 nanoparticles on aerogel's surface (Fig. 11a). SEM analysis has shown that the nanocomposite polymer aerogel obtained had 3D porous structures piled up by sheet fibers (Fig. 11b). This m-CPMCA aerogel had high porosity and low

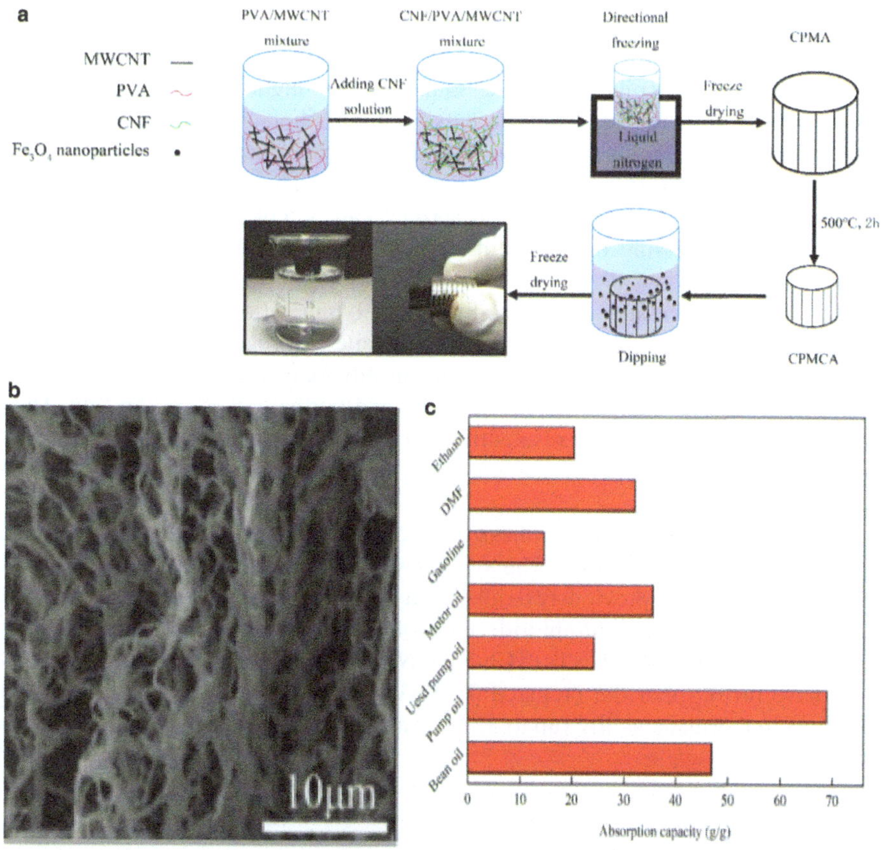

Fig. 11 **a** Complete process for the synthesis of magnetic cellulose nanofiber/polyvinyl alcohol/MWCNT carbon aerogel (m-CPMCA) (the picture also illustrate the ability of m-CPMA to float and trap by a magnet); **b** SEM image of m-CPMCA and **c** absorption performance of m-CPMCA for diverse organic liquids and oils (reprinted with permission from [104])

density. From the water contact angle analysis, Li and colleagues demonstrated that m-CPMCA nanocomposite was super hydrophobic and can always float on the water surface as well as adsorb pollutants. They were also able to demonstrate m-CPMCA's adsorption capacity toward different types of oil or organic liquids commonly found in contaminated water or industrial wastewater (Fig. 11c). Thus, among the organic liquids used for testing, m-CPMA efficiently removed pump oil from contaminated water. They have shown that m-CPMCA can be regenerated by combustion and be reused more than five times [104].

Xu et al. [105] have conducted a study using bamboo powder as a raw material to prepare CNF/MWCNTs carbon aerogel by a simple dipping and carbonization process. From their results obtained, CNF/MWCNTs carbon aerogel (CMCA) had low density, high porosity, and surface morphology similar to 3D sheet-like skeleton porous structure. Xu et al. have proven that CMCA could absorb oils and organic liquids at 35–110 times its weight. They have reported that this high absorption capacity for different oils or organic liquids' types might be due to the CMCA surface's properties (density, surface tension, hydrophobicity, high porosity) and oils properties. In addition, they have shown that CMCA can be regenerated using distillation and combustion; then reused over 50 cycles. They have also demonstrated the developed CNF/MWCNTs carbon aerogel can be considered as a promising candidate for addressing the problems arising from the spill of oily compounds [105].

Moreover, by grafting and cross-linking polymerization, Khoerunnisa et al. [106] have developed a polymer nanocomposite hydrogel, cellulose/polyvinyl alcohol/CNTs. They have shown that the introduction of CNTs in the polymer matrix significantly changed the morphology and porosity of the composite. They have also depicted from their results that polymer nanocomposite hydrogel (cellulose/polyvinyl alcohol/CNTs) had improved mechanical strength and absorption capacity [106].

Liu and Li [107] have also synthesized a polymer nanocomposite hydrogel made of SWCNTs and tea cellulose. The synthesis was achieved by dispersing a small quantity of SWCNTs (previously oxidized) in a mixture solution of tea cellulose and 1-allyl-3-methylimidazolium chloride, then following by casting techniques to obtain tea cellulose hydrogel-SWCNTs (TCh-SWCNTs) polymer nanocomposite (Fig. 12). They have proved that the prepared TCh-SWCNTs polymer nanocomposite had improved in thermal stability (up to 323 °C) and adsorption capacity for methylene blue [107].

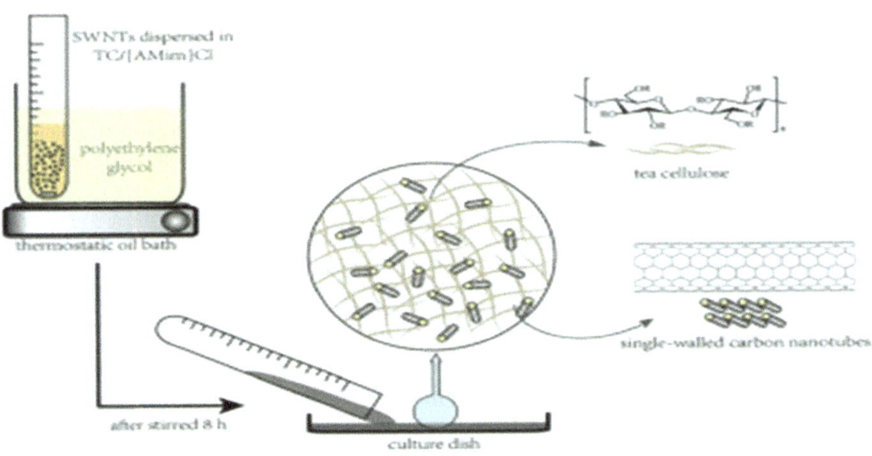

Fig. 12 Illustration of the synthesis process of Tea cellulose hydrogel-SWCNTs (TCh-SWCNTs) polymer nanocomposite (reprinted with permission from [107])

5 Cost and Economy Value of Carbon Nanotubes Modified with Natural Polymeric Materials used as Nano-sorbents for Water Purification

Carbon nanotubes modified with natural polymeric materials are economically valuable because they are environmentally friendly, recyclable, and highly degradable materials that can be considered for water purification. In addition, these green CNTs-based nanocomposite natural polymeric materials are made from renewable resources, and they favor sustainability development because they can be recycled and recovered after every water treatment cycle [34, 95]. The utilization of these CNTs-based nanocomposite natural polymeric materials will also help to avoid secondary pollution because, after many repeating cycles of water purification, they are not the source of waste generation, the spent nano-biosorbents can be applied in other applications such as latent fingerprint detection, biogas, cement, and particleboard production [34, 95, 108, 109]. These CNTs-based nanocomposite natural polymeric nano-biosorbent materials have been demonstrated to possess different shapes, structures, and multifunctional properties that can be applied in other various fields such as biomedicine, electronics, energy, tissue engineering, and drug delivery [34, 95].

On the other hand, to evaluate the total cost from the synthesis of CNTs-based nanocomposite natural polymeric nano-sorbents to their application in water purification is a challenging task; in fact, such evaluation is not often reported. This is because there are many factors to be considered. For instance, Adewale Adewuyi has reported on the cost evaluation of biosorbents and biosorption [109], and Leudjo taka et al. have reported on the cost and the practical aspect of the modification of cyclodextrin nanosponge polyurethanes with CNTs and metal nanoparticles [2]. Therefore, based on their reports one can say that the factors involved in the cost estimation could be the synthesis process including pretreatment, optimization of the process, energy consumption, time required, transportation, maintenance, type of water purification used, desorption process, regeneration process and maybe disposal of the spent nano-biosorbent (unless reused in other application) [2, 109]. Besides, literature has demonstrated that only a very small weight percent of CNTs and/or other nanoparticles were used to reinforced the polymeric materials. Then, the benefits resulting from this reinforcement can reduce some costs. Additionally, if the water purification process used is adsorption, this will help alleviate the overall because the adsorption process is not costly; and it is simple to operate compared to other water purification processes such as membrane filtration, photocatalysis, and reverse osmosis, just to cite a few.

6 Conclusion

To this end, research studies reviewed have proven that carbon nanotubes reinforced with natural polymer possess excellent thermal stability and multifunctional properties (e.g., organic, inorganic antimicrobial properties, mechanical properties) arising from the combination of each component in the polymer nanocomposites. Each of these components present in the polymer nanocomposites has been shown to play an essential role during the water purification process. These CNTs-based natural polymer nanocomposites have been demonstrated to be reused after several cycles of adsorption and desorption without any significant loss in their water treatment performance efficiency. Thus, these CNTs-based natural polymer nanocomposites can be considered as a cost-effective nano-sorbent for water purification. However, it will also be necessary to investigate the nanotoxicity of these CNTs-based natural polymer nanocomposites. Furthermore, the results obtained from these nanotoxicity studies will be crucial to the environment, humans, and the future of nanotechnology for water purification.

Acknowledgements The authors are grateful for the valuable support of the National Research Foundation (NRF, Grant Number 120740), the Vaal University of Technology, and the University of Johannesburg. The God Almighty for the knowledge and wisdom.

Funding National Research Foundation (NRF) "This work is based on the research supported wholly/in part by the National Research Foundation of South Africa (Grant Numbers: 120740)". This work is also partly supported by the Vaal University of Technology.

References

1. Alonso A, Macanás J, Davies GL, Gun YK (2011) Environmentally-safe polymer-metal nanocomposites with most favorable distribution of catalytically active and biocide nanoparticles. In: Hashim DA (ed) Advances in nanocomposite technology, July 2011. InTech, pp 175–200
2. Leudjo Taka A, Pillay K, Mbianda Yangkou X (2017) Nanosponge cyclodextrin polyurethanes and their modification with nanomaterials for the removal of pollutants from waste water : a review. Carbohydr Polym 159:94–107
3. Aslam MMA, Kuo HW, Den W, Usman M, Sultan M, Ashraf H (2021) Functionalized carbon nanotubes (Cnts) for water and wastewater treatment: preparation to application. Sustain 13:1–54
4. Qu X, Alvarez PJJ, Li Q (2013) Applications of nanotechnology in water and wastewater treatment. Water Res 47:3931–3946
5. Goh K, Karahan HE, Wei L, Bae T-H, Fane AG, Wang R, Chen Y (2016) Carbon nanomaterials for advancing separation membranes: a strategic perspective. Carbon N Y 109:694–710
6. Goh PS, Ismail AF, Hilal N (2016) Nano-enabled membranes technology: sustainable and revolutionary solutions for membrane desalination? Desalination 380:100–104
7. Goh PS, Matsuura T, Ismail AF, Hilal N (2016) Recent trends in membranes and membrane processes for desalination. Desalination 391:43–60
8. Gupta VK, Agarwal S, Saleh TA (2011) Synthesis and characterization of alumina-coated carbon nanotubes and their application for lead removal. J Hazard Mater 185:17–23

9. Santhosh C, Velmurugan V, Jacob G, Jeong SK, Grace AN, Bhatnagar A (2016) Role of nanomaterials in water treatment applications: a review. Chem Eng J 306:1116–1137
10. Lijima S (1991) Helical microtubules of graphitic carbon. Nature 354:56–58
11. Leudjo Taka A (2014) Metal-decorated carbon nanostructures for photocatalytic reduction of CO2. University of Johannesburg
12. Stefanescu EA, Daranga C, Stefanescu C (2009) Insight into the broad field of polymer nanocomposites: from carbon nanotubes to clay nanoplatelets, via metal nanoparticles. Materials (Basel) 2:2095–2153
13. Lee J, Jeong S, Liu Z (2016) Progress and challenges of carbon nanotube membrane in water treatment. Crit Rev Environ Sci Technol 46:999–1046
14. Leudjo Taka A (2018) Phosphorylated carbon nanotube-cyclodextrin/silver-doped titania nanobiocomposites for water purification. University of Johannesburg
15. He H, Pham-Huy LA, Dramou P, Xiao D, Zuo P, Pham-Huy C (2013) Carbon nanotubes: applications in pharmacy and medicine. Biomed Res Int 2013:1–12
16. Ihsanullah AA, Al-Amer AM, Laoui T, Al-Marri MJ, Nasser MS, Khraisheh M, Atieh MA (2016) Heavy metal removal from aqueous solution by advanced carbon nanotubes: critical review of adsorption applications. Sep Purif Technol 157:141–161
17. Ihsanullah A-K, Abu-Sharkh B, Abulkibash AM, Qureshi MI, Laoui T, Atieh MA (2016) Effect of acid modification on adsorption of hexavalent chromium (Cr(VI)) from aqueous solution by activated carbon and carbon nanotubes. Desalin Water Treat 57:7232–7244
18. Ji L, Zhou L, Bai X, Shao Y, Zhao G, Qu Y, Wang C, Li Y (2012) Facile synthesis of multiwall carbon nanotubes/iron oxides for removal of tetrabromobisphenol A and Pb(II). J Mater Chem 22:15853
19. Miyashiro D, Hamano R, Umemura K (2020) A review of applications using mixed materials of cellulose, nanocellulose and carbon nanotubes. Nanomaterials 10:1–23
20. Leudjo Taka A, Klink MJ, Yangkou Mbianda X, Naidoo EB (2020) Chitosan nanocomposites for water treatment by fixed-bed continuous flow column adsorption: a review. Carbohydr Polym 255:117398
21. Leudjo Taka A, Fosso-Kankeu E, Naidoo EB, Yangkou Mbianda X (2021) Recent development in antimicrobial activity of biopolymer-inorganic nanoparticle composites with water disinfection potential: a comprehensive review. Environ Sci Pollut Res 28:26252–26268
22. Olivera S, Muralidhara HB, Venkatesh K, Guna VK, Gopalakrishna K, Kumar KY (2016) Potential applications of cellulose and chitosan nanoparticles/composites in wastewater treatment: a review. Carbohydr Polym 153:600–618
23. Shariatnia S, Kumar AV, Kaynan O, Asadi A, Asadi A, Asadi A (2020) Hybrid cellulose nanocrystal-bonded carbon nanotubes/carbon fiber polymer composites for structural applications. ACS Appl Nano Mater 3:5421–5436
24. Hersam MC (2008) Progress towards monodisperse single-walled carbon nanotubes. Nat Nanotechnol 3:387
25. Punetha VD, Rana S, Yoo HJ, Chaurasia A, McLeskey JT, Ramasamy MS, Sahoo NG, Cho JW (2017) Functionalization of carbon nanomaterials for advanced polymer nanocomposites: a comparison study between CNT and graphene. Prog Polym Sci 67:1–47
26. Fujigaya T, Nakashima N (2015) Non-covalent polymer wrapping of carbon nanotubes and the role of wrapped polymers as functional dispersants. Sci Technol Adv Mater 16:024802
27. Bekyarova E, Thostenson ET, Yu A, Kim H, Gao J, Tang J, Hahn HT, Chou TW, Itkis ME, Haddon RC (2007) Multiscale carbon nanotube-carbon fiber reinforcement for advanced epoxy composites. Langmuir 23:3970–3974
28. Zhou H, Du X, Liu H-Y, Zhou H, Zhang Y, Mai Y-W (2017) Delamination toughening of carbon fiber/epoxy laminates by hierarchical carbon nanotube-short carbon fiber interleaves. Compos Sci Technol 140:46–53
29. White KL, Sue H-J (2012) Delamination toughness of fiberreinforced composites containing a carbon nanotube/polyamide-12 epoxy thin film interlayer. Polymer (Guildf) 53:37–42
30. Motchelaho MAM, Xiong H, Moyo M, Jewell LL, Coville NJ (2011) Effect of acid treatment on the surface of multiwalled carbon nanotubes prepared from Fe-Co supported on $CaCO_3$: correlation with Fischer-Tropsch catalyst activity. J Mol Catal A Chem 335:189–198

31. Zhao H, Wang R, Deng H, Zhang L, Gao L, Zhang L, Jiao T (2021) Facile preparation of self-assembled chitosan-based POSS-CNTs-CS composite as highly efficient dye absorbent for wastewater treatment. ACS Omega 6:294–300
32. Zhang S, Zhang F, Pan Y, Jin L, Liu B, Mao Y, Huang J (2018) Multiwall-carbon-nanotube/cellulose composite fibers with enhanced mechanical and electrical properties by cellulose grafting. RSC Adv 8:5678–5684
33. Leudjo Taka A, Pillay K, Mbianda XY (2018) Synthesis and characterization of a novel bio nanosponge filter (pMWCNT-CD/TiO_2-Ag) as potential adsorbent for water purification. In: Ramasami P, Bhowon MG, Laulloo SJ, Li H, Wah K (eds) Emerging trends in chemical sciences. © Springer International Publishing AG, pp 313–343
34. Mallakpour S, Khodadadzadeh L (2018) Biocompatible and biodegradable Chitosan nanocomposites loaded with carbon nanotubes. In: Biodegradable and biocompatible polymer composites. Elsevier Ltd, pp 187–221
35. Mhlanga SD, Mamba BB, Krause RW, Malefetse TJ (2007) Removal of organic contaminants from water using nanosponge cyclodextrin polyurethanes Sabelo. J Chem Technol Biotechnol 82:382–388
36. Li D, Ma M (1998) Cyclodextrin polymer separation materials. Patent WO 9822197
37. Bambo MF (2007) Synthesis, characterization and application of nanoporous cyclodextrin polymers of nanoporous cyclodextrin polymers. University of Johannesburg
38. Villiers A (1891) Sur la fermentation de la fécule par l'action du ferment butyrique. Comptes Rendus l'Académie des Sci 112:536–538
39. Del Valle EMM (2004) Cyclodextrins and their uses: a review. Process Biochem 39:1033–1046
40. Tejashri G, Amrita B, Darshana J (2013) Cyclodextrin based nanosponges for pharmaceutical use: a review. Acta Pharm 63:335–358
41. Buschmann HJ, Schollmeyer E (2002) Applications of cyclodextrins in cosmetic products: a review. J Cosmet Sci 53:575–592
42. Li D, Ma M (2000) Nanosponges for water purification. Clean Prod Process 2:0112–0116
43. Krause RW, Mamba BB, Bambo FM, Malefetse TJ (2010) Cyclodextrin polymers: Synthesis and application in water treatment. In: Hu J (ed) Cyclodextrins: chemistry and physics. Transworld Research Network, Kerala, India, pp 1–25
44. Trotta F, Tumiatti W (2003) Cross-linked polymers based on cyclodextrins for removing polluting agents. WO 03/085002 A1
45. Trotta F, Tumiatti W (2005) Cross-linked polymers based on cyclodextrins for removing polluting agents. Patent US 2005/0154198 A1. 1:1–3
46. Salipira KL, Mamba BB, Krause RW, Malefetse TJ, Durbach SH (2007) Carbon nanotubes and cyclodextrin polymers for removing organic pollutants from water. Environ Chem Lett 5:13–17
47. Salipira KL, Krause RW, Mamba BB, Malefetse TJ, Cele LM, Durbach SH (2008) Cyclodextrin polyurethanes polymerized with multi-walled carbon nanotubes: synthesis and characterization. Mater Chem Phys 111:218–224
48. Mamba B, Krause R, Malefetse T, Sithole S (2015) Degradation studies of β-cyclodextrin polyurethane polymers using soil burial experiments. South African J Chem 61:133–140
49. Mamba BB, Krause RW, Malefetse TJ, Gericke G, Sithole SP (2008) Cyclodextrin nanosponges in the removal of organic matter to produce water for power generation. Water SA 34:657–660
50. Mamba B, Krause R, Malefetse T, Sithole S, Nkambule T (2009) Humic acid as a model for natural organic matter (NOM) in the removal of odorants from water by cyclodextrin polyurethanes. Water SA 35:117–120
51. Salipira KL, Mamba BB, Krause RW, Malefetse TJ, Durbach SH (2008) Cyclodextrin polyurethanes polymerised with carbon nanotubes for the removal of organic pollutants in water. Water SA 34:113–118
52. Mamba G, Mbianda XY, Govender PP (2013) Phosphorylated multiwalled carbon nanotube-cyclodextrin polymer: Synthesis, characterisation and potential application in water purification. Carbohydr Polym 98:470–476

53. Lukhele LP, Krause RWM, Mamba BB, Momba MNB (2010) Synthesis of silver impregnated carbon nanotubes and cyclodextrin polyurethanes for the disinfection of water. Water SA 36:433–436
54. Lukhele L, Krause R, Mamba B Application of silver impregnated carbon nanotubes and cyclodextrin polymers, for the destruction of bacteria in water. In: Water Inst. SA
55. Mamba G, Mbianda XY, Govender PP, Mamba BB, Krause RW (2010) Application of multi-walled carbon nanotube-cyclodextrin polymers in the removal of heavy metals from water. J Appl Sci 10:940–949
56. Leudjo Taka A, Fosso-Kankeu E, Pillay K, Yangkou Mbianda X (2020) Metal nanoparticles decorated phosphorylated carbon nanotube/cyclodextrin nanosponge for trichloroethylene and Congo red dye adsorption from wastewater. J Environ Chem Eng 8:1–11
57. Leudjo Taka A, Doyle BP, Carleschi E, Youmbi Fonkui T, Erasmus R, Fosso-Kankeu E, Pillay K, Mbianda XY (2020) Spectroscopic characterization and antimicrobial activity of nanoparticle doped cyclodextrin polyurethane bionanosponge. Mater Sci Eng C 115:1–15
58. Leudjo Taka A, Fosso-kankeu E, Pillay K, Mbianda XY (2018) Removal of cobalt and lead ions from wastewater samples using an insoluble nanosponge biopolymer composite: adsorption isotherm, kinetic, thermodynamic, and regeneration studies. Environ Sci Pollut Res 25:21752–21767
59. Escudero-Oñate C, Elena Martínez-Francés (2016) A review of chitosan-based materials for the removal of organic pollution from water and bioaugmentation. In: Dongre RS (ed) Chitin-Chitosan—myriad functionalities in science and technology. IntechOpen, pp 71–87
60. Rizeq BR, Younes NN, Rasool K, Nasrallah GK (2019) Synthesis, bioapplications, and toxicity evaluation of chitosan-based nanoparticles. Int J Mol Sci 20:1–24
61. Shukla SK, Mishra AK, Arotiba OA, Mamba BB (2013) Chitosan-based nanomaterials: a state-of-the-art review. Int J Biol Macromol 59:46–58
62. Vafakish W (2019) Surface-modified chitosan: an adsorption study of a "Tweezer-Like" biopolymer with fluorescein. Surfaces 2:468–484
63. Kumar S, Ye F, Dobretsov S, Dutta J (2019) Chitosan nanocomposite coatings for food, paints, and water treatment applications. Appl Sci 9:1–27
64. Wang J, Chen C (2014) Chitosan-based biosorbents: modification and application for biosorption of heavy metals and radionuclides. Bioresour Technol 160:129–141
65. Casettari L, Vllasaliu D, Castagnino E, Stolnik S, Howdle S, Illum L (2012) PEGylated chitosan derivatives: Synthesis, characterizations and pharmaceutical applications. Prog Polym Sci 37:659–685
66. Zhao Y, Guo L, Shen W, An Q, Xiao Z, Wang H, Cai W, Zhai S, Li Z (2020) Function integrated chitosan-based beads with throughout sorption sites and inherent diffusion network for efficient phosphate removal. Carbohydr Polym 230:1–9
67. Chu C-C, Zhong C (2009) Organo-soluble chitosan salts and chitosan-derived biomaterials prepared thereof. pp 1–72
68. Fiorani G, Saoncella O, Kaner P, Altinkaya SA, Figoli A, Bonchio M, Carraro M (2014) Chitosan-polyoxometalate nanocomposites: synthesis, characterization and application as antimicrobial agents. J Clust Sci 25:839–854
69. Gagnon J, Binette A (2008) Chitosan salts, methods of manufacture and uses thereof. pp 50–80
70. Dhanavel S, Manivannan N, Mathivanan N, Kumar V, Narayanan V, Stephen A (2018) Preparation and characterization of cross-linked chitosan/palladium nanocomposites for catalytic and antibacterial activity. J Mol Liq 257:32–41
71. Kemp MM, Kumar A, Mousa S, Park TJ, Ajayan P, Kubotera N, Mousa SA, Linhardt RJ (2009) Synthesis of gold and silver nanoparticles stabilized with glycosaminoglycans having distinctive biological activities. Biomacromol 10:589–595
72. Kemp M, Kumar A, Clement D, Ajayan P, Mousa S, Linhardt RJ (2009) Hyaluronanand heparin-reduced silver nanoparticles with antimicrobial properties. Nanomedicine 4:421–429
73. Travan A, Marisch E, Donati I, Benincasa M, Giazzon M, Felisari L, Paoletti S (2011) Silver-polysaccharide nanocomposite antimicrobial coatings for methacrylic thermosets. Acta Biomater 7:337–346

74. Sambhy V, Macbride MM, Peterson BR, Sen A, Park UV, Pennsyl V (2006) Silver bromide nanoparticle/polymer composites: dual action tunable antimicrobial materials. J Am Chem Soc 128:9798–9808
75. Nthunya LN, Masheane M, Malinga S, Nxumalo E (2017) Greener approach to prepare electrospun antibacterial β-cyclodextrin/cellulose acetate nanofibers for removal of bacteria from water. Sustain Chem Eng 5:153–160
76. Zheng Y, Monty J, Linhardt RJ (2015) Polysaccharide-based nanocomposites and their applications. Carbohydr Res 405:23–32. https://doi.org/10.1016/j.carres.2014.07.016
77. Nthunya LN, Masheane ML, Malinga SP, Edward N, Mamba BB, Mhlanga SD, Nthunya LN, Masheane ML, Malinga SP (2017) Thermally and mechanically stable β -cyclodextrin/cellulose acetate nanofibers synthesized using an environmentally benign procedure. Int J Smart Nano Mater 8:1–19
78. Singh PP, Ambika (2018) Environmental remediation by nanoadsorbents-based polymer nanocomposite. In: New polymer nanocomposites for environmental remediation. Elsevier Inc., pp 223–241
79. Campelo JM, Luna D, Luque R, Marinas JM, Romero AA (2009) Sustainable preparation of supported metal nanoparticles and their applications in catalysis. Chemsuschem 2:18–45
80. Palza H (2015) Antimicrobial polymers with metal nanoparticles. Int J Mol Sci 16:2099–2116
81. Triebel C, Vasylyev S, Damm C, Stara H, Özpınar C, Hausmann S, Peukert W, Münstedt H (2011) Polyurethane/silver-nanocomposites with enhanced silver ion release using multifunctional invertible polyesters. J Mater Chem 21:4377–4383
82. Bui VKH, Park D, Lee YC (2017) Chitosan combined with ZnO, TiO_2 and Ag nanoparticles for antimicrobialwound healing applications: a mini review of the research trends. Polymers (Basel) 9:1–24
83. Masheane ML, Nthunya LN, Sambaza SS, Malinga SP, Nxumalo EN, Mamba BB, Mhlanga SD (2017) Chitosan-based nanocomposite beads for drinking water production. IOP Conf Ser Mater Sci Eng 195:1–8
84. Byrne JA, Fernandez-ibanez PA, Dunlop PSM, Alrousan DMA, Hamilton JWJ (2010) Photocatalytic enhancement for solar disinfection of water: a review. Int J Photoenergy 2011:1–13
85. Yu Z, Day D, Ciston S (2012) Improving solar water disinfection (SODIS) with a photoreactive TiO2/SWCNT composite on plastic PET bottles. J Exp Second Sci 1:40–43
86. ZabihiSahebi A, Koushkbaghi S, Pishnamazi M, Askari A, Khosravi R, Irani M (2019) Synthesis of cellulose acetate/chitosan/SWCNT/Fe_3O_4/TiO_2 composite nanofibers for the removal of Cr(VI), As(V), Methylene blue and Congo red from aqueous solutions. Int J Biol Macromol 140:1296–1304
87. Salam MA, Makki MSI, Abdelaal MYA (2011) Preparation and characterization of multi-walled carbon nanotubes/chitosan nanocomposite and its application for the removal of heavy metals from aqueous solution. J Alloys Compd 509:2582–2587
88. Zhang L, Zeng Y, Cheng Z (2016) Removal of heavy metal ions using chitosan and modified chitosan: a review. J Mol Liq 214:175–191
89. Keshvardoostchokami M, Babaei S, Piri F, Zamani A (2017) Nitrate removal from aqueous solutions by ZnO nanoparticles and chitosan-polystyrene–Zn nanocomposite: Kinetic, isotherm, batch and fixed-bed studies. Int J Biol Macromol 101:922–930
90. Moridi Z, Mottaghitalab V, Haghi AK (2011) A Detailed review of recent progress in carbon nanotube/ chitosan nanocomposites. Cellul Chem Technol 45:549–563
91. Beheshti H, Irani M, Hosseini L, Rahimi A, Aliabadi M (2016) Removal of Cr (VI) from aqueous solutions using chitosan/MWCNT/Fe_3O_4 composite nanofibers-batch and column studies. Chem Eng J 284:557–564
92. Alsuhybani M, Alshahrani A, Haidyrah AS (2020) Synthesis, characterization, and evaluation of evaporated casting MWCNT/chitosan composite membranes for water desalination. J Chem 2020:1–9
93. Shawky HA, El-Aassar AHM, Abo-Zeid DE (2012) Chitosan/carbon nanotube composite beads: preparation, characterization, and cost evaluation for mercury removal from wastewater of some industrial cities in Egypt. J Appl Polym Sci 125:E93–E101

94. Rania E Morsi, Alsabagh AM, Nasr SA, Zaki MM (2017) Multifunctional nanocomposites of chitosan, silver nanoparticles, copper nanoparticles and carbon nanotubes for water treatment: Antimicrobial characteristics. Int J Biol Macromol 97:264–269
95. Dong YD, Zhang H, Zhong GJ, Yao G, Lai B (2021) Cellulose/carbon composites and their applications in water treatment—a review. Chem Eng J 405:126980
96. Tapia-Orozco N, Ibarra-Cabrera R, Tecante A, Gimeno M, Parra R, Garcia-Arrazola R (2016) Removal strategies for endocrine disrupting chemicals using cellulose-based materials as adsorbents: a review. J Environ Chem Eng 4:3122–3142
97. Bejoy T, Midhun CR, Athira KB, Rubiyah MH, Jithin JA, MooresGlenna LD, Clément S (2018) Nanocellulose, a versatile green platform: from biosources to materials and their applications. Chem Rev 118:11575–11625
98. Eyley S, Thielemans W (2014) Surface Modification of Cellulose Nanocrystals. Nanoscale 6:7764–7779
99. Chen P, Kim HS, Kwon SM, Yun YS, Jin HJ (2009) Regenerated bacterial cellulose/multi-walled carbon nanotubes composite fibers prepared by wet-spinning. Curr Appl Phys 9:e96–e99
100. Lu P, Hsieh Y-L (2010) Multiwalled Carbon Nanotube (MWCNT) reinforced cellulose fibers by electrospinning. ACS Appl Mater Interfaces 2:2413–2420
101. Silva MA, Hilliou L, de Amorim MTP (2020) Fabrication of pristine-multiwalled carbon nanotubes/cellulose acetate composites for removal of methylene blue. Polym Bull 77:623–653
102. Sabir A, Shafiq M, Islam A, Sarwar A, Dilshad MR, Shafeeq A, Butt MTZ, Jamil T (2015) Fabrication of tethered carbon nanotubes in cellulose acetate/polyethylene glycol-400 composite membranes for reverse osmosis. Carbohydr Polym 132:589–597
103. Nouran El Badawi, Ramadan AR, Esawi AMK, El-Morsi M (2014) Novel carbon nanotube–cellulose acetate nanocomposite membranes for water filtration applications. Desalination 344:79–85
104. Li J, Zhou L, Jiang X, Tan S, Chen P, Zhou H, Xu Z (2019) Directional preparation of superhydrophobic magnetic CNF/PVA/MWCNT carbon aerogel. IET Nanobiotechnol 13:565–570
105. Xu Z, Jiang X, Tan S, Wu W, Shi J, Zhou H, Chen P (2018) Preparation and characterisation of CNF/MWCNT carbon aerogel as efficient adsorbents. IET Nanobiotechnol 12:500–504
106. F. Khoerunnisa, Sonjaya Y. Hendrawan ODP (2015) Superabsorbent hydrogel composite based on copolymer cellulose/poly(Vinyl Alcohol)/CNT. In: Mart T, Triyono D (eds) International symposium on current progress in mathematics and sciences. AIP Conference Proceedings, p 17292016
107. Liu Z, Li D (2021) Preparation and characterization of cellulose composite hydrogels from tea residue and single-walled carbon nanotube oxides and its potential applications. Front Chem 9:1–8
108. Fouda-Mbanga BG, Prabakaran E, Pillay K (2020) Synthesis and characterization of CDs/Al_2O_3 nanofibers nanocomposite for Pb^{2+} ions adsorption and reuse for latent fingerprint detection. Arab J Chem 13:6762–6781
109. Adewuyi A (2020) Chemically modified biosorbents and their role in the removal of emerging pharmaceutical waste in the water system. Water (Switzerland) 12:1–31

Chitosan Based Nano-Hybrid Materials for Water Purification

Niladri Sarkar and Gyanaranjan Sahoo

Abstract Now a days, water contamination, caused by the presence of synthetic dyes and heavy metal pollutants, is recognized as the potential threat to the human civilization. These toxic chemicals are sufficient to attribute severe health issues even in a very low consumption due to their carcinogenic and mutagenic behaviors. Moreover, these chemicals are nonbiodegradable with high tendency of bioaccumulation and hence, need to be removed from the main-stream water bodies prior to entering the food chain. Among various separation/removal techniques, adsorption through the polymer nanocomposites, is regarded as the easiest and cost-effective way to deal with the water pollution. Chitosan is a pseudo-natural cationic polysaccharide, obtained largely from various natural sources. Because of its high surface functionality (amine and hydroxyl groups), biocompatibility and non-toxic nature, this gel-forming polymer is highly preferred in preparing nanohybrid adsorbent. This book chapter mainly provides an insight into the chitosan nanohybrid structures used for the removal of a variety of different dyes and heavy-metal pollutants.

Keywords Chitosan nanohybrid · Water-pollution · Synthetic-dyes · Heavy-metals

1 Introduction

With the ever-growing industrialization, huge use of synthetic fertilizers and pesticides in agriculture and various domestic activities, life-saving water became polluted and turn out to be the recent concern of applied research [1]. In this connection, dyes, heavy metals, phenols, detergents, insecticides, and pesticides, etc. are found to be

N. Sarkar (✉)
Department of Chemistry, Centurion University of Technology and Management, Odisha Jatni, India
e-mail: nil.sarkar88@gmail.com

School of Chemistry, University of Hyderabad, Gachibowli, Telangana 500046, India

G. Sahoo
Department of Basic Science and Humanities, Majhighariani Institute of Technology and Science, Bhujabala, Rayagada, Odisha 765017, India

S. K. Swain (ed.), *Nanohybrid Materials for Water Purification*, Composites Science and Technology, https://doi.org/10.1007/978-981-19-2332-6_10

largely responsible as they are toxic as well as having mutagenic attributes. Among these various pollutants, colored-dyes and heavy metals are on top of the list to create potential threats for the sustainable livelihood [2]. This problem is associated with the huge demands of synthetic colorants or dyes which have complex structures like several aryl rings to delocalize electrons and showing bright colors. These are readily available and cheap in cost and hence, widely used in textile and printing industries instead of natural pigments. Moreover, these toxic chemicals are also used in food and cosmetics along with some other additives and binders. The major environmental contamination through dyes is happened due to the discharge of these toxic dyes as textile sewage from textile industries. Although, the content of dyes, contaminated with the main-stream water bodies is very low, but this low amount is sufficient to create fatal diseases in animals [3, 4]. Textile dyes with nitro- and azo-functionalities are proven to be more dangerous in creating long- term health issues [5, 6]. This type of dyes can deteriorate into carcinogenic amines in anaerobic aquatic environment and contributes to tumor growth, skin inflammation, allergic dermatitis, and mutagenesis with structurally degraded secondary pollutants. The presence of dyes not only hampers the photosynthesis in aquatic plants, but also leads to the chromosomal aberration. The actual pathway of these dyes to encounter the human organs is the food chain. Initially, these toxic dyes are aggregated or bioaccumulated in some tissues of aquatic animals and thereafter, entering the human body. Two well-known reactive dyes, Red1 and Red 9 which have their huge use in textile industries are already established as potential mutagenic chemicals. Another, hazardous textile dye, Azur-B is reported [7] to be fused with plasma membrane via intercalation with helical-DNA structures and contributes cytotoxicity by acting as intracellular enzyme (monoamine oxidase A). Dye-contaminated wastewater is also related with the issue of soil contamination and eventually degrades the quality of soil and therefore, plant growth. In the same way, heavy metal pollution also leads the degradation of water quality. The prime source of heavy-metal pollution is the discharge of metallurgical waste from various mining companies. Other sources include geologic-weathering, agricultural or municipal discharge. Bioaccumulation and biomagnification of heavy metals also create health issues like allergies, headache, and pancreas problems along with long-term chronic disease like cancer due to carcinogenic nature of these heavy metals. Therefore, there are some rules and guidelines which have been issued by the world health organization (WHO), US-environmental protection agency (USEPA), and central pollution control board (CPCB), etc., related to the permissible limits of these metals in drinking water [8–10]. Heavy metals include lead (Pb), cadmium (Cd), mercury (Hg), Arsenic (As), Chromium (Cr), Copper (Cu), Zinc (Zn), iron (Fe), Manganese (Mn), cobalt (Co), nickel (Ni), etc. and these are not biodegradable in nature. Among these, some metals like Fe, Cu, Zn, Mn are essentials to maintain some important biological processes, but the requirement is very low. When these are in excess amount, create potential toxicity in both plants, and animals. In India, CPCB and other environmental regulatory authority, have already listed almost 43 industrial groups as the potential source of environmental pollution as per the report on 2009. State-wise survey in India has also been performed by the CPCB, and some states like Orissa, West Bengal and Jharkhand have been declared as the highly polluted areas. As per the report, some heavy metals like Fe, Mn, As, Cr, Ni and Cd have been found in greater quantity [11, 12] than the permissible limit as suggested by the USEPA which seems to be very dangerous situation for our upcoming future. Not only in India, but this situation also persists everywhere in the globe. Case study shows that Bangladesh [13] and Nigeria [14] have the metals like Cr, Co and Ni in main-stream water bodies, much higher than the acceptable range of USEPA(2009). Same situation is also observed for Germany [11] with high contamination of Ni and Mn. In some urban areas of China [15] like Shaoguan, Zhuzhou, Chenzhou and Kunming, heavy-metal contamination is very high and beyond the permissible limits. Few years ago, in 2018, Qu and his research team [16] published a paper on the high-level heavy metal contamination of the Wen-Rui Tang River which includes various toxic metals like As, Cr, Pb, Cd, and Hg. The team highlighted the nearby electroplating and leather tanning industries as the potential source of heavy metal pollution. Another report showed that the China exceeds the permissible limits of As, Fe, Ni, Mn and Co as instructed by WHO-2017 and USEPA-2009. The Hg-level were reported to be much higher in Greece [17], whereas Ni and Cr levels were high in South Africa and Iran [18]. In United-state, Cd, Cr and Ni were found in high content. Pakistan was found to be extremely polluted with Mn, Cd, Ni, Cr, Co and Cu (Khanpur lake), having higher values than USEPA and WHO limits. Severe heavy metal contamination was also found in Phewa and Gosainkunda lake in Nepal [19].

However, a widespread perception concerning heavy metals in water bodies is yet to be set up to a tremendous level and is restrained to literature. Therefore, it is an urgent need to remove synthetic dyes and heavy-metal pollutants from the water bodies to avoid unnecessary health issues and diseases. On this basis of the characters of pollutants, numerous strategies, ranging from coagulation to adsorption and electrochemical techniques, had been studied.

2 Various Techniques Available to Remove Water Pollutants

Nowadays, the most common processes in water treatment utilized by municipal water systems (mostly surface water treatment) involve coagulation and flocculation [20], sedimentation, filtration and disinfect [21]. Numerous methods have been utilized to eliminate dyes from contaminated media in both industrial wastewater and water resources such as electrochemical and membrane processes, biological treatment, coagulation, chemical oxidation, adsorption, and aerobic microbial degradation. Water remedy technology for heavy metals and dyes removal is depending on the nature of contaminant to be eliminated or basis for the treatment, whereas water treatment methods that are ideal for dye removal may not be suitable for getting rid of heavy metals. For example, chemical precipitation is used to put off heavy

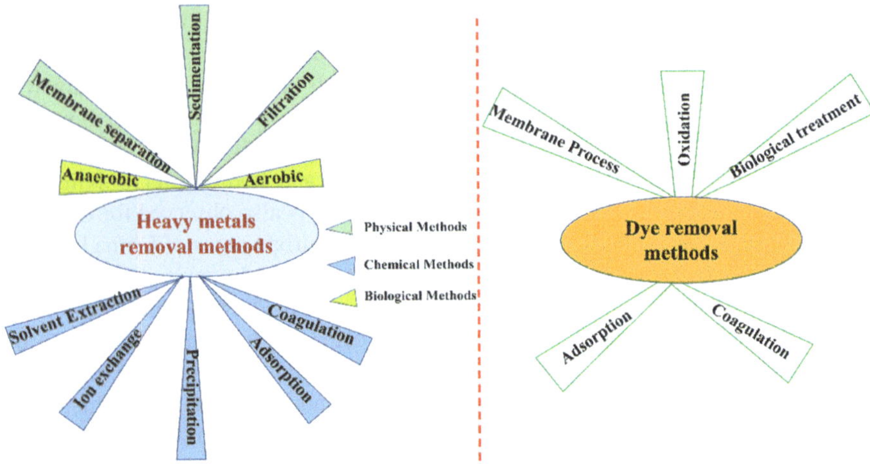

Fig. 1 Various technologies used for removal of dye pollutants and heavy metal

metals, even though it isn't always proposed for dye removal from water [22]. Treatment strategies had been employed to cognizance at the removal of harmful dyes from wastewater and aquatic water. On the opposite hand, the adsorption approach is extensively used because of its several advantages, such as monetary feasibility, improved efficiency in putting off the dyes, and simplicity of processing in addition to low cost for dye elimination from aquatic water and wastewater. The adsorption procedure consists of numerous steps: (i) dissolving dyes in solution, (ii) exterior dye diffusion to the adsorbents' surrounds, (iii) intra/internal-particle diffusion (filling nanoparticle pores with dye materials), and (iv) desorption or adsorption on interior sites. Fig. 1 shows the various techniques used for removing dye and heavy metal pollutants. The strengths, weaknesses, opportunities, and threats of persistent removal of dye pollutants and heavy metals, based on adsorption techniques are effectively discussed on the next section.

3 Chitosan and Its Functional Modifications

3.1 Overview of Chitin and Chitosan as Biopolymer

Polymers are large molecular-weight macromolecules with wide applicability from biomedical to engineering fields. Polymers based on their origin, can be categorized as synthetic (polyethylene, polypropylene, polystyrene, polyethylene terethalate, etc.), semi-synthetic (vulcanized rubber, cellulose nitrate, cellulose acetate, and other cellulose derivatives) and biopolymers (cellulose, starch, chitin, chitosan, sodium alginate, silk fibroin, keratin, etc.). Among all these categories of polymers, biopolymers are preferred over other polymers due to their low cost, excellent biodegradability, and non-cytotoxicity nature [23]. The choice of biopolymer is very high in case of preparing bio-based capsules for targeted and sustained release of drug molecules, wound healing applications and tissue engineering, etc. On the other hand, cellulose and chitosan [24] are the most abundant biopolymers, extracted from the important biomass resources like plant and sea-food creatures like shrimp, crab, etc. In practice, chitin is regarded as the raw source of the most important biopolymer, chitosan. Over the past few decades, the upward push of Asian affluence has brought about a growth in the demand for seafood and a surge within the international seafood processing enterprise, causing an exponential increase in the production of marine products from crabs, prawns, shrimps, and krills. Approximately, eight million tons of seafood waste is generated each year. The heads, tails, and shells of shrimps are undesirable to the processing industries. The surge in shrimp waste has motivated the researcher for chitin extraction. Structurally, chitin is characterized with N-acetyl-D-glucosamine units linked with one another to form a long heteropolymer via β-(1–4) glycosidic bonds [25]. Morphologically, chitin is a nanofibrous component, connected as a matrix to the proteins, lipids and minerals like calcium carbonate/phosphate. Various crystalline forms of chitin are known to us such as, α, β, and γ with varying % of acetylation [25]. Among these, α form has the high occurrence with antiparallel arrangement of the nanofibrils and shows high crystallinity index (80%). On the other hand, β chitins are formed by the parallel arrangement of nanofibrils and offer a crystallinity index of 70%. In discussion to the particular γ form of chitin, it is in between of α and β forms. The extraction chitin from the shrimp shell is generally performed with the process of demineralization and deproteination. In chemical approach, after demineralization, alkali is generally used to avoid the acid-susceptible breaking of glycosidic linkages. After chemical treatment, it is very tough to remove unreacted chemicals and therefore, some researchers are preferred for bio-enzymatic extraction of chitin from sea-food wastes. Enzymes are highly specific and imparts catalytic activity in the process of deproteination. Moreover, it reduces the risk of toxicity as it is more in the chemical process with the highly reactive chemicals. Generally, two important enzymes, namely lipases and proteases are used for deproteination purpose. In a report [26], immobilized lipases and proteases from sunflower seeds was acted like biocatalyst and utilized in chitin extraction. Chitin was reported to be extracted with the help of a protein/enzyme mixture, obtained from*Loligo vulgaris*, Alacase® [27]. On the other side, crude microbial proteases is widely used in extracting chitin from shrimp shells. Various living organisms of ocean annually produce 10^{12}–10^{14} tons of chitin. Arthropods in freshwater and marine environments generate about 2.3×10^{19} and 1.08×10^{11} kg chitin in a month, respectively. These quantities are enough for the extraction of chitosan and commercialize this polymer with low cost. Actually, crystalline chitin is more viable than chitosan, but cannot be used in many cases due to its insoluble nature in green solvents. Therefore, chitosan has been selected as the prime choice of biopolymer in various applications. Basically, the extraction of chitosan from the direct sea-food wastes can be performed with three major steps

Fig. 2 Schematic presentation of chitosan preparation

like demineralization, deproteination, and most importantly, deacetylation. These steps are generally performed either through the chemical route or biological route (enzyme-based treatment/fermentation). The Process of deacetylation is generally performed with concentrated NaOH solution for a period of 3 h and then immediately transferred in crushed ice [28]. The ultimate outcome is the soluble chitosan with a degree of deacetylation, ranging between 48% and 55%. Figure 2 depicts the schematic representation of chitosan preparation from shrimp shell.

3.2 Structural, Physical, and Chemical Aspects of Chitosan

Structurally, chitosan is a polymer of glucosamine and N-acetyl glucosamine, connected via β(1–4) glycosidic linkages [29]. This polysaccharide can be extracted from the low cost chitin via the route of deacetylation. The random distribution of acetyl and deacetyl-units are happened during the process of deacetylation [29]. Depending on the condition of deacetylation, various attributes like physical (solubility, tensile strength, surface area, flexibility, porosity, conductivity, and viscosity), chemical and biological (absorptivity, antioxidant, biodegradability, biocompatibility, and bioavailability) behaviors have been changed. Normally, chitosan is almost insoluble in aqueous medium with neutral pH due to intermolecular hydrogen bonding between chitosan chains, but in presence of organic acids like glacial acetic acid and formic acid; amine functionality gets protonated, and it shows high solubility. In case of processing any polymer in liquid medium, viscosity is a very important parameter. Actually, the viscosity depends on various parameters like degree of deacetylation, molecular weight of the polymer and pH, ionic strength of the medium. The viscosity of chitosan solution also depends on the temperature in an inverse manner, just like other solution. Use of radiation to degrade biological polymer is the unique technique to control the molecular weight of oligomers without affecting the environment [30]. In a study, Choi et al. [31] used the radiation tool to obtain chitosan oligomer from the solution of chitosan in acetic acid. On the other hand, Jia and Shen [32] applied the irradiation method in presence of 85% phosphoric acid at different temperatures to successfully obtain the low molecular weight chitosan (7.1×10^4–21.4×10^4) and observed a decrease in the viscosity value. Low molecular-weight chitosan molecules are highly soluble and hence, the application of ultrasound wave and irradiation in presence of NaOH based deacetylation were performed to derive high yield of acid soluble chitosan [33]. This also improves various physical properties. As per the report, deacetylation with 600 W microwave irradiation, improved the antimicrobial properties of the processed chitosan film via glow-discharge plasma technique along with other benefits like improved tensile strength, thermal stability, and water vapour permanent [34]. The degree of deacetylation of chitosan can be understood via the comparison of FTIR spectra of chitosan and chitin. Undoubtedly, chitosan has aunique physiochemical property and hence, widely used in environmental and biomedical applications. Chitosan with varying molecular weight shows distinct physical properties. Chitosan is widely used in preparing antimicrobial packaging application. Chitosan shows very low antioxidant behavior due to lack of hydrogen donor atoms. Moreover, limited reactivity, very high hydrophilicity, rigidity, and brittleness are some weak points of raw chitosan. But these issues can be easily handled by functional modifications via easily available low-cost chemical reagents. These include carboxyalkylation, alkylation, quaternization, acylation, sulfation, phosphorylation (Fig. 3), and hydroxyalkylation [35].

As stated earlier, chitosan-based hybrid and nanohybrid systems are extremely useful in various applications like corrosion protection, heavy metal ions and dye removals, antimicrobial packaging and many more. In this connection, pure chitosan has been modified via blending [36] with other polymers like polyaniline, polymethyl methacrylate, polystyrene, polycarbonate, etc. Blending improves the mechanical strength, bioactive properties, surface morphology, adsorption behaviors and photoluminescence. Apart from blending, chitosan nanostructures and nanohybris are also prepared for various technological applications, particularly removal of dye pollutants and heavy metal ions. Nanostructured and nanohybrid chitosan shows high surface area, improved surface functionality and morphology and most importantly, adsorption capabilities. In this book chapter we basically have focused on the different chitosan nanohybrids for the removal application of dye and heavy metal pollutants.

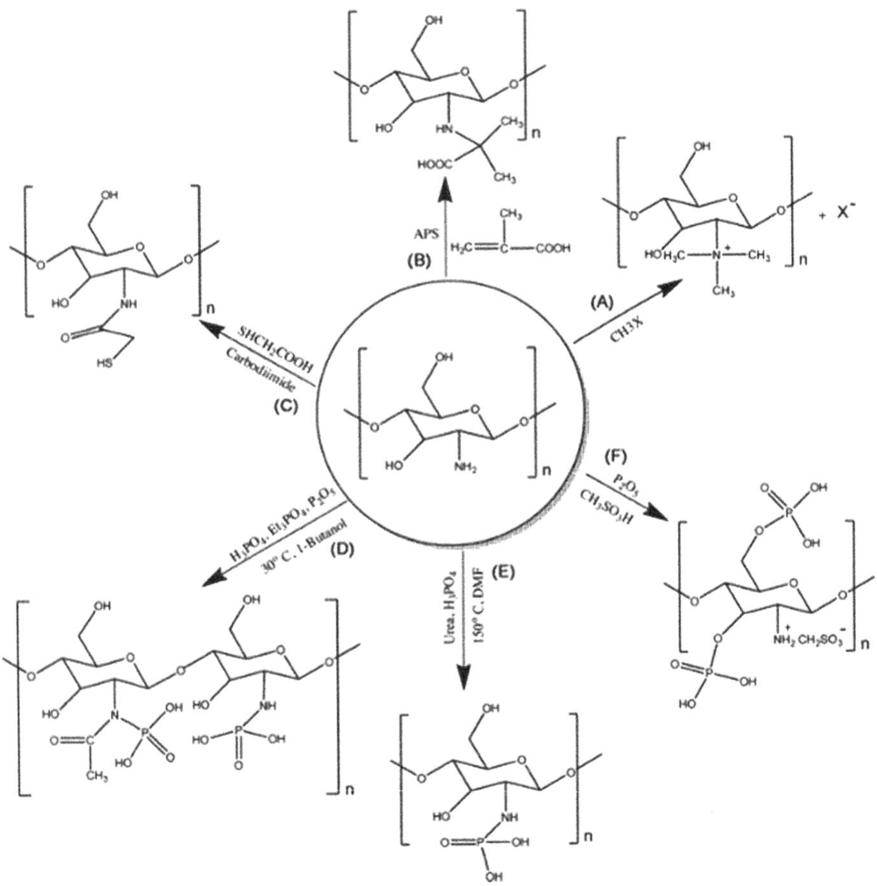

Fig. 3 Functional modification of chitosan through various reagents (Reproduced with permission from Elsevier [35])

4 Chitosan Nanohybrid Materials for Adsorptive Removal of Dyes

Although chitosan-based hybrid and nanohybrid systems are widely used for multifunctional applications like drug delivery, wound healing, tissue engineering, antimicrobial packaging films, but their applicability in wastewater treatment by removing dye-stuffs and heavy metal pollutants are highly investigated due to their non-toxic eco-friendly nature. In 2011, Zhou et al. [37] published an article on the preparation of ethylenediamine-modified magnetic chitosan nanohybrids to remove acid orange 7 and acid orange 10 from the aqueous solutions. Adsorption analysis showed that the maximum adsorption behavior for acid orange 7 was happened at pH 4.0, whereas for the maximum adsorption of acid orange 10 was observed at pH 3.0.

Simple solution casting route was adopted for the inclusion of maghemite (γ-Fe_2O_3) nanoparticles into the chitosan matrix and utilized for quick adsorption of sulfonic acid dyes via batch adsorption experiment. Particle size distribution showed the very narrow size distribution (15 nm–20 nm) of the prepared chitosan nanohybrid and therefore, indicated a very high surface area. The prepared material also exhibited good reusability after the desorption experiment with NH_4Cl/NH_4OH (pH 10) solution. The maximum adsorption values at 298 K were found 3.47 mmol/g for acid orange 7 and 2.25 mmol/g for acid orange 10, respectively. The model adsorption isotherm is well-fitted with the Langmuir model. In another report, Khataee et al. [38] prepared a chitosan based bio-adsorbent for the removal of acid red 88 from aqueous solution and observed the maximum adsorption capacity was about 25.84 mg g^{-1}. The high adsorption capacity was contributed from the chitosan matrix as well as incorporated bio silica, having diameter near about 20 nm. The biopolymer chitosan and biosilica was mixed in solution stage and allowed to form bubble free homogeneous mixture with the process of magnetic stirring, only. After then the solution mixture was drop-wise added to the NaOH(15%)/EtOH (95%) solution to form the biosilica incorporated chitosan nanohybrid beads. The adsorption kinetic was well established with the pseudo 2nd order kinetic along with some correlation of intra-particle diffusion model, whereas, the adsorption isotherm was well-finked with the Langmuir adsorption isotherm. Recently, we had designed chitosan nanohybrid hydrogel with incorporations of organically modified nanoclay and magnetite (Fe_3O_4) nanoparticles into the hybrid chitosan matrix. Hybrid chitosan matrix was developed with the combination of chitosan, polyvinyl alcohol as agar–agar. Chitosan nanohybrid was formulated with simple sulfate (SO_4^{2-}) induced gelation route [39]. The prepared magnetic chitosan nanohybrid hydrogels were morphologically and structurally characterized with FESEM, TEM, XRD and FTIR.

The scanning electron micrograph of (Fig. 4) revealed the three-dimensional tubular shaped surface morphology of the as-prepared chitosan based nanohybrid hydrogel. These tubular architectures were observed to form the bundle structure which had a close resemblance to the wrapped rice straws. TEM images were in well correlation with the SEM images and reflected both tubular structures and particles structures, where both of them were seen to have thin layer coverage of hybrid hydrogel network. Particles were appeared as polymer wrapped cluster at the bottom part of the tubular phase without much affecting the crystalline behaviours of Fe_3O_4 nanostructures. The highly rough surface (136.2 nm) of chitosan nanohybrid hydrogel depicted the increased surface area in nanoscale dimension, a useful criterion for good adsorbent. The variation of solution pH referred the improved dye adsorption behavior at basic region due to exploration of high negative charge contribution via rolled nanoclay platelets. The adsorption of MB/RhB onto the surface of magnetic chitosan nanohybrid hydrogel followed the pseudo 2nd order kinetics along with Langmuir adsorption isotherm model. The significant adsorption capacity of chitosan based nanohybrid hydrogel hydrogel towards MB (877 mg/g) and RhB (893 mg/g) along with high regeneration capability, low cost, effective magnetization value (38.16 emu/g) was found to enable the hydrogel material towards industrial application. In a recent report, Wu et al. [40] reported the preparation of chitosan

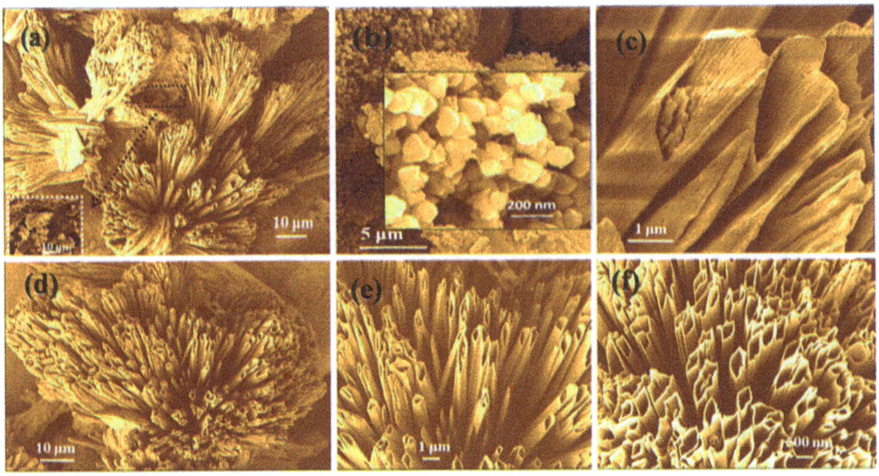

Fig. 4 SEM images of (**a**, **b**, **c**, **d**, **e**, **f**) magnetic chitosan nanohybrid hydrogels with different magnification (Reproduced with permission from American Chemical Society [39])

nanohybrid with incorporation of nanostructural graphene oxide for the adsorption of cationic dye, methylene blue from water through column experiment. Hydrogel was formulated with the electrostatic interactions between the negatively charged graphene oxides with positively charged chitosan at pH 5. Freeze drying technique was utilized to dry the samples. SEM analysis showed that the graphene oxide layers were completely exfoliated within the chitosan matrix at a higher loading of chitosan, whereas with lower loading of chitosan higher adsorption behavior was achieved. Chen et al. [41] had carefully incorporated the concept of self-assembly to prepare the physically cross-linked chitosan based nanocomposite hydrogel with incorporation of graphene oxide and applied the prepared CS–GO nanocomposite hydrogel towards removal of organic (Methylene blue, Eosin-Y) as well as heavy metal pollutants [Cu (II) and Pb (II)] from the aqueous environment.

Two different synthesis routes were adopted by Chen et al. to design self-assembled CS–GO nanocomposite hydrogels (Fig. 5a). In first route of synthesis, chitosan–graphene oxide mixture was vigorously shaken to achieve gel structure, whereas in second route of synthesis *in situ* column technique was performed with aid of NaOH solution to produce the gel structure. The prepared CS–GO nanocomposite hydrogel was structurally characterized by XRD which confirmed the stacking of GO sheets along with the destruction of crystalline phases of chitosan chains. This fact was also supported by the SEM micrographs (Fig. 5b) which showed the loosely cross-linked porous morphology, having pore diameter ranging from several micrometers to ten micrometers without any traces of agglomerated chitosan chains. The prepared CS–GO (10) nanocomposite hydrogel showed the excellent adsorption capacity for methylene blue and Eosin Y (>300 mg/g) which was shown to be varied with changing the content of graphene oxide and chitosan. Lower chitosan and higher GO in hydrogel structure was found to be effective for methylene blue

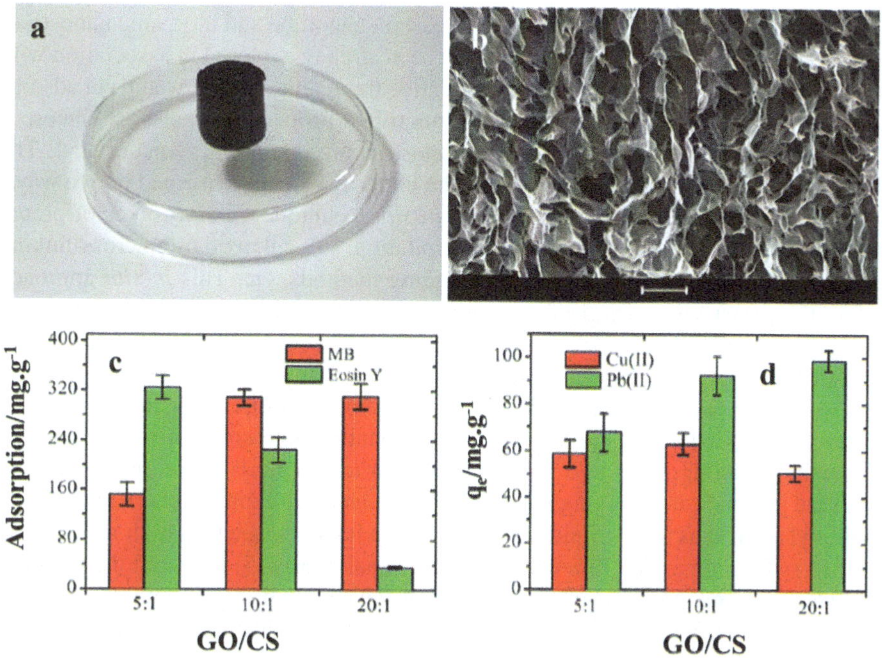

Fig. 5 Digital image of the GO/CS10 hydrogel (**a**) SEM micrograph of the lyophilized GO–CS10 hydrogel (**b**), Adsorption of MB and Eosin Y on the different GO/CS hydrogels. (The initial concentration of the dyes is 80 mg/L) (**c**) and adsorption of Cu2+ and Pb2+ on different GO/CS composite hydrogels. (The initial concentration of Cu2+ and Pb2+ is 100 mg/L) (**d**). (Reproduced with permission from Royal Society of Chemistry [41])

removal, whereas; reverse trend was reported to be effective for removal of Eosin Y. The CS–GO (10) nanocomposite hydrogel was estimated for their adsorption capacity towards Cu(II) and Pb(II) ions and the values were found to be 70 mg/g and 90 mg/g, respectively (Fig. 5c, d). Cationic chitosan shows a very good complexation with multivalent, non-toxic tripolyphosphate (TPP) ions and therefore capable to form nanostructured chitosan [42]. During complexation, the charge distribution of chitosan and tripolyphosphate can be tuned by tuning the pH of the solution. In 2016, Linda et al. [43] developed an efficient adsorbent of chitosan nanohybrid with incorporation of hybrid nanostructure, CdO/NiO. Semiconductor hybrid nanostructures played the role of photocatalytic dye degrader, whereas the chitosan matrix was acting as the good adsorbent site for Rhodamin B. Similar, type of approach was also adopted by Saad et al. [44]. They prepared the chitosan based nanohybrid systems with inclusions of ZnO and cerium doped zinc oxide (Ce-ZnO), respectively for the advanced high performance catalyst under visible light irradiation for the photocatalytic degradation of malachite green. The percentages of dye-degradation were found 54% and 87% for chitosan/ZnO and chitosan/Ce-ZnO nanocomposites, respectively. Due to attractive physical and chemical properties, chitosan is always getting the priority as matrix component. Various carbon-based nanostructures like graphene oxide, reduced graphene oxide, carbon nanotube and inorganic nanostructures like ZnO, TiO_2, Fe_3O_4, Ag, etc. have been already explored in association with chitosan as matrix component for the effective treatment of wastewater via adsorption, photocatalytic degradation and antimicrobial properties. Therefore, incorporations of carbon and inorganic-based nanostructures are highly encouraged. The association of nanostructural materials with chitosan can be performed in two ways. One is *ex-situ* technique and the other is *in-situ* technique. *In-situ* routes of preparation mostly includes impregnation, precipitation, sol–gel, oxidation, cross-linking polymerization, reduction and electrospinning methods, etc. This *in-situ* approach refers the direct preparation of the metal or metal-oxide nanoparticles within the polymer solution. In most of time, *in-situ* approach provides the uniform distribution of the so-formed nanostructures over polymer matrix. Contrariwise, the ex-situ technique, which refers to as physical methods, requires first the preparation of the NPs, followed by the addition of this nanoparticle solution to the polymer dispersion. Among these two strategies, in-situ technique is more preferred as in this case polymer, itself acting as both reducing and stabilizing agent. Chitosan has hydroxyl as well as amine functionalities, hence it can be acted as both reducing as well as stabilizing agent. In a recent report, Bahal et al. [45] prepared the titanium dioxide (TiO_2) nanoparticle incorporated chitosan nanohybrids via both physical and chemical routes. They used the polymer grafting method with acrylic acid to introduce carboxyl functionalities on to the chitosan chains. *In-situ* polymerization route was performed for the uniform dispersion of TiO_2 nanoparticles within the copolymeric matrix. The prepared chitosan nanohybrid was applied for the photocatalytic degradation of malachite green under visible light irradiation. In another literature [46], chitosan was modified with cellulose acetate and various nanostructures like single walled carbon nanotube, iron oxide (ferrite) and titanium oxide to prepare the adsorbent for removal of dye pollutants (azo-dyes) and heavy metal ions. This chitosan based nanocomposite fiber was fabricated with the combination of three processes such as impregnation, sonication and electrospinning methods. Incorporation of nanostructured materials improved the wettability of the nanocomposite fibers and therefore, showed high adsorption behaviors with an optimum loading of the nanostructured materials.

5 Chitosan Nanohybrid Materials for Removal of Heavy Metals

Chitosan on chemical modification shows the improvement in physical and chemical properties and therefore, chosen as effective adsorbent for heavy metal pollutants also. The use of various cross-linking agents like formaldehyde, glutaraldehyde, isocyanates, ethylene glycol, diglycidyl ether, etc. improves the stability of the chitosan nanostructures. The adsorption behavior of chitosan based nanohybrid system is different as compared to the other forms of chitosan. In terms of publication, a large increase was experienced since 2001 on chitosan based articles which might be due to the global concern on environmental pollution and searching for biodegradable alternative. In past 20 years, total number of publication on chitosan based material is increased to 30,172 research articles (Pubmed information) which clearly shows an inclination towards the urge of searching biodegradable materials with improved properties. More than 76,000 articles in Scopus were published related to chitosan, whereas, the article numbers with title of "Chitosan for removal of heavy metals" has the noticeable value (approximately, 7000 papers). As per the information on pubmed, in 2019 only 500 research articles were published related to chitosan on heavy metal ion removal. Charlet and his workers [47] reported the preparation of a novel chitosan nanohybrid with incorporation of iron-oxyhydroxide for the adsorptive removal of As(III) and As(V) from the water system in a pH range of 5 to 9. Here, nanostructural iron oxy-hydroxide was identified as goethite and dispersed well in chitosan matrix. Chitosan/goethite nanohybrid was formulated in the bead form for easy handling. XRF study revealed that diffusion of As(V) from the water system to the pores of chitosan nanohybrid was much faster as compared to As(III). Bead was observed to purify As(V)-spiked water system (0.5 mg/L) to drinking water system ($\leq$0.01 mg/L). Anjali Paul and her coworkers [48] modified the chitosan with anionic surfactant, sodium dodecyl sulfate (SDS) by forming a bilayer over chitosan matrix. The prepared nanoadsorbent was applied for the removal of toxic Cd^{+2} from the water system. It showed maximum adsorption capacity of 125 mg/g. The adsorption model was well fitted with Langmuir adsorption isotherm, whereas the adsorption kinetic was correlated with pseudo 2nd order kinetics. The SDS modified chitosan nanohybrid showed excellent removal efficiency (100%) with the adsorbent dosage of 0.45 g/L for the spiked water samples with Cd^{+2} concentration ranging from 10 to 30 mg/L in 10 h. In another report [49], chitin nanofibers were surface modified with thiol-functionalities and utilized for the adsorption of arsenic from water samples. Chitosan nanohybrid with nanoclay and magnetite nanoparticles [50] showed an excellent adsorption behaviors for both, As (V) and Cu (II). The adsorption kinetics was correlated with pseudo 2nd order kinetic and intra-particle diffusion model for both the cases. Organoclay containing chitosan [51] also showed good adsorptivity of Cu(II) [176 mg/g] and Ni(II) [144 mg/g]. Kolodynska et al. [52] used chitosan to modify zeolite and then applied that material as the adsorbent for removal of methylene blue dye and copper(II). The adsorption process was found spontaneous and endothermic in nature. In a report, chitosan nanohybrid was prepared with inclusion of clinoptilotile and magnetite nanoparticles and used for the removal of Pb(II). The material showed a maximum theoretical adsorption of 136.99 mg/g at pH 6. Li et al. [53] developed a porous poly lactic acid/chitosan based nanofibers via electrospinning route followed by acetone treatment and utilized as potential adsorbent for removal of copper(II) ions from solution. Actually, the chitosan was included in the structures after the formation of polylactic acid nanofibers via direct immersion coating method. The prepared material showed a very fast second order diffusion kinetics and Langmuir adsorption isotherm for the adsorption of copper ions. Prepared porous nanofiber demonstrated the very high adsorption capacity (111.66

± 3.22 mg/g) for copper ions at pH 7. In a recent literature, Wei et al. [54] incorporated the zeolite imidazolate framework (ZIF-67) to the hybrid polymeric matrix, formed by the combination of bacterial cellulose and chitosan and processed as hybrid aerogel for the removal of Cu^{2+} and Cr^{+6} ions from water samples. Distribution of ZIF-67 to the hybrid polymeric matrix, increased the surface roughness value to 268.7 m^2/g as compared to the pure hybrid matrix. The adsorption values for Cu^{2+} and Cr^{+6} ions were found to be 200.6 mg/ g and 152.1 mg /g, respectively. In a recent report, Debnath et al. [55] prepared a chitosan/clay nanohybrid beads for removal of heavy metal ions and dyes from industrial effluent. In this work, prawn shell derived chitosan was mixed with kaolinite rich modified clay and processed as bead with treatment of NaOH/EtOH solution. The prepared bionanohybrid showed excellent adsorption capacity for Cr(VI) and Pb(II) from water sample. From tannery effluent, 50.90% of Cr(VI) and 39.50% of Pb(II) were removed by the bionanohybrid, rich with chitosan. Dinh et al. [56] preapared the chitosan based nanohybrid by dispersing MnO_2 nanoparticles on to the chitosan matrix in order to enhance the adsorption and removal of Pb(II). The monolayer adsorption capacity of this adsorbent towards Pb(II) was reported as 126.1 mg/g which was interpreted as the result of "ions-hole" mechanism. Wei et al. [57] prepared an ion imprinted reversible-thermosensitive chitosan derivative for the adsorptive removal of Cu^{+2} ions from water sample. The adsorption (at 45 °C) and desorption (at 20 °C) processes were controlled just by tuning the temperature. An adsorption capacity of 88.94 mg/g was achieved along with high desorption (98.90%) at proper condition. Adsorbent was designed with Cu^{2+} as template paired with N-isopropyl acrylamide monomer. The adsorption kinetics and isotherms were well-fitted with pseudo 2nd order kinetic and Langmuir adsorption isotherm. In another literature, shanget al. [58] prepared the Pb-ion imprinted thermosensitive polymer with incorporation of multi-walled carbon nanotube, N-isopropyl acrylamide, and methyl methacrylate. The maximum adsorption for Pb^{+2} was found 18.09 mg/g. Wang and his coworkers [59] prepared a chitosan-based magnetic nanohybrid with incorporation of magnetite and bentonite nanostructures for the effective removal of Cr^{+6} ions from water sample. The maximum adsorption capacity was observed 62.1 mg/g for Cr^{+6} ions at 25 °C (pH 2). The adsorption kinetic was well-demonstrated with the pseudo-2nd order kinetics, whereas adsorption isotherm was closely related to the Langmuir adsorption isotherm model. The polyelectrolyte type nanostructured chitosan [60] was prepared by the combination of carboxymethyl cellulose and chitosan for removal of Cd(II) and Cr(VI) ions from water. Itshowed excellent regeneration and recyclability. Magnetic Loofah biochar-modified chitosan was reported [61] to remove Cr(VI) [30.14 mg/g] and Cu(II) [54.68 mg/g] ions from water sample. The Freundlich model best described the isotherm kinetics of the composite. Even after three cycles, the absorptivity was relatively high, which might be contributed from the ion exchange and surface complexation mechanisms. Sudha et al. [62] adopted the graft copolymerization of acrylonitrile with chitosan to prepare the adsorbent for removal of copper and chromium from water. The optimum adsorption was achieved at pH 5.0. In another report, Ali et al. [63] prepared the chitosan based biocomposites with acrylonitrile in presence of carbon disulfide and successfully utilized for removal of lead (2.54 mmol/g) and copper (2.02 mmol/g) from water. The combination of chitosan, CS_2 and acrylonitrile was performed to increase the Lewis base centers. Moreover, thiocarbamate moiety was introduced to the position amine functionality of chitosan to improve sorption capacity most. The Langmuir model is the best fitted model for its isotherm kinetics. In a report, chitosan based nanohybrid fiber was prepared by Beheshti et al. [64]. In this approach, chitosan was mixed with multiwalled carbon nanotube and magnetite nanoparticles and processed as hybrid nanofiber via electrospinning technique to impart improved adsorption behavior for Cr^{+6} ions from water medium. They observed that with a loading of hybrid nanofiller (MWCNT/Fe_3O_4) up to 2 wt%, adsorptions towards Cr^{+6} ions were significantly increased which might be the result of the improved adsorption sites with high surface area of chitosan-based adsorbent. On the other hand, higher loading of nanohybrid filler in chitosan matrix provided an agglomerated phase and therefore, showed less adsorption behavior. The kinetic and equilibrium data were well corresponded to the pseudo 2nd order kinetics and Langmuir adsorption isotherm model. Sometimes, anions and some toxic pathogens also strongly contributed towards the pollution of main-stream water bodies. Keshvardoostchokami et al. [65] dispersed the zinc oxide nanostructures into the hybrid chitosan matrix, formed by the combination of chitosan and polystyrene for adsorptive removal of nitrate ions from water samples. Wrinkled surface morphology of the adsorbent due to strong coordination between Zn and hybrid polymer matrix helped to create various porous channels which in turn was responsible for quick diffusion of the nitrate ions into the adsorbent. In another report [66], chitosan/polyvinylidene chloride matrix was designed with the incorporation metallic nanosilver. Simple chemical reduction method was employed for the preparation of this chitosan nanohybrid system and successfully utilized for the uptake of various heavy metals like Pb, Cd, Fe from the water system. Moreover, this material showed excellent antibacterial behavior towards *E. coli* and *G. Bacillus*. As we know, chemical cross-linking approach helps to create water insoluble three-dimensional network of chitosan, many of the researchers tried this to obtain recyclable chitosan nanohybrid adsorbents for the treatment of wastewater. In some cases, cross-linking agent may act as harmful reagent like epichorohydrin in presence of water release toxic components and therefore, such cross-linking agent is not preferred. Magnetic chitosan hydrogel [67] was prepared with the cross-linking between chitosan and glyoxal. Incorporation of magnetite nanostructures helped to create high surface area and adsorption sites for Cr^{+6} ions along with magnetic nature for easy separation. In a study, conducted by Salam et al. [68], chitosan/multi-walled carbon nanotube based three dimensional nanohybrid structure was formulated via cross-linking with glutaraldehyde and successfully employed for the removal of various heavy metal pollutants like Cu, Zn, Cd and Ni. Moreno Lopez et al. [69] prepared a biosorbent for removal of Cr^{+6} ions from water sample. In this work, cross-linked chitosan and chitosan xanthate were coated on to agave to prepare the composite. The maximum adsorption capacity of cross-linked chitosan xanthane based composite towards Cr^{+6} ions was reported 284.7 mg/gat pH 4 which was much larger value than the agave fiber composites, prepared by chitosan only. This demonstrated the formation of high surface area and active sites due to cross-linking of chitosan with adipoyl chloride. Most of the isotherm models follow pseudo second order kinetics, indicating monolayer coverage and a rate-limiting chemisorption step.

6 Conclusion and Future Scope

This book chapter includes the detailing of chitosan in terms of its surface functionality, ease of surface modification by means of physical and chemical methods, and recent works related to preparing hybrid nanostructures for the cost-effective remediation of wastewater. Chitosan based nanohybrids with inclusion of various carbon nanostructured and inorganic particles have been proven as potential candidates in wastewater remediation. On the other hand, polymeric chitosan is well-known to capture heavy metals like As, Cu, Cd, and Pb, due to coordinating surface functionalities such as amine and hydroxyl groups. These functionalities also help to modify and functionalize chitosan macromers. Proper modification/cross-linking leads to the improvement in physical and mechanical properties of the chitosan nanohybrid structures as adsorbent. Chitosan nanohybrids can replace the commercial adsorbents in terms of high stability, reusability, high selectivity, and efficiency. Most heavy metal adsorptions follow the pseudo-second order kinetics and Langmuir isotherm, suggesting unique chemisorption behaviour. The optimization of adsorption through modelling systems, such as RSM, should be conducted to explore adsorption mechanisms. Thus, researchers should focus on other kinetic models for data fitting. The prepared chitosan nanohybrids in both the cases like removal of dyes and heavy metal pollutants have been tested with synthetic wastewater solution. Real time application of these chitosan nanohybrids should be more focused for the ultimate goal of commercialization. On the other hand, extraction of costly heavy metals like Au and Pt needs to be more focused on desorption for its recovery. Development of multi-metal sensors could be a breakthrough in research.

References

1. Oelkers EH, Hering JG, Zhu C (2011) Water: is there a global crisis? Elements 7(3):157–162
2. Basha SA, Rajaganesh K (2014) Microbial bioremediation of heavy metals from textile industry dye effluents using isolated bacterial strains. Int J Curr Microbiol Appl Sci 3:785–794
3. Khan S, Malik A (2014) Environmental and health effects of textile industry wastewater. In: Environmental deterioration and human health. Springer, Dordrecht, pp 55–71
4. Wani KA, Jangid NK, Bhat AR (eds) (2019) Impact of textile dyes on public health and the environment. IGI Global
5. Chequer FD, De Oliveira GR, Ferraz EA, Cardoso JC, Zanoni MB, de Oliveira DP (2013) Textile dyes: dyeing process and environmental impact. Eco-Friendly TextE Dye Finish 6(6):151–176
6. Chung KT (2016) Azo dyes and human health: a review. J Environ Sci Health C 34(4):233–261

7. Paul P, Kumar GS (2013) Spectroscopic studies on the binding interaction of phenothiazinium dyes toluidine blue O, azure A and azure B to DNA. Spectrochim Acta Part A Mol Biomol Spectrosc 107:303–310
8. Edition F (2011) Guidelines for drinking-water quality. WHO chronicle 38:104–108
9. Liu G, Yu Y, Hou J, Xue W, Liu X, Liu Y, Liu Z (2014) An ecological risk assessment of heavy metal pollution of the agricultural ecosystem near a lead-acid battery factory. Ecol Ind 47:210–218
10. Mohankumar K, Hariharan V, Rao NP (2016) Heavy metal contamination in groundwater around industrial estate vs residential areas in Coimbatore, India. J Clin Diagn Res 10:BC05
11. Kumar V, Parihar RD, Sharma A, Bakshi P, Sidhu GPS, Bali AS, Karaouzas L, Bhardwaj R, Thukral AK, Gyasi-Agyei Y, Rodrigo-Comino J (2019) Global evaluation of heavy metal content in surface water bodies: a meta-analysis using heavy metal pollution indices and multivariate statistical analyses. Chemosphere 236:124364
12. Kumar V, Sharma A, Minakshi BR, Thukral AK (2018) Temporal distribution, source apportionment, and pollution assessment of metals in the sediments of Beas river, India. Hum Ecol Risk Assess Int J 24(8):2162–2181
13. Bhuyan MS, Islam MS (2017) A critical review of heavy metal pollution and its effects in Bangladesh. Environ Energy Econ 2:12–25
14. Olubunmi FE, Olorunsola OE (2010) Evaluation of the status of heavy metal pollution of sediment of Agbabu bitumen deposit area, Nigeria. Eur J Sci Res 41(3):373–382
15. Hu H, Jin Q, Kavan P (2014) A study of heavy metal pollution in China: current status, pollution-control policies and countermeasures. Sustainability 6(9):5820–5838
16. Qu L, Huang H, Xia F, Liu Y, Dahlgren RA, Zhang M, Mei K (2018) Risk analysis of heavy metal concentration in surface waters across the rural-urban interface of the Wen-Rui Tang River, China. Environ Pollut 237:639–649
17. Karaouzas I, Kapetanaki N, Mentzafou A, Kanellopoulos TD, Skoulikidis N (2021) Heavy metal contamination status in Greek surface waters: a review with application and evaluation of pollution indices. Chemosphere 263:128192
18. Karimian S, Chamani A, Shams M (2020) Evaluation of heavy metal pollution in the Zayandeh-Rud River as the only permanent river in the central plateau of Iran. Environ Monit Assess 192(5):1–13
19. Rupakheti D, Tripathee L, Kang S, Sharma CM, Paudyal R, Sillanpää M (2017) Assessment of water quality and health risks for toxic trace elements in urban Phewa and remote Gosainkunda lakes, Nepal. Hum Ecol Risk Assess: Int J 23(5):959–973
20. Riera-Torres M, Gutiérrez-Bouzán C, Crespi M (2010) Combination of coagulation–flocculation and nanofiltration techniques for dye removal and water reuse in textile effluents. Desalination 252(1–3):53–59
21. Katheresan V, Kansedo J, Lau SY (2018) Efficiency of various recent wastewater dye removal methods: a review. J Environ Chem Eng 6(4):4676–4697
22. Gunatilake SK (2015) Methods of removing heavy metals from industrial wastewater. Methods 1(1):14
23. Gowthaman NSK, Lim HN, Sreeraj TR, Amalraj A, Gopi S (2021) Advantages of biopolymers over synthetic polymers: social, economic, and environmental aspects. In: Biopolymers and their Industrial Applications. Elsevier, pp 351–372
24. Thakur VK, Voicu SI (2016) Recent advances in cellulose and chitosan-based membranes for water purification: a concise review. Carbohyd Polym 146:148–165
25. Kumirska J, Czerwicka M, Kaczyński Z, Bychowska A, Brzozowski K, Thöming J, Stepnowski P (2010) Application of spectroscopic methods for structural analysis of chitin and chitosan. Mar Drugs 8(5):1567–1636
26. Mehdi WA, Mehde AA, Özacar M, Özacar Z (2018) Characterization and immobilization of protease and lipase on chitin-starch material as a novel matrix. Int J Biol Macromol 117:947–958
27. Abdelmalek BE, Sila A, Haddar A, Bougatef A, Ayadi MA (2017) β-Chitin and chitosan from squid gladius: biological activities of chitosan and its application as clarifying agent for apple juice. Int J Biol Macromol 104:953–962

28. Venkatesan J, Kim SK (2010) Chitosan composites for bone tissue engineering—an overview. Mar Drugs 8(8):2252–2266
29. Ali A, Ahmed S (2018) A review on chitosan and its nanocomposites in drug delivery. Int J Biol Macromol 109:273–286
30. Kundu CK, Wang X, Hou Y, Hu Y (2018) Construction of flame retardant coating on polyamide 6.6 via UV grafting of phosphorylated chitosan and sol–gel process of organo-silane. Carbohyd Polym 181:833–840
31. Choi WS, Ahn KJ, Lee DW, Byun MW, Park HJ (2002) Preparation of chitosan oligomers by irradiation. Polym Degrad Stab 78(3):533–538
32. Jia Z, Shen D (2002) Effect of reaction temperature and reaction time on the preparation of low-molecular-weight chitosan using phosphoric acid. Carbohyd Polym 49(4):393–396
33. Hajji S, Younes I, Ghorbel-Bellaaj O, Hajji R, Rinaudo M, Nasri M, Jellouli K (2014) Structural differences between chitin and chitosan extracted from three different marine sources. Int J Biol Macromol 65:298–306
34. Alves HJ, Vieceli M, Alves C, Muñiz GI, de Oliveira CL, Feroldi M, Arantes MK (2018) Chitosan depolymerization and nanochitosan production using a single physical procedure. J Polym Environ 26(9):3913–3923
35. Taka AL, Klink MJ, Mbianda XY, Naidoo EB (2021) Chitosan nanocomposites for water treatment by fixed-bed continuous flow column adsorption: a review. Carbohyd Polym 255:117398
36. El-Hefian EA, Nasef MM and Yahaya AH (2014) Chitosan-based polymer blends: current status and applications. J Chem Soc Pak 36(1)
37. Zhou L, Jin J, Liu Z, Liang X, Shang C (2011) Adsorption of acid dyes from aqueous solutions by the ethylenediamine-modified magnetic chitosan nanoparticles. J Hazard Mater 185(2–3):1045–1052
38. Soltani RDC, Khataee AR, Safari M, Joo SW (2013) Preparation of bio-silica/chitosan nanocomposite for adsorption of a textile dye in aqueous solutions. Int Biodeterior Biodegradation 85:383–391
39. Sarkar N, Sahoo G, Das R, Swain SK (2018) Three-dimensional rice straw-structured magnetic nanoclay-decorated tripolymeric nanohydrogels as superadsorbent of dye pollutants. ACS Applied Nano Materials 1(3):1188–1203
40. Qi C, Zhao L, Lin Y, Wu D (2018) Graphene oxide/chitosan sponge as a novel filtering material for the removal of dye from water. J Colloid Interface Sci 517:18–27
41. Chen Y, Chen L, Bai H, Li L (2013) Graphene oxide–chitosan composite hydrogels as broad-spectrum adsorbents for water purification. J Mater Chem A 1(6):1992–2001
42. Zhao LM, Shi LE, Zhang ZL, Chen JM, Shi DD, Yang J, Tang ZX (2011) Preparation and application of chitosan nanoparticles and nanofibers. Braz J Chem Eng 28:353–362
43. Linda T, Muthupoongodi S, Shajan XS, Balakumar S (2016) Fabrication and characterization of chitosan templated CdO/NiO nano composite for dye degradation. Optik 127(20):8287–8293
44. Saad AM, Abukhadra MR, Ahmed SAK, Elzanaty AM, Mady AH, Betiha MA, Rabie AM (2020) Photocatalytic degradation of malachite green dye using chitosan supported ZnO and Ce–ZnO nano-flowers under visible light. J Environ Manage 258:110043
45. Bahal M, Kaur N, Sharotri N, Sud D (2019) Investigations on amphoteric chitosan/TiO2 bionanocomposites for application in visible light induced photocatalytic degradation. Adv Polym Technol 2019:1–9
46. ZabihiSahebi A, Koushkbaghi S, Pishnamazi M, Askari A, Khosravi R, Irani M (2019) Synthesis of cellulose acetate/chitosan/SWCNT/Fe3O4/TiO2 composite nanofibers for the removal of Cr (VI), As (V), Methylene blue and Congo red from aqueous solutions. Int J Biol Macromol 140:1296–1304
47. He J, Bardelli F, Gehin A, Silvester E, Charlet L (2016) Novel chitosan goethite bionanocomposite beads for arsenic remediation. Water Res 101:1–9
48. Pal P, Pal A (2019) Dye removal using waste beads: efficient utilization of surface-modified chitosan beads generated after lead adsorption process. J Water Process Eng 31:100882

49. Yang R, Su Y, Aubrecht KB, Wang X, Ma H, Grubbs RB, Chu B (2015) Thiol-functionalized chitin nanofibers for As(III) adsorption. Polymer 60:9–17
50. Cho DW, Jeon BH, Chon CM, Kim Y, Schwartz FW, Lee ES, Song H (2012) A novel chitosan/clay/magnetite composite for adsorption of Cu(II) and As(V). Chem Eng J 200:654–662
51. Malayoglu U (2018) Removal of heavy metals by biopolymer (chitosan)/nanoclay composites. Sep Sci Technol 53(17):2741–2749
52. Kołodyńska D, Hałas P, Franus M, Hubicki Z (2017) Zeolite properties improvement by chitosan modification—Sorption studies. J Ind Eng Chem 52:187–196
53. Javanbakht V, Ghoreishi SM, Habibi N, Javanbakht M (2016) A novel magnetic chitosan/clinoptilolite/magnetite nanocomposite for highly efficient removal of Pb (II) ions from aqueous solution. Powder Technol 302:372–383
54. Li D, Tian X, Wang Z, Guan Z, Li X, Qiao H, Ke H, Luo L, Wei Q (2020) Multifunctional adsorbent based on metal-organic framework modified bacterial cellulose/chitosan composite aerogel for highly efficient removal of heavy metal ion and organic pollutant. Chem Eng J 383:123127
55. Biswas S, Rashid TU, Debnath T, Haque P, Rahman MM (2020) Application of chitosan–clay biocomposite beads for removal of heavy metal and dye from industrial effluent. J Compos Sci 4(1):16
56. Dinh VP, Nguyen MD, Nguyen QH, Luu TT, Luu AT, Tap TD, Ho TH, Phan TP, Nguyen TD, Tan LV (2020) Chitosan-MnO2 nanocomposite for effective removal of Cr (VI) from aqueous solution. Chemosphere 257:127147
57. Han P, Li Z, Wei X, Tang L, Li M, Liang Z, Wei S (2020) Ion-imprinted thermosensitive chitosan derivative for heavy metal remediation. Carbohyd Polym 248:116732
58. Wang H, Shang H, Sun X, Hou L, Wen M, Qiao Y (2020) Preparation of thermo-sensitive surface ion-imprinted polymers based on multi-walled carbon nanotube composites for selective adsorption of lead (II) ion. Colloids Surf, A 585:124139
59. Feng G, Ma J, Zhang X, Zhang Q, Xiao Y, Ma Q and Wang S (2019) Magnetic natural composite Fe3O4-chitosan@ bentonite for removal of heavy metals from acid mine drainage. J Colloid Interface Sci 538:132–141
60. Borsagli FGLM, Borsagli A (2019) Chemically modified chitosan bio-sorbents for the competitive complexation of heavy metals ions: a potential model for the treatment of wastewaters and industrial spills. J Polym Environ 27(7):1542–1556
61. Xiao F, Cheng J, Cao W, Yang C, Chen J, Luo Z (2019) Removal of heavy metals from aqueous solution using chitosan-combined magnetic biochars. J Colloid Interface Sci 54:579–584
62. Ramya R, Sankar P, Anbalagan S, Sudha PN (2011) Adsorption of Cu (II) and Ni (II) ions from metal solution using crosslinked chitosan-g-acrylonitrile copolymer. Int J Environ Sci 1(6):1323–1338
63. Ali N, Khan A, Bilal M, Malik S, Badshah S, Iqbal H (2020) Chitosan-based bio-composite modified with thiocarbamate moiety for decontamination of cations from the aqueous media. Molecules 25(1):226
64. Beheshti H, Irani M, Hosseini L, Rahimi A, Aliabadi M (2016) Removal of Cr(VI) from aqueous solutions using chitosan/MWCNT/Fe3O4 composite nanofibers-batch and column studies. Chem Eng J 284:557–564
65. Keshvardoostchokami M, Babaei S, Piri F, Zamani A (2017) Nitrate removal from aqueous solutions by ZnO nanoparticles and chitosan-polystyrene–Zn nanocomposite: Kinetic, isotherm, batch and fixed-bed studies. Int J Biol Macromol 101:922–930
66. Al-Sherbini ASA, Ghannam HE, El-Ghanam GM, El-Ella AA, Youssef AM (2019) Utilization of chitosan/Ag bionanocomposites as eco-friendly photocatalytic reactor for Bactericidal effect and heavy metals removal. Heliyon 5(6):e01980
67. Mirabedini M, Kassaee MZ, Poorsadeghi S (2017) Novel magnetic chitosan hydrogel film, cross-linked with glyoxal as an efficient adsorbent for removal of toxic Cr(VI) from water. Arab J Sci Eng 42(1):115–124

68. Salam MA, Makki MS, Abdelaal MY (2011) Preparation and characterization of multi-walled carbon nanotubes/chitosan nanocomposite and its application for the removal of heavy metals from aqueous solution. J Alloy Compd 509(5):2582–2587
69. López-Cervantes J, Sánchez-Machado DI, Sánchez-Duarte RG, Correa-Murrieta MA (2018) Study of a fixed-bed column in the adsorption of an azo dye from an aqueous medium using a chitosan–glutaraldehyde biosorbent. Adsorpt Sci Technol 36(1–2):215–232

Metal Oxide Nanoparticles for Water Decontamination

Preeti Jain, Mridula Guin, and N. B. Singh

Abstract Clean water is very important for living being and other activities. However, water is continually being polluted and become harmful. Number of techniques is being used for purification of water and out of that adsorption is found to be the most affordable and fast technique. In recent years, nanotechnology has played an important role in water purification and decontamination. Nanomaterials (NMs) have proved to be a very good adsorbent for the removal of organic and inorganic pollutants and heavy metals from water and also kill microorganisms in the wastewater. Due to electronic structure, electronic, optical, and magnetic properties, metal oxide nanoparticle (NPs) are found to be unique materials for water remediation. Metal oxide-based NMs, such as zinc oxides, iron oxides, manganese oxides, titanium oxides, aluminum oxides, magnesium oxides, cerium oxides, zirconium oxides, etc. have shown their effectiveness for water remediation. Nanosized metal oxides possess many exceptional properties, such as a high removal capacity and selectivity towards heavy metals and organic compounds. Thus, they have great potential as promising adsorbents for heavy metals, dyes and other pollutants. In this chapter synthesis of number of metal oxide NMs and their applications for water decontaminations have been discussed in detail.

Keyword Nanomaterial · Metal oxide · Iron oxide · Zinc oxide · Magnesium oxide · Photodegradation

P. Jain · M. Guin (✉) · N. B. Singh (✉)
Department of Chemistry and Biochemistry, Sharda University, Greater Noida, India
e-mail: mridula.guin@sharda.ac.in

N. B. Singh
e-mail: n.b.singh@sharda.ac.in

N. B. Singh
Research and Development Cell, Sharda University, Greater Noida, India

S. K. Swain (ed.), *Nanohybrid Materials for Water Purification*, Composites Science and Technology, https://doi.org/10.1007/978-981-19-2332-6_11

1 Introduction

Water is the most important component for various activities on this planet. The distribution of water on the earth is shown in Fig. 1. 97.5% water is in sea and only 2.5% is available in other areas for various activities.

Water is the most important component for various activities on this planet and therefore, clean and pure water is required for development as well as for the survival of living organisms. However, the quality of water is continually deteriorating due to rapid urbanization and industrialization [1–5]. The harmful chemicals going into water bodies are heavy metal ions, inorganic cations, dyes and organic compounds. Pollutants mainly come from different sources (Fig. 2) [6].

Toxic metal ions, from different sources go into water body and make them injurious (Fig. 3) [1].

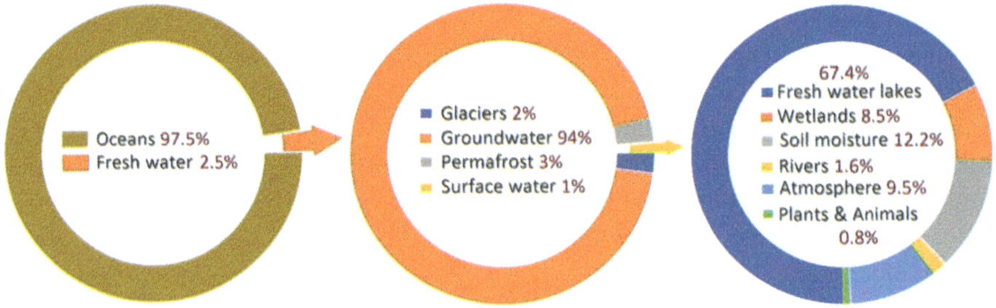

Fig. 1 Distribution of water on earth

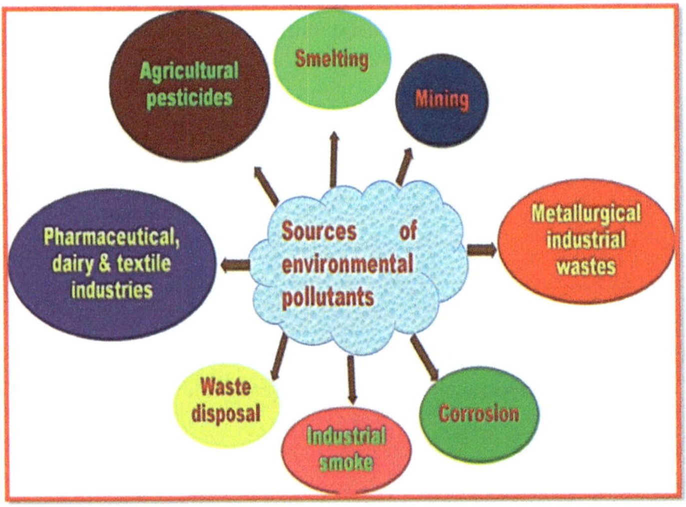

Fig. 2 Different sources for pollutants (Reproduced with permission from Elsevier [6])

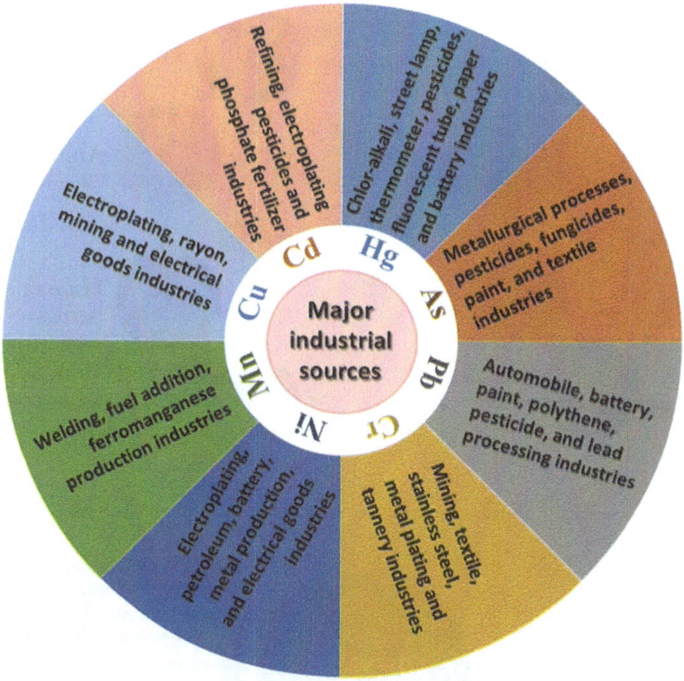

Fig. 3 Major sources for generation of toxic heavy ions (Reproduced with permission from Elsevier [1])

When pollutants are present in water, they are very injurious. Different type of pollutants and their effects are shown in Fig. 4 [7].

These pollutants if increase beyond certain limit becomes harmful to living system and environment. Therefore, these pollutants are to be removed in an environmentally friendly and economical ways [8–10]. There are number of methods used for the removal of pollutants from water and are given in Fig. 5. Out of all, adsorption technique is one of the most convenient techniques for removal of pollutants from water. For this purpose, a suitable adsorbent is needed. Amongst different adsorbents, nanoadsorbents have been reported to be the most effective adsorbents. Some of the adsorbents are given in Fig. 6 [11].

In recent years, it is reported that metal oxide NP and many other NMs are found to be as an effective adsorbent and photocatalyst for the removal of pollutants from water (Fig. 7)[12].

In this chapter removal of pollutants from water using different nano metal oxides as adsorbents a photocatalyst have been discussed.

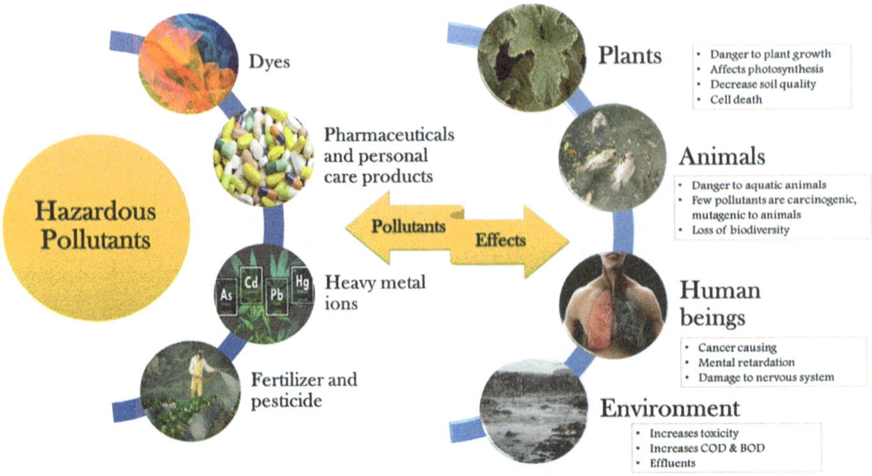

Fig. 4 Pollutants and their impact (Reproduced with permission from Elsevier [7])

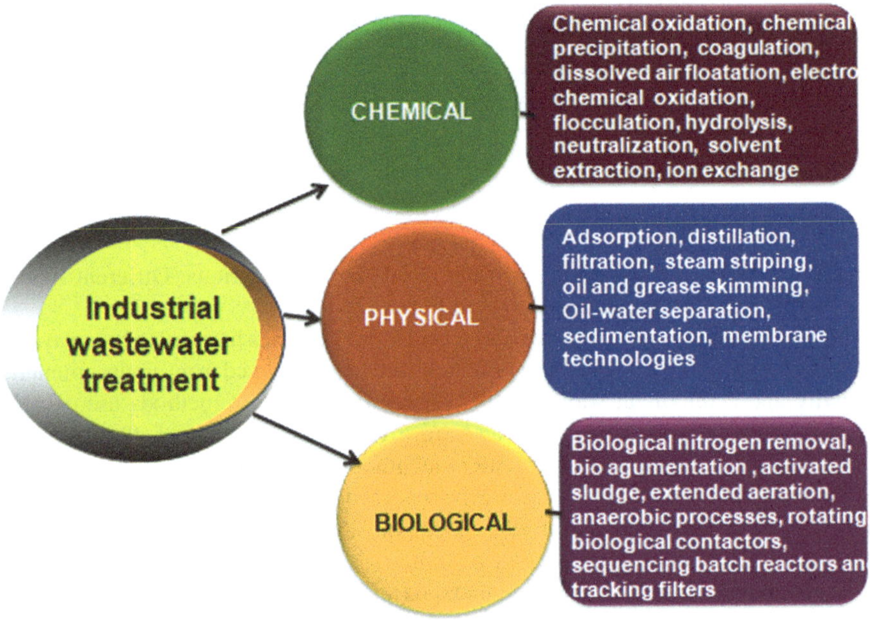

Fig. 5 Water purification methods

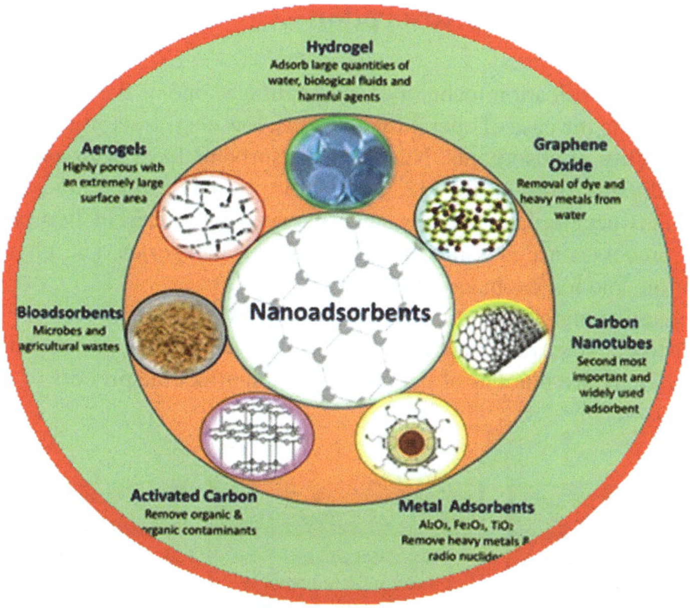

Fig. 6 Nanoadsorbents (Reproduced with permission from Elsevier [11])

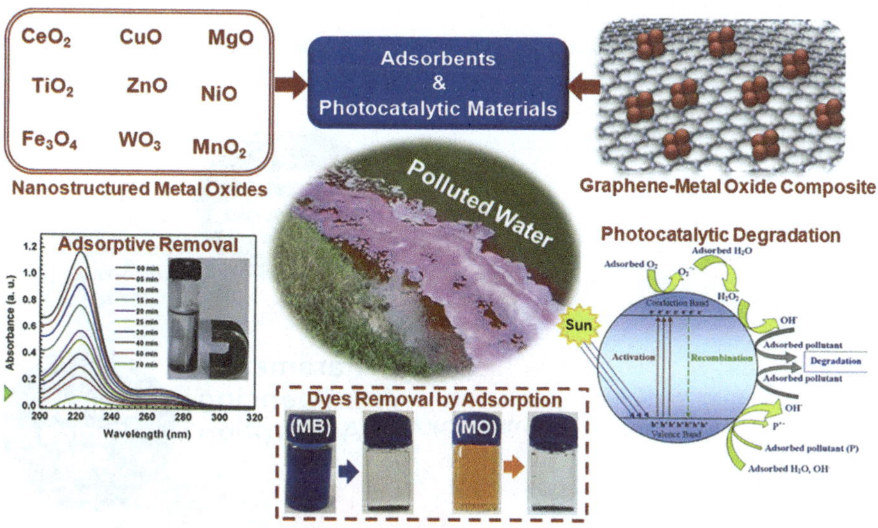

Fig. 7 Adsorption of pollutants by different type of NMs (Reproduced with permission from Elsevier [12])

2 Decontamination by Adsorption

Out of different purification techniques, adsorption is one of the most useful technique, considering the ease of operational aspects, low cost, scalability, high efficacy, and regenerability of adsorbents. Number of adsorbents have been used to remove different pollutants from water [12, 13]. Adsorption is a mass transfer process, which includes accumulating a substance between interfaces of two phases. The adsorption processes are classified into three major categories, i.e., physisorption, chemisorption, and ion exchange (Fig. 8) [3].

The qualitative and quantitative aspects of the adsorption process are evaluated by using different adsorption isotherm models, kinetic models, and thermodynamic parameters. There are number of parameters which affect the process of adsorption (Fig. 9).

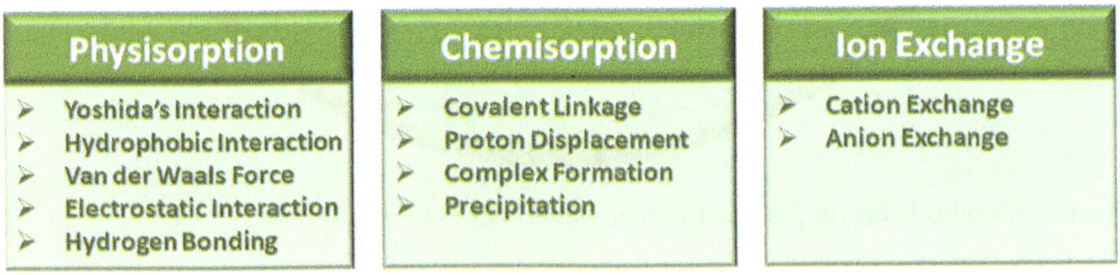

Fig. 8 Different type of adsorption (Reproduced with permission from Elsevier [3])

Fig. 9 Parameters affecting adsorption

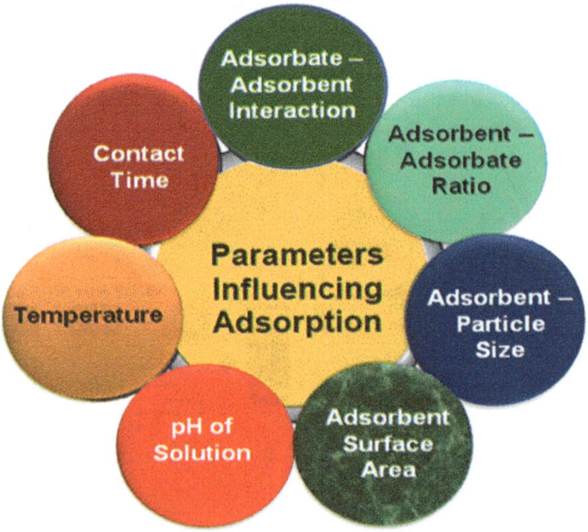

3 Photocatalytic Degradation of Pollutants

Prof. Fujishima of Tokyo University, Japan, accidently in 1967, discovered that in presence of TiO_2, water splitted evolving oxygen. This phenomenon was named as photocatalysis. After that various effects of photocalysis was studied with industrial applications. Number of semiconductors with nanodimensions are being used as photocatalysts. Out of all, transition metal oxides are found to be most effective photocatalysts. Many nano metal oxides have been used as photo catalysts for degradation of dyes and organic compounds contaminated with water. Metal oxide NMs are semiconductors with valence band and conduction band separated by band energy of about 3.0 eV and act as photocatalyst (PC). When light of appropriate wavelength and energy hv falls, electron from the valence band jump to conduction band leaving a positive hole (h_{VB}^+) and trapped electron in the conduction band, as given by Eq. (1) [14]

$$PC + hv \rightarrow e_{CB}^- + h_{VB}^+ \quad (1)$$

The electrons liberated through irradiation could be trapped by O_2 absorbed on the surface of the catalyst and give superoxide radicals ($\boldsymbol{O_2^-}$):

$$e_{CB}^- + O_2 \rightarrow O_2^- \quad (2)$$

$O_2^{\cdot -}$ obtained reacts with H_2O to form hydroperoxy ($HO_2^{\cdot}$) and hydroxyl radicals ($OH^{\cdot}$). These radicals are strong oxidizing agents and decompose the organic molecule, dyes and other type of organic contaminants in water as given by Eq. (3).

$$O_2^- + H_2O \rightarrow HO_2 + OH \quad (3)$$

Simultaneously, the photoinduced holes could be trapped by surface hydroxyl groups (or H_2O) on the photocatalyst surface to give hydroxyl radicals ($OH^{\cdot}$):

$$\begin{aligned} h_{VB}^+ + OH^- &\rightarrow OH_{ad} \\ h_{VB}^+ + H_2O &\rightarrow OH + H^+ \end{aligned} \quad (4)$$

Finally the organic molecules will be oxidized to yield carbon dioxide and water as follows:

$$OH + \text{organic molecules} + O_2 \rightarrow \text{products } (CO_2 \text{and } H_2O) \quad (5)$$

Meanwhile, recombination of positive hole and electron could take place which could reduce the photocatalytic activity of prepared nanocatalyst:

$$e_{CB}^- + h_{VB}^+ \rightarrow PC \quad (6)$$

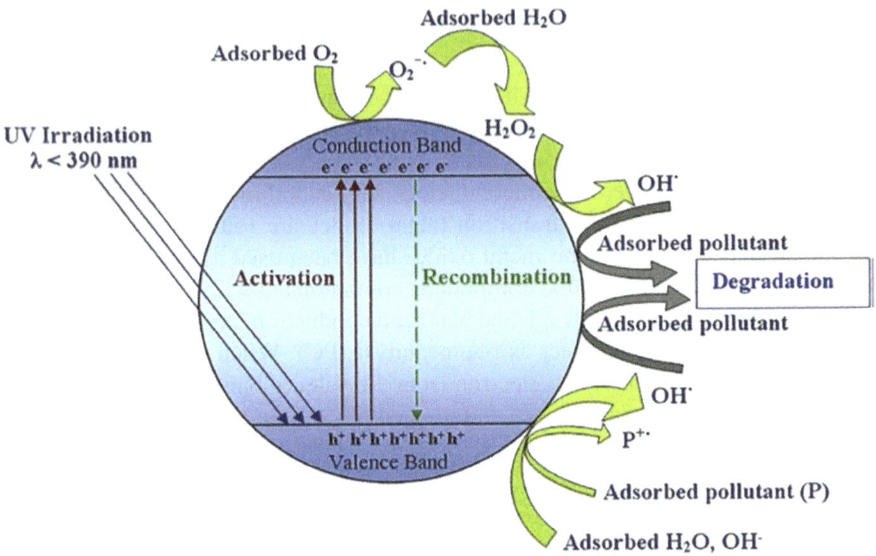

Fig. 10 General mechanism of photodegradation of pollutants by semiconducting metal oxide NM (Reproduced with permission from Elsevier [12])

A wide range of metal oxides such as *tungsten oxides*, copper oxides titanium oxides, zinc oxides, *iron oxides*, magnesium oxide, metal oxides composites, and graphene-metal oxides composites have been studied for photocatalytic degradation and adsorptive removal of organic pollutants viz. *phenolic compounds*, dyes, pesticides, and so on. Figure 10 represents the photocatalytic degradation of organic pollutants, along with the role of the photogenerated hole and electron pairs [12].

Photocatalytic degradation of some organic pollutants in water by titanium dioxides-based NMs and their composites are given in Table 1 [12].

In general photocatalytic degradation is fast and ecofriendly as compared to adsorption technique.

4 Synthesis of Metal Oxide NP

The purpose of synthesizing metal oxide NP is to change the properties of corresponding metal NP. For example oxidizing iron NP are converted into iron oxide NP, which increase the reactivity. In the past ten years, effective synthesis methods to obtain metal/metal oxide NMs with controllable shape, high stability and monodispersion have been extensively studied [15]. In general metal oxide NPs can be synthesized by using Chemical Precipitation, Sol –Gel, Hydrothermal, Chemical Vapour Deposition methods. However, synthesis is divided into three major categories to

Table 1 Titanium dioxide *NM* and their composites as photocatalyst for degradation of *organic pollutants* from *wastewater* (Reproduced with permission from Elsevier [12])

No.	Photocatalyst	Light source	Targeted pollutant	Degradation efficiency (%)
1	Surface-fluorinated TiO_2 assembled on GO	UV	Methylene blue	96
2	TiO_2 NP	UV	Malachite green	99.9
3	Phosphorous acid-modified Degussa P25 TiO_2	Visible	Rhodamine B	~100
4	Glass coated TiO_2 thin films	UV–VIS	Methylene blue	90.3
5	TiO_2/Fe_2O_3 nanocomposite	Visible	Diazinon	95.1
6	Zn^{2+}-doped TiO_2 NP	UV	Malathion	98
7	In and S co-doped TiO_2@rGO	Visible	Atrazine	99.5

understand the difference in methodology, advantages/disadvantages: (i) solution-based methods (ii) gas phase methods and (iii) biological methods (Fig. 11). This classification is based on the type of medium in which the oxidation reaction occurs. The physical and chemical properties of NMs like size, shape, dispersibility, morphology, internal/external defects and crystal structure are generally influenced by the choice of synthesis method which ultimately affects their applications. For example, nano Mg doped ZnO (ZnMgO) fabricated via three different synthesis methods were found different in geometry. Where regular cubic structure was obtained by CVS method, mixture of cubes and tetrapods for metal combustion method and irregular nano rods

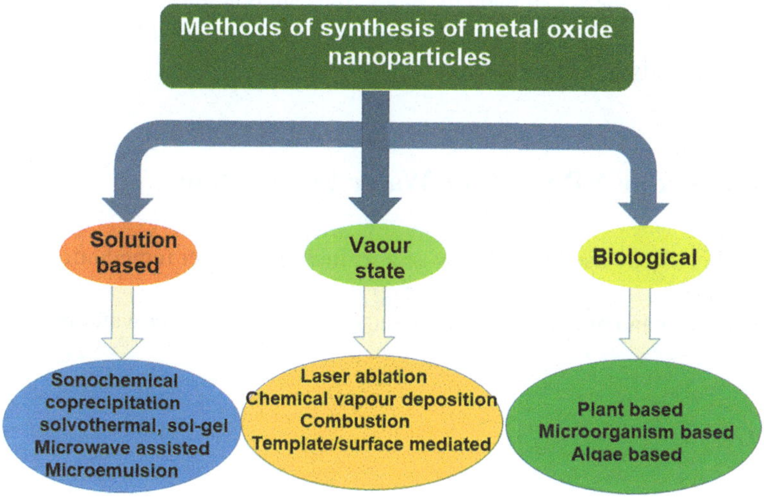

Fig. 11 Synthetic methods of metal oxide NMs

by sol-gel method [16, 17]. To understand the difference in various methods they are briefly discussed in Table 2 [18].

Due to the vast and varying applications of these nanostructures, various synthetic methods have been utilized to synthesize them as discussed in the Table 2. All the described methods provide high quality metal oxide nanocrystals with definite size and shape except the biological method. It is very difficult to control all the required features of NMs like size, shape yield, purity, cost etc. in most of the methods. This problem is more common in the case of multi-metallic oxide NPs. The most effective method with respect to high crystal purity is chemical vapour deposition method [19]. This method is also very useful to give stability to otherwise unstable crystal phase. For example, Zinc oxide NPs in cubic crystal form can only be achieved at very high pressure but chemical vapour synthesis method allows c-ZnO NPs to be dispersed on MgO surface [20].

On the other hand, sonochemical method has been successfully applied to get enhanced photocatalytic performance of TiO_2 NPs [21] and varying magnetic properties of iron oxide NPs [18, 21]. The sol-gel method has been widely used for synthesizing almost all kind of metal oxide NPs. This method is also very useful in doping group 5 oxides, which is often a challenge, for example Co doped Hf-oxide NPs [22]. This method has been utilized by researchers with certain modifications, for example Corr et al. have reported an improved one-step sol-gel aqueous method for the synthesis of iron oxide-silica NPs [23]. To avoid the oxidation of the products at very high temperature, use of ultrasonic conditions is also reported. Some solution-based manufacturing techniques use surfactants [24], which, in addition to affecting particle size, also tend to reduce the degree of aggregation between particles. In addition, solution-based technology combats pollution problems in the resulting metal oxide products. Most of the solution-based methods suffer with the problem of the contamination in the products specially contamination of anions of precursor salt [25]. Biological method is suitable for biomedical applications due to its biocompatibility but face the problem of contamination and composition of NPs also cannot be defined completely [18].

5 Metal Oxide NP Used for Water Purification

Metal oxide NPs are used in different sectors including water remediation as shown in Fig. 12 [26].

Numbers of nano metal oxides discussed below were used for water remediation.

5.1 TiO_2

TiO_2 NPs have become the most widely used NMs for water remediation due to their high photosensitivity, availability, non-toxicity, cost-effectiveness and environmental

Table 2 Summary of most common synthetic methods with their advantages/disadvantages

Category	Method	Specific Reaction conditions	Advantages	Disadvantages	Example
Solution based methods					
Sonochemical method	The solution of metal salt is subjected to a strong ultrasonic vibration flow, which breaks the chemical bond of the compound, resulting in alternating compression and relaxation. which leads to acoustic cavitation. It results in formation, growth, and implosion of bubbles in the liquid	Too high cooling rate will affect the formation and crystallization of the resulting product	Uniform particle size with higher surface area, faster reaction time and improved phase purity of Nps	It requires high energies and pressures in short time	ZnO, MoO_3, CeO_2 V_2O_5, In_2O_3, TiO_2, In_2O_3, $PbWO_4$, $BiPO_4$,$ZnAl_2O_4$, $ZnFe_2O_4$, Eu/Dy-doped
Coprecipitation	Use of a precipitation medium to precipitate out the oxyhydroxide form from a solution of a salt precursor (metal salt) in a solvent (such as water)	pH of medium, nature of alkali andspeed of its addition and the drying modality of synthesized powders affect the size, magnetic properties and degree of agglomeration of the synthesized NMs	Low cost, mild reaction conditions, possibility to perform direct synthesis in water, ease of scale-up, flexibility in modulation of core and surface properties, possibility of using surfactants to improve the surface properties	Low removal efficiency, high energy consumption, and production of toxic sludge	ZnO, $BiVO_4$, MnO_2,MgO,SnO_2, Ni1-xZnxFe_2O_4, $MgFe_2O_4$, Ni–$CeZrO_2$ and Y_2O_3:Eu^{+3}, Cu-doped ZnO,

(continued)

Table 2 (continued)

Category	Method	Specific Reaction conditions	Advantages	Disadvantages	Example
Solvothermal	By dispersing the solution of starting material in a suitable solvent and exposed it to high temperature and pressure conditions	Composition and concentration of the reactants, ratio of solvent/ reducing agent and various thermodynamic parameters like temperature, pressure and reaction time affect the final particle formation	Nanostructures with different morphologies—such as, nanocubes could be produced and use of surfactants to avoid agglomeration is also possible	Impossibility of observing the reaction process, need for expensive autoclaves	TiO_2, $MnFe_2O_4$, Nb_2O_5, MgO, $CoFe_2O_4$, Fe_3O_4, Ag-Fe_3O_4, Graphene-TiO_2, $CoFe_2O_4$@$BaTiO_3$, Fe_3O_4@NiO and Fe_3O_4@Co_3O_4)
Sol–gel method	Include the hydrolysis of precursor organometallic compound, like alcoxysilane to produce corresponding oxohydroxide, followed by condensation to form a network of the metal hydroxide	Super critical liquids can also be used. it provides the particle size to be tuned by simply varying the gelation time	More bioactive, high quality material, promising in doping of Group 5 oxides, clean to synthesize, excellent control over the texture and surface properties of the materials		ZnO, TiO_2, CuO, MgO, ZrO_2 and Nb_2O_5 Cu-ZnO, Ce-doped ZrO_2 CuO/Cu_2O, oxides of Hf, Ta and Nb, $LiCoO_2$ thin film,

(continued)

Table 2 (continued)

Category	Method	Specific Reaction conditions	Advantages	Disadvantages	Example
Microwave assisted method	It involvesuniform and quick heating of the reaction medium with no temperature gradients using two mechanisms: dipolar polarization and ionic conduction	High heating rates favor rapid nucleation and formation of small and highly monodisperse particles	Very short reaction time without compromising the particle purity or size	Unsuitable to scale up and reaction monitoring is not feasible	CeO_2, Cr_2O_3, ZnO MnO, Fe_3O_4, Mn_2O_3 CaO, CoO, and MgO, $BaTiO_3$,
Microemulsion method	Two immiscible phases (oil and water) interfaced by surfactant molecules form two binary systems. At CMC, reversed micelles are formed. This core act as a nanorecator where variety of chemical reactions take place	Reducing agent/oxidizing agent/precipitating agent is added under vigorous stirring to precipitate the NP	The morphological properties of NMs can be modified by affecting the various self-assembled structures formed in the binary systems	Controlling the deposit parameters is difficult to achieve	ZnO,Fe_2O_3, NiO, TiO_2, CeO_2, CuO, and NPs like $BaAlO_2$, alumina NP doped with iron-oxide

(continued)

Table 2 (continued)

Category	Method	Specific Reaction conditions	Advantages	Disadvantages	Example
Vapour state method					
Laser ablation method	Irradiation of different metal salts submerged in solution by high power laser beam, condense a plasma to produce NP	By increasing the laser fluence and changing the liquid media, the thickness of molten layer can be changed which affect the size of NMs		Chances of nanoparticle agglomeration leads tolack of long-term stabilization	ZnO, Al_2O_3 NiO, ZrO_2, SnO_2, iron-oxide, Au-SnO_2, Cu/Cu_2O
Chemical vapor-based method	Precursor materials in the gaseous state is deposited onto a heated substrate surface at which it reacts or decompose to form NMs	Choice of nature and concentration of the oxidizing agent affect the nucleation process and therefore affects the average size of the NMs	Uniform and contamination-free metal oxide NP and films with flexibility in product composition can be produced	Volatility of the reactants and degree of agglomeration use of toxic, corrosive, flammable, and/or explosive precursor gases	Defect free ZnO nanowires, films, Doped-ZnO, nano cubes and nanospheres of magnetite, SrO, SnO_2, Cu_2O, MgO, CoO and Co_3O_4, CaO.,Eu doped YO, Li-doped MgO
Combustion method	Pure metallic precursor is heated to get it evaporated into a background gas then mixed with second reactant i.e., oxidizing agent	Partial pressure of oxidizing agent can be controlled which determines the nucleation growth and affect the particle size to some extent	Control over phases and morphologies of NMs, successful commercially	Complex process that is rather difficult to control	ZnO, CuO, FeO, Mn_2O_3, CdO, MgO and Co_3O_4 or Ag-MgO composite, Co_3O_4@CuO, La0.82Sr0.18MnO_3

(continued)

Table 2 (continued)

Category	Method	Specific Reaction conditions	Advantages	Disadvantages	Example
Template/surface-mediated synthesis	Fabrication of the desired NMI within the pores or channels of a nonporous template depending on the properties of the template		Can be used to synthesize self-assembly systems with tubular and fibrillary andhighly monodisperse nanostructures with small diameters, enhanced activities, uniform morphology with high surface area		Various morphologies of metal oxide NMI such as rods, tubules and fibrils
Biological method					
Plant based synthesis	By using Plant parts (Enzymatic reaction) and their extract as reducing and capping agents to convert metal salts to metal oxide NPs	It is somewhat possible to control size of particle by controlling the pH, temperature, solvent conditions	Consumes lesser energy, environment-friendly and eliminates the production of toxic wastes, NMs have higher catalytic reactivity, biocompatible, attractive for biomedical applications	Low yield, inability to obtain desired size and/or shape of NP, Slow and time taking	ZnO, CuO, Fe_3O_4, Ag_2O

(continued)

Table 2 (continued)

Category	Method	Specific Reaction conditions	Advantages	Disadvantages	Example
Microorganism based synthesis	By using Microorganisms like bacteria, fungi and molds as reducing and capping agents	By varying parameters like microorganism strain, culture growth medium, size and shape of NPs can be controlled			ZnO, TiO_2, MgO NP by using fungus
Algae based synthesis	Using algae as reducing agent	By varying the species type, the enzymatic reaction may change and affect the properties of NPs			ZnO, CuO, Fe_3O_4, Ag_2O

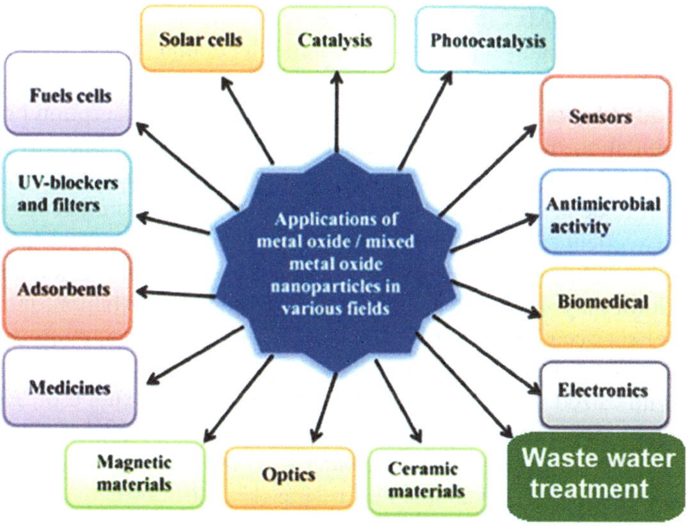

Fig. 12 Applications of nano metal oxides/mixed metal oxides in different areas (Reproduced with permission from Elsevier [26])

friendliness [27]. These NMs have been widely used in the oxidation and reduction conversion of organic and inorganic pollutants in air and water, such as phenolic compounds, metal ethylene diamine tetraacetate complexes, microorganisms in the air and odorous chemicals, halogenated compounds degradation, dye removal, metal and metal removal, etc. [28]. Photodegradation leads to complete oxidation and reduction of organic and inorganic pollutants and converts them into carbon dioxide, water and inorganic acids [29]. Its large band gap energy (3.2 eV) requires ultraviolet excitation to induce charge separation within the particles [30]. TiO_2 and TiO_2 films have been successfully used to degrade atrazine and organochlorine pesticides in water, respectively [27]. Photocatalytic degradation of methyl orange using ZnO/TiO_2 composites has been studied [31]. Non metal elements like N, F, C and S can improve the photocatalytic activity of TiO_2 NMs by narrowing its band gap. This is achieved by the substitution of oxygen by these nonmetals in the TiO_2 lattice [32]. Doping with transition metals like Fe, Co and Cu has also been proved to improve photocatalytic activity of TiO_2 NPs under UV irradiation However, noble metals like silver, have received much attention for this purpose [33].

5.2 *Iron Oxides*

In recent years, the synthesis of iron oxide NMs with modified properties and their applications have gained widespread attention due to their high porosity and surface-to-volume ratio, low cost, strong adsorption capacity, easy magnetic separation response. Iron oxide NMs can act as immobilized carrier to remove contaminants or can also act as photocatalyst/catalyst to degrade the contaminants. Magnetic separation is a unique property of iron oxide NMs which is a challenge due to small size of nanoadsorbents [34]. Therefore, the combination of adsorption process and magnetic separation has been widely used in water treatment and environmental purification. Strong paramagnetic characters of Fe_2O_3 NMs make them effective for the removal of toxic heavy metals like Cd(II), Pb(II) etc. Super magnetic Fe_3O_4 NPs have shown excellent catalytic activity for dye degradation in waste water to convert them in less toxic form. According to reports, the preparation method and surface coating medium play a key role in determining the size distribution, morphology, magnetic and surface chemistry of NMs in the form of NP, nanoellipses nanobelts and nanorings or other nanostructures [35].

Green synthesized iron oxide (Fe_3O_4) NP using an extract of *Excoecaria cochinchinensis* leaves were found much effective for the removal of a contaminant antibiotic (rifampicin) from aqueous media. This was found much more effective as compared to commercially available Fe_3O_4 (Fig. 13) [36].

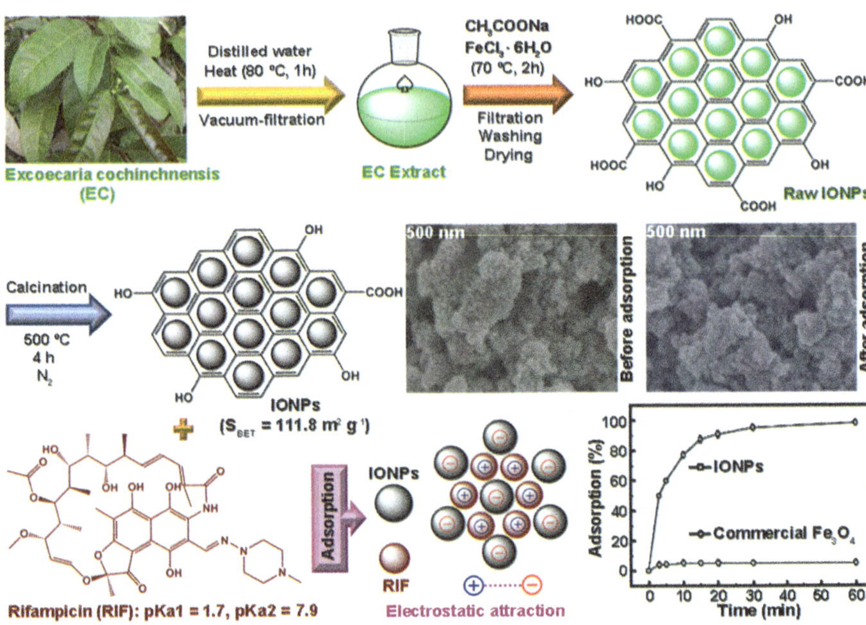

Fig. 13 Green synthesis of Fe_3O_4 and removal of rifampicin (Reproduced with permission from Elsevier [36])

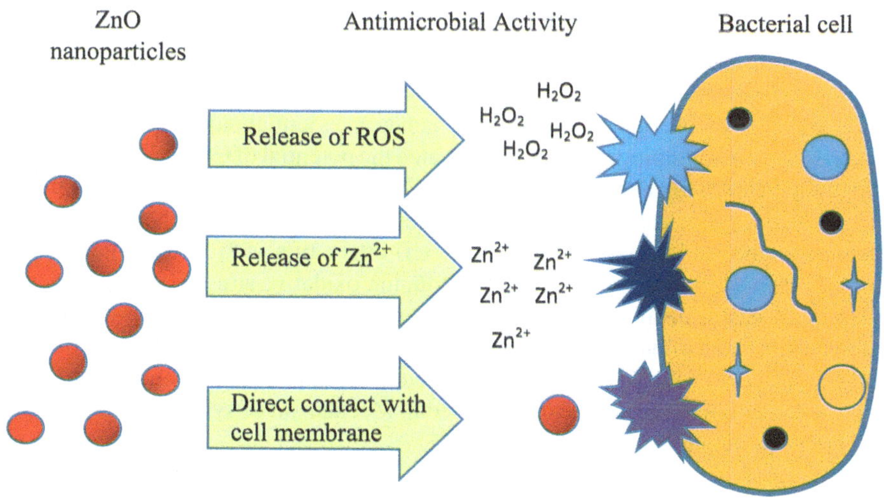

Fig. 14 ZnO disinfection mechanisms (Reproduced with permission from Elsevier [38])

5.3 *Zinc Oxide*

Zinc oxide is another metal oxide NMs based photo catalyst which shows most promising water treatment due to its high chemical stability and excellent photocatalytic activity. Large number of research groups across the globe have already reported potential applications of ZnO NMs along with their variable morphology and structural characteristics including Nano sheets, nanowires, Nano rods, nanoribbons and complex hybrid structures [37, 38]. ZnO has a wide band gap (3.37 eV), and the excitation binding energy (60 meV) is also large at room temperature which makes it an excellent photo catalyst. In addition, easy availability, low toxicity and antibacterial efficiency of ZnO NMs make them ideal for water treatment. Hollow spheres in these nanostructures are of great interest due to their light trapping efficiency and highly enhanced photocatalytic activity, as well as their high surface area, low density, and good surface permeability [37].

Nano Zno acts as an antibacterial agent and the mechanism of its action is given in Fig. 14 [38].

5.4 *Copper Oxide*

Due to its inherent compatibility, low-cost manufacturing and excellent electrochemical performance, research on copper oxide NMs has grown significantly in the recent years. It is reported that copper oxide NMs show a little photocatalytic activity which can be significantly enhanced by activating it with H_2O_2 [39]. More than a photo catalyst, CuO NMs are used as a good adsorbent of water pollutants such as congo reed, malachite green, methylene blue, ciprofloxacin, methyl orange dyes and many heavy metals like Pb (II), Hg(II) As (III). The main application of CuO NM in water treatment is its antimicrobial efficiency. Bacterial disinfections, one of the main applications of copper-based NM, which is an essential requirement of potable water. Many biotechnologists have investigated the potential of CuO NMs to disinfect water with respect to microorganisms [9]. Scientific reports suggest few mechanisms behind it (i) Cu is released from copper oxide NPs which damages the bacterial cell membrane and lead to bacterial cell death (ii) interaction with DNA molecule and disorder its helical structure (iii) and by inducing oxidative stress [18].

In many cases metal oxide composites were found more effective in removal of pollutants. Extract of pine needle was used for the synthesis of nano composite of iron and copper oxides (Fe/Cu oxides) and was found an efficient adsorbent for ofloxacin and norfloxacin removal from aqueous media. Mechanism of synthesis and removal of organic pollutant is given in Fig. 15 [36].

5.5 Silver Oxide

Silver oxide exists in many nanostructural forms which includes, NPs, nanohorns, nanorods, and nanopyramids. Silver oxide NM exhibits excellent antibacterial properties which has been already used in many commercial products [40]. A few research groups have reported the photocatalytic activity of these NMs for the degradation of dyes like methylene blue and methyl orange present in water along with their antimicrobial activity [22]. Silver oxide NMs are syntheisezd be various synthetic routes which includes direct precipitation, sol gel, hydrothermal and biological route [41, 42]. These are mostly spherical particles (20–80 nm) with high surface area (10–50 m^2/g) and good magnetic properties.

5.6 Manganese Oxide

Different forms of manganese oxide NP such as MnO, MnO_2, Mn_2O_3, Mn_3O_4 are tested for removel of heavy metals in water decontamination process. They are structurally flexible and display novel physical and chemical properties. The primary benefits of managanese oxide NMs come from their low cost, high activity and non-toxic nature. A large number of heavy metals including Cu(II), Cd(II), Pb(II), As(III), As(V), U(VI) and organic contaminants are successfully removed by MnO_2 and its NPs. The mutual interference of Zn(II), Cd(II) and Pb(II) ions with various nanostructures of MnO_2 e.g. nanoparticle, nanotube and nanobowl are investigated by Zhang et al. [43]. Nanoflakes of MnO_2 are reported for the detection and removal of Cr(III) ion [44]. Differential pulse voltammetric method was used for the detection of Hg(II) ion by MnO_2 nanotubes [45].

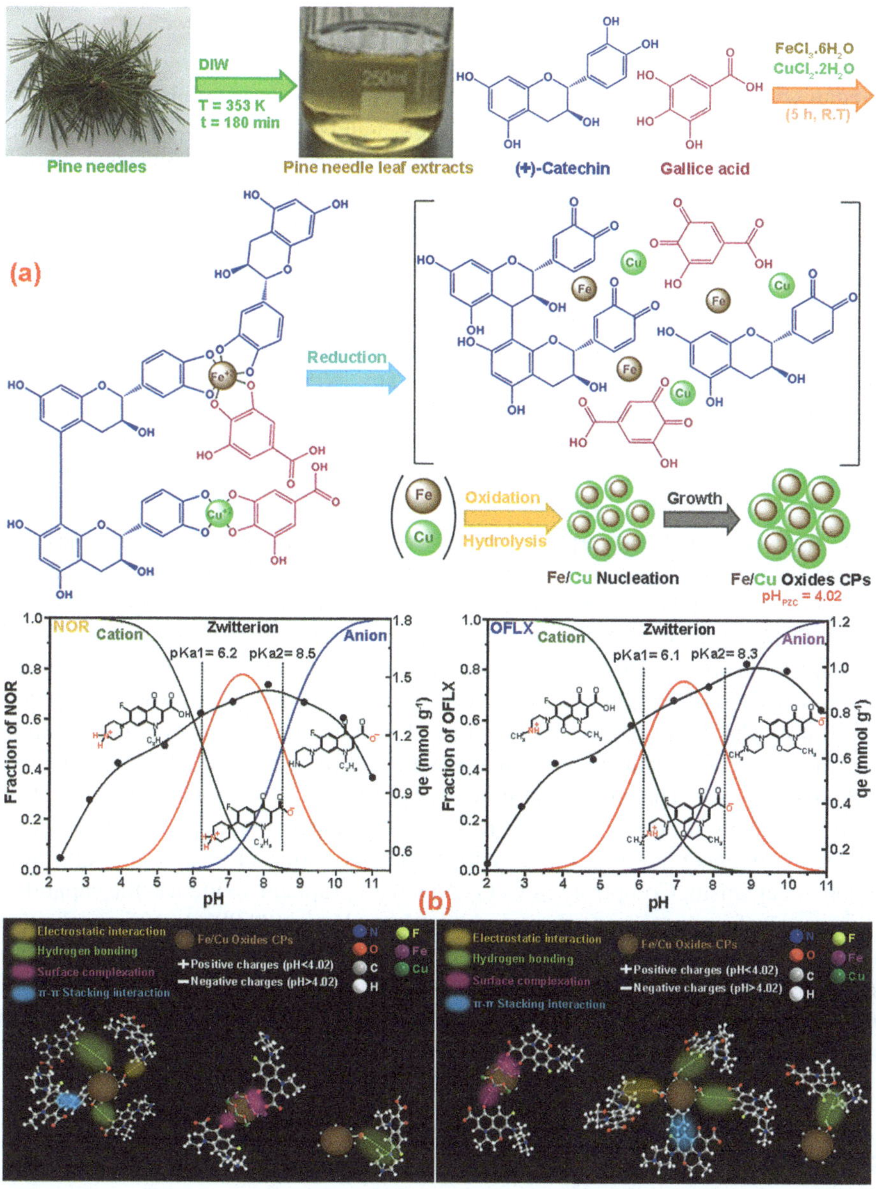

Fig. 15 Synthesis of Fe/Cu oxide (Reproduced with permission from Elsevier [36])

5.7 Cerium Oxide

Cerium oxide (CeO_2), a non-toxic rare earth metal oxide is gaining attention for application as UV-blocking agent, sensing agent and in water remediation. Recently nanoscale CeO_2 is investigated for their applications in removal of heavy metals from water [46]. The properties of nanocrystalline CeO_2 are found to be effective for removal of inorganic heavy metals. Low ionic potential and high basicity leads to dissociation of hydroxy group into hydroxyl ions. The size, porosity, surface area, bulk density etc. are in favor of their selectivity, stability and activity during adsorption process. Recillas et al. reported removal of Cr(VI) metal ions using 12 nm average sized CeO_2 NPs [47]. Their results indicate that low concentration of Cr(VI) (80 mg/L) can be effectively removed by CeO_2 NPs with maximum adsorption capacity of 121.95 mg/g. Arsenic metal in the form of As(III) and As(V) has successfully been removed from water by CeO_2 NPs by Mishra et al. [48]. In their work, the BET surface area of 3–5 nm sized CeO_2 NPs was 257 m^2/g and the adsorption capacity of As(III) and As(V) ions were 71.9 and 36.8 mg/g^{-1} respectively. It is observed that the adsorption capacity of CeO_2 NPs reduced in the presence of some anions such as sulpahte, bicarbonate, dihydrogen phosphate etc. Further, CeO_2 NPs are found to be compatible with other metal oxides for treatment of heavy metals from water [49, 50]. Recently, Meepho et al. have synthesized samaria doped cerium nanopowder (SDC) by doping samaria with different morphologies of cerium nanopowder [51]. The samaria doped cerium nanopowder (SDC) was used for the removal of Cu(II), Zn(II) and Pb(II) ions. The outcome of the investigation indicates that spherically shaped SDC nanopowder was more effective than the plate like SDC nanopowder. The surface modifications of CeO_2 NPs enhance the adsorption of heavy metals in terms of material stability and selectivity. The hydrous CeO_2 NPs with adequate hydoxyl group help in the adsorption of arsenic through inner sphere mechanism. Composite of CeO_2 NPs with graphene oxide has the capability of removal of arsenate and arsenite almost completely (99.99%). A cost-effective adsorbent is developed by supporting CeO_2 NP over carbon nanotube (CNT) for removing AS(V) ions [52]. The only drawback of ceria in water remediation is its high cost which can be taken care by the surface modification or composite formation of ceria.

5.8 Magnesium Oxide

Magnesium oxide NP have high potential in removing pollutants from water. MgO NP are associated with exceptionally high absorption ability, abundantly available, non toxic and inexpensive [53]. These unique properties make it one of the sought-after metal oxides NMs for removing heavy metals from water. MgO NP also displays superb antibacterial activity for both gram-positive bacteria, gram-negative bacteria and spore cells [54]. Reported literature also indicated the effect of size of MgO NP in

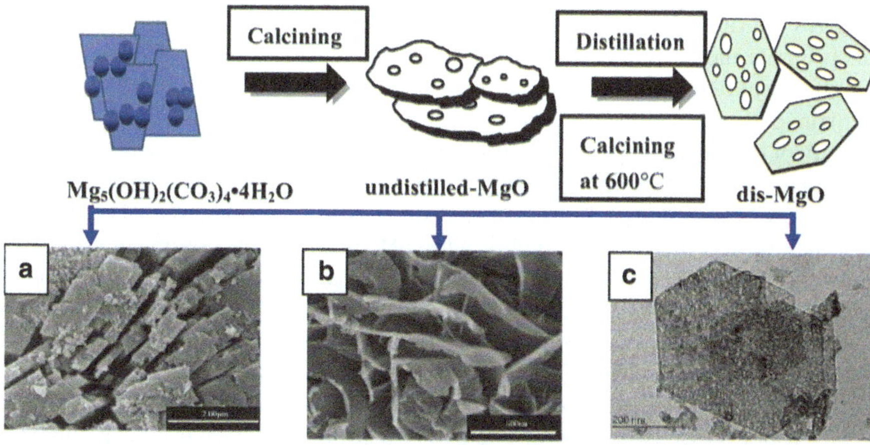

Fig. 16 Schematic illustration for the formation of mesoporous MgO nanosheets. (Reproduced with permission from Elsevier [58])

its bactericidal properties. Cai et al. reported simultaneous removal of heavy metals Cd(II) and Pb(II) and *Escherichia coli* bacteria by MgO NP [53]. Three different nano metal oxide e.g. TiO_2, MgO and Al_2O_3 was investigated for elimination of heavy metals Cd(II), Cu(II), Ni(II) and Pb(II) ions form water [55]. It was observed that the efficiency of MgO NP was better than the other two metal oxide NP. MgO NP follows adsorption and precipitation mechanism for the removal of heavy metals while TiO_2 and Al_2O_3 were via adsorption mechanism only. Interesting work by Madzokere et al. revealed that MgO NP are capable of removing 96% Cu(II) ion compared to the 15% removal ability of commercial MgO [56]. A batch adsorption experiment performed by Xiong et al. indicated excellent adsorption capacity of MgO NP [57]. Langmuir model was used by Jing et al. to establish the remarkably high adsorption of Ni(II) ion over mesoporous MgO nanosheets (Fig. 16) [58]. All these works suggest MgO NP as a very promising material for the removal of heavy metals from water.

5.9 *Zirconium Oxide*

Among metal oxide NMs, zironia or zirconium oxides also exhibited remarkable potential in removing water pollutants specially the heavy metal ions [59, 60]. They have high thermal and chemical stability, less toxicity and biocompatibility. Zirconia display high resistivity against acids and alkalis. Presence of large number of -OH groups over the surface leads to high surface area which in turn makes zirconia a good adsorbent. Both nanoscale zirconia and hydrous zirconia are excellent for removing heavy metals like Cd(II), Zn(II), Pb(II), arsenate and arsenite ions. Silicate ions adsorb strongly over zirconia surface, thus hampering the adsorption of arsenic ions using zirconia adsorbent. The presence of alkaline earth metals e.g. Ca(II) and Mg(II) ions promote the adsorption of arsenic pollutants by reacting with the silicate ions [61]. It is reported that simultaneous adsorption of arsenate and arsenite is possible over nanocomposite of hydrated zirconia-graphene oxide sheet [62]. The adsorption capacity of this nanocomposite was higher compared to the pristine nano zirconia. In addition to that it is recyclable up to five times. Removal of Cd(II) was investigated by a composite of polystyrene supported nanosized hydrous zirconia [63]. The removal efficiency of this material lies within wide pH range. Further promising result of removal of Pb(II) and Cd(II) ions are observed by nanocomposite based on hydrous zirconium (IV) oxide [64].

Another composite of zirconia with γ-Fe_2O_3 is investigated for the removal of arsenic from leach out water of gold cyanidation industry [65]. The iron oxide core helps in improving the recyclability of the adsorbent by easy separation. To improve the adsorption capacity towards arsenate ions at strong acidic environment, zirconia is encapsulated in D201 (polystyrene anion exchanger [66]. The electrostatic interaction between arsenate ions and D201 and inner sphere complexation explain the mechanism of the adsorption. Presence of sulphate ions restricts the electrostatic interaction and in turn reduces the adsorption capacity.

An interesting report on removal of Cr(VI) by a series of mesoporous transition metal oxides suggests that ZrO_2 as the most attractive adsorbent among other nano metal oxides e.g. TiO_2, HfO_2 and NbO_2 [67]. A hybrid nanocomposite made from ZrO_2/B_2O_3 displayed satisfactory results in removal of Cu(II), Cd(II) and Cu(II) ion [68].

5.9.1 Aluminium Oxide

Aluminium oxide NMs are inexpensive and can be prepared easily. Alumina adsorbents have high efficiency in removing heavy metal ions [69]. Many research works is reported on the application of alumina for the adsorption of several heavy metals. Among several crystalline forms of aluminium oxide, γ-alumina is the most effective for decontamination purpose because of its high surface area [70]. In addition to that γ-alumina has high mechanical strength, excellent thermal stability and high adsorption capacity. Tabesh et al. has reported 97% and 87% removal of Pb(II) and Cd(II) ion respectively by γ-alumina NP [71]. It is observed that adsorption of Zn(II) and Cd(II) ions by alumina become more enhanced in presence of phosphate ions and humic acid while presence of citrate ion reduces the adsorption of Zn(II) ion [72]. Moreover heavy metals ions such as As(III), Hg(II), Ni(II), Cu(II), Cr(VI) are also reported to be removed by alumina NP [73–75].

Applications of some selected nano metal oxides in removal of pathogens, dyes and heavy metals are listed in Tables 3, 4 and 5 respectively.

Table 3 Nanometal oxide in pathogen removal for water purification

Nano metal oxide	Targeted pathogen	Mechanism	References
TiO_2	Escherichia coli and human pathogens	Antibacterial activity	[76]
Ag_2O	Bacillus subtilis, Staphylococcus aureus, Psedomonas aeruginosa, Esherichia coli, Canadida albicans and Aspergillus niger	//	[77]
	Pseudomonas aeruginosa, Staphylococcus aureus, Bacillus subtilis, Escherichia coli	//	[78]
	Streptococcus mutans and Lactobacilli sp.	//	[79]
CuO	Staphylococcus aureus, Escherichia coli, Bacillus licheniformis and Pseudomonas aeruginosa	//	[80]
	Vibrio anguillarum, Proteus mirabilis, Bacillus cereus, Edwardsiella tarda, Staphylococcus aureus, Aeromonas hydrophila, and Aeromonas caviae	//	[81]
	Staphylococcus aureus and Escherichia coli	//	[82]
	Escherichia coli and Salmonella typhimurium	//	[83]
ZnO	Escherichia coli and Enterococcus faecalis	//	[84]
	Escherichia coli and Salmonella typhimurium	//	[85]
	Enterobacter aerogenes and Bacillus subtilis	//	[86]
	Escherichia coli and Bacillus subtilis	//	[87]

6 Challenges

The metal oxide NPs are extensively studied for their application in water purification technology. But the validation and development of nanotechnology for purification of water at mass scale is full of challenges. The toxicity of the nanometal oxides is of primary concern. When nano metal oxides are used for water purification, consumers are exposed to the toxicity of these nano materials. Numerous research works have been performed on toxicity analysis of these materials both in vitro and in vivo [115, 116]. Various factors control the level of toxicity of these engineered NMs. Size of the NMs, dose, administration mode and exposure duration are important factors that controls the toxicity levels. It is reported that large TiO_2 NP with size more than 100 nm are non toxic in nature. Concentration of TiO_2 nanoparticle in the range of 1000–2000 $\mu g/g$ is found to be toxic [116]. The health issues from TiO_2 NP primarily

Table 4 Nanometal oxide in dye removal for water purification

Nano metal oxide	Targeted dye	Mechanism	References
CuO	Methylene blue	Photocatalytic degradation	[88]
	Basic red 14, Basic violet 16	Adsorption	[89]
	Melachite green oxalate, Methyl orange	Adsorption	[90]
TiO_2	Methylene blue, Rhodamine B	Photocatalytic activity	[91]
	Methyl orange	Photocatalytic activity	[92]
Ag_2O	malachite green	Adsorption	[93]
	Methyl orange	Photocatalytic activity	[94]
	Methylene blue	//	[95]
	AO8	//	[96]
ZnO	Methylene blue	Photocatalytic degradation	[85]
	Azo dye	Adsorption	[97]
	Reactive blue 19, Acid Black 210	Adsorption	[98]

Table 5 Nanometal oxide in heavy metal removal for water purification

Nano metal oxide	Targeted heavy metal	Mechanism	References
TiO_2	Cu, Zn, Pb, Cd, Ni	Adsorption	[99]
	Zn(II),Sr(II)	//	[100]
	Cr(VI), Cr(III)	//	[101]
CuO	As(V)	//	[102]
	Pb(II)	//	[103]
ZnO	Cr(VI)	//	[104]
	Pb(II)	//	[105]
	As(III)	//	[106]
	As(V)	//	[107]
ZrO2	As(III), As(V)	//	[108]
Al2O3	As(V)	//	[109]
MgO	As(III)	//	[110]
MgO	As(V)	//	[111]
$MnFe_2O_4$	As(III), As(V)	//	[112]
Ce-Mn mixed oxide	As(III)	//	[113]
Ce-Fe Mixed oxide	As(V)	//	[114]

come from inhalation not from ingestion with water. Thus toxicity of TiO_2 is not a serious concern. Oral administration of high dose (2.5 mg/g body weight) of ZnO NP is known to be accumulated in different body parts e.g. lung, kidney, liver and spleen. A detailed in vitro toxicity study on ZnO NM is reported by Vandebriel and Jong [117]. The toxicity of silver oxide nanoparticle is found to be more compared to other nano metal oxides. Silver can interact with most of the biomolecules and impart toxicity which in turn leads to cellular apoptosis [115, 118]. Magnetic iron oxide nano particles used for purification of water has insignificant toxic effect and are not serious issue [119]. Thus, technological advancement on nanometal oxide purification system is possible after addressing the toxicity issues convincingly.

Next the economic viability is another challenge that needs to be sorted out. To make the nano metal oxide based water purification technology acceptable it must be affordable. In this regard development of highly effective filtration membrane with multifunctional capabilities is extremely necessary to reduce the cost of the membrane-based purification technology.

In addition to the above issues, the aggregation and dispersion properties of nanometal oxides make the operational conditions critical. Mixing of nano metal oxides in water gets accumulated and forms aggregate. The surface immobilization of the nano metal oxides is used for killing various water borne microbes and pathogens. However, leaching of NMs beyond their acceptable limit is a serious threat for human and other living beings. Report of aggregation of TiO_2 NPs as waste from industry and consumer products in water is well documented [120]. One important strategy to reduce the leaching of nano metal oxide in water is to sediment or coagulates the NPs before supply to the consumers. This method has been successfully applied for TiO_2 and silver oxide NPs [121, 122]. More technological innovations are needed in these directions to make the nano metal oxide-based water purification in large scale.

Thus, to assure the safety of the consumers for the use of nano metal oxides-based purification technology, regulatory board must be formed [123]. In China, the use of NM and its issues are taken care by NSCNN (National Steering Committee for Nanoscience and Nanotechnology) which work closely with National Nanotechnology Standardization Technical Committee [124]. Similarly in Europe REACH63 (Registration, Evaluation, Authorization and Restriction of Chemicals) controls the use of NMs and their impact on health and environment is monitored [125]. Few other developed countries are in process of bringing regulatory law to control the usage of NM based technology products. Till now in India there is no such organization for governing the usage and legal constraints of NMs [126].

7 Conclusions

Water is the most important element on this planet for living things and plants. However, the water is contaminated with different type of toxic materials. The major cause of this pollution is industrial waste going into water bodies. Numbers of techniques have been used for remediation but the adsorption technique is found to be the most effective. Nanomaterials have been considered to be the most important adsorbent. Because of various specific properties, numbers of nano metal oxides and their composites have been found to be a suitable adsorbent for removal of pollutants. Synthesis of nano metal oxides and their applications for water remediation have been discussed. These metal oxides have also been used as photocatalysts. Considering the advantages and disadvantages, further research is needed.

References

1. Dhiman V, Kondal N (2021) ZnO Nanoadsorbents: A potent material for removal of heavy metal ions from wastewater. Colloid Interf Sci Commun 41:100380
2. Singh NB, Nagpal G, Agrawal S, Rachna (2018) Water purification by using adsorbents: a review. Environ Technol Innov 11:187–240
3. Gupta K, Joshi P, Gusain R, Khatri OP (2021) Recent advances in adsorptive removal of heavy metal and metalloid ions by metal oxide-based NMs. Coordination Chem Rev 445:214100
4. Muralikrishna IV, Manickam V (2017) Chapter one—introduction. Environ Manage 1–4
5. Agasti N (2021) Decontamination of heavy metal ions from water by composites prepared from waste. Curr Res Green Sustain Chem 4:100088
6. Yadav N, Garg VK, Chhillar AK, Rana JS (2021) Detection and remediation of pollutants to maintain ecosustainability employing nanotechnology: a review. Chemosphere 280:130792
7. Senthil Rathi B, Senthil Kumar P, Dai-Viet N Vo (2021) Critical review on hazardous pollutants in water environment: occurrence, monitoring, fate, removal technologies and risk assessment. Sci Total Environ 797:149134
8. Naseem T, Durrani T (2021) The role of some important metal oxide NP for wastewater and antibacterial applications: a review. Environ Chem Ecotoxicol 3:59–75
9. de Mendonça VR, Mourão HAJL, Malagutti AR, Ribeiro C (2019) The role of the relative dye/photocatalyst concentration in TiO_2 assisted photodegradation process. Photochem Photobiol 90(1):66–72. https://doi.org/10.1111/php.12175
10. Stoller M, Azizova G, Mammadova A, Vilardi G, Di Palma L, Chianese A (2016) Treatment of olive oil processing wastewater by ultrafiltration, nanofiltration, reverse osmosis and biofiltration. Chem Eng Trans 47:409–414. https://doi.org/10.3303/CET1647069
11. Jain A, Kumari S, Agarwal S (2021) SuphiyaKhan, Water purification via novel nano-adsorbents and their regeneration strategies. Process Saf Environ Prot 152:441–454
12. Gusain R, Gupta K, Joshi P, Khatri OP (2019) Adsorptive removal and photocatalytic degradation of organic pollutants using metal oxides and their composites: a comprehensive review. Adv Colloid Interf Sci 272:102009
13. De Gisi S, Lofrano G, Grassi M, Notarnicola M (2016) Characteristics and adsorption capacities of low-cost sorbents for wastewater treatment: A review. Sustain Mater Technol 9:10–40
14. Gnanaprakasam A, Sivakumar VM, Thirumarimurugan M (2015) Influencing parameters in the photocatalytic degradation of organic effluent via nanometal oxide catalyst: a review. Indian J Mater Sci Article ID 601827, 16 p. Hindawi Publishing Corporation.https://doi.org/10.1155/2015/601827

15. Taghipour S, Hosseini SM, Ataie-Ashtiani B (2019) Engineering NMs for water and wastewater treatment: review of classifications, properties and applications. New J Chem 43(21):7902–7927
16. Khin M, Nair AS, Babu VJ, Murugan R, Ramakrishna S (2012) A review on NMs for environmental remediation. Energy Environ Sci 5(8):8075–8109
17. Guo T, Yao MS, Lin YH, Nan CW (2015) A comprehensive review on synthesis methods for transition-metal oxide nanostructures. CrystEngComm 17(19):3551–3585
18. Stankic S, Suman S, Haque F, Vidic J (2016) Pure and multi metal oxide NP: synthesis, antibacterial and cytotoxic properties. J Nanobiotechnol 14(1):1–20
19. Nikam AV, Prasad BLV, Kulkarni AA (2018) Wet chemical synthesis of metal oxide NP: a review. CrystEngComm 20(35):5091–5107
20. Stankic S, Sternig A, Finocchi F, Bernardi J, Diwald O (2010) Zinc oxide scaffolds on MgO nanocubes. Nanotechnology 21(35):355603
21. El Nemr A, Helmy ET, Gomaa EA, Eldafrawy S, Mousa M (2019) Photocatalytic and biological activities of undoped and doped TiO_2 prepared by Green method for water treatment. J Environ Chem Eng 7(5):103385
22. Buha J, Arčon D, Niederberger M, Djerdj I (2010) Solvothermal and surfactant free synthesis of crystalline Nb_2O_5, Ta_2O_5, HfO_2, and Co-doped HfO 2 NP. PhysChemChemPhys 12:15537–15543
23. Corr SA, Gun'ko YK, Douvalis AP, Venkatesan M, Gunning RD, Nellist PD (2008) From nanocrystals to nanorods: new iron oxide−silica NPs from metallorganic precursors. J Phys Chem C 112(4):1008–1018
24. Parashar M, Shukla VK, Singh R (2020) Metal oxides NP via sol–gel method: a review on synthesis, characterization and applications. J Mater Sci Mater Electron 31(5):3729–3749
25. Ghasemzadeh G, Momenpour M, Omidi F, Hosseini MR, Ahani M, Barzegari A (2014) Applications of NMs in water treatment and environmental remediation. Front Environ Sci Eng 8(4):471–482
26. Singh L, Sharma R, Singh N, Kumar A, Mahato DK, Lee Y, Bechelany M, Mandal KD (2021) Semi-wet growth and characterization of multi-functional nano-engineered mixed metal oxides for industrial application. Progr Cryst Growth Charact Mater 67:100542
27. Zhang L, Kanki T, Sano N, Toyoda A (2003) Development of TiO_2 photocatalyst reaction for water purification. Sep Purif Technol 31(1):105–110
28. Xiong Z, Ma J, Ng WJ, Waite TD, Zhao XS (2011) Silver-modified mesoporous TiO_2 photocatalyst for water purification. Water Res 45(5):2095–2103
29. Konstantinou IK, Albanis TA (2003) Photocatalytic transformation of pesticides in aqueous titanium dioxide suspensions using artificial and solar light: intermediates and degradation pathways. Appl Catal B 42(4):319–335
30. Shahid M, McDonagh A, Kim JH, Shon HK (2015) Magnetised titanium dioxide (TiO_2) for water purification: preparation, characterization and application. Desalin Water Treat 54(4–5):979–1002
31. Ge M, Guo C, Ma L, Han Z, Hu W, Wang Y (2009) Photocatalytic degradation of methyl orange using ZnO/TiO_2 composites. Front Environ Sci Eng China 3(3):271–280
32. Dong F, Zhao W, Wu Z (2008) Characterization and photocatalytic activities of C, N and S co-doped TiO_2 with 1D nanostructure prepared by the nano-confinement effect. Nanotechnology 19(36):365607
33. Zhang R, Wang Q, Liang J, Li Q, Dai QJ, Li W (2012) Optical properties of N and transition metal R (R= V, Cr, Mn, Fe Co, Ni, Cu, and Zn) codoped anatase TiO2. Phys B 407(14):2709–2715
34. Ullah S, Faiz P, Leng S (2020) Synthesis, mechanism, and performance assessment of zero-valent iron for metal-contaminated water remediation: a review, CLEAN–soil, air. Water 48(9):2000080
35. Nassar NN (2012) Iron oxide nanoadsorbents for removal of various pollutants from wastewater: an overview. Appl Adsorb Water Poll Control 81–118

36. Fallah Z, Zare EN, Ghomi M, Ahmadijokani F, Amini M, Tajbakhsh M, Arjmand M, Sharma G, Ali H, Ahmad A, Makvandi P, Lichtfouse E, Sillanpa M, Varma RS (2021) Toxicity and remediation of pharmaceuticals and pesticides using metal oxides and carbon nanomaterials. Chemosphere 275:130055
37. Vinayagam R, Pai S, Murugesan G, Varadavenkatesan T, Selvaraj R (2021) Synthesis of photocatalytic zinc oxide nanoflowers using Peltophorumpterocarpum pod extract and their characterization. Appl Nanosci 1–11
38. Alsharaeh E, Ahmed F, Arshi N, Alturki M (2017) Metal oxide nanophotocatalysts for water purification. Renew Energy Technol Water Desalinat 57–72
39. Tortella GR, Pieretti JC, Rubilar O, Fernández-Baldo M, Benavides-Mendoza A, Diez MC, Seabra AB (2021) Silver, copper and copper oxide NP in the fight against human viruses: progress and perspectives. Crit Rev Biotechnol 1–19
40. Abbasi BA, Iqbal J, Nasir JA, Zahra SA, Shahbaz A, Uddin S, Mahmood T (2020) Environmentally friendly green approach for the fabrication of silver oxide NP: characterization and diverse biomedical applications. Microsc Res Tech 83(11):1308–1320
41. Vinay SP, Nagaraju G, Chandrappa CP, Chandrasekhar N (2019) Novel Gomutra (cow urine) mediated synthesis of silver oxide NP and their enhanced photocatalytic, photoluminescence and antibacterial studies. J Sci Adv Mater Dev 4(3):392–399
42. Davis AS, Prakash P, Thamaraiselvi K (2017) Nano bioremediation technologies for sustainable environment. In: Bioremediation and sustainable technologies for cleaner environment 13–33
43. Zhang Q-X, Wen H, Peng D, Fu Q, Huang X-J (2015) Interesting interference evidences of electrochemical detection of Zn(II), Cd(II) and Pb(II) on three different morphologies of MnO_2 nanocrystals. J Electroanal Chem 739:89e96
44. Salimi A, Pourbahram B, Mansouri-Majd S, Hallaj R (2015) Manganese oxide nanoflakes/multi-walled carbon nanotubes/chitosan nanocomposite modified glassy carbon electrode as a novel electrochemical sensor for chromium (III) detection. Electrochim Acta 156:207e215. https://doi.org/10.1016/j.electacta.2014.12.146
45. Fayazi M, Taher MA, Afzali D, Mostafavi A (2016) Fe_3O_4 and MnO_2 assembled on haloysite nanotubes: a highly efficient solid-phase extractant for electrochemical detection of mercury(II) ions. Sensor Actuator B Chem 228:1e9. https://doi.org/10.1016/j.snb.2015.12.107
46. Recillas S, García A, González E, Casals E, Puntes V, Sánchez A (2011) Use of CeO_2, TiO_2 and Fe_3O_4 NP for the removal of lead from water toxicity of NP and derived compounds. Desalination 277:213–220
47. Recillas S, Colón J, Casals E, González E, Puntes V, Sánchez A (2010) Chromium VI adsorption on cerium oxide NP and morphology changes during the process. J Hazard Mater 184:425–431
48. Mishra PK, Saxena A, Rawat AS, Dixit PK, Rai PK (2018) Surfactant-free one-pot synthesis of low-density cerium oxide np for adsorptive removal of arsenic species. Environ Prog Sustain Energy 37:221–231
49. Ayawanna J, Teoh WT, Niratisairak S, Sato K (2015) Gadolinia-modified ceria photocatalyst for removal of lead(II) ions from aqueous solutions. Mater Sci Semicond Process 40:136–139
50. Ayawanna J, Sato K (2017) Photoelectrodeposition effect of lanthanum oxide-modified ceria particles on the removal of lead(II) ions from water. Catal Today 321–322:128–134
51. Meepho M, Sirimongkol W, Ayawanna J (2018) Samaria-doped ceria nanopowders for heavy metal removal from aqueous solution. Mater Chem Phys 214:56–65
52. Peng XJ, Luan ZK, Ding J, Di ZC, Li YH, Tian BH (2005) Ceria NP supported nanotubes for the removal of arsenate from water. Mater Lett 59:399–403
53. Cai Y, Li C, Dan W, Wei W, Tan F, Wang X, Wong PK, Qiao X (2016) Highly active MgO NP for simultaneous bacterial inactivation and heavy metal removal from aqueous solution. Chem Eng J 312:158–166
54. Stoimenov PK, Klinger RL, Marchin GL, Klabunde KJ (2002) Metal oxide NP as bactericidal agents. Langmuir 18:6679–6686

55. Mahdavi S, Jalali M, Afkhami A (2013) Heavy metals removal from aqueous solutions using TiO_2, MgO, Al_2O_3 NP. Chem Eng Commun 200:448–470
56. Madzokere TC, Karthigeyan A (2017) Heavy metal ion effluent discharge containment using magnesium oxide (MgO) NP. Mater Today 4:9–18
57. Xiong C, Wang W, Tan F, Luo F, Chen J, Qiao X (2015) Investigation on the efficiency and mechanism of Cd(II) and Pb(II) removal from aqueous solutions using MgO NP. J Hazard Mater 299:664–674
58. Jing F, Zou L, Wang Y, Li B, He X, Fan Z, Ren Y, Lv Y, Zhang M, Dan C (2015) Synthesis of high surface area, mesoporous MgO nanosheets with excellent adsorption capability for Ni(II) via a distillation treating. J Colloid Interf Sci 438:259–267
59. Hang C, Li Q, Gao S, Shang JK (2011) As(III) and As(V) adsorption by hydrous zirconium oxide NP synthesized by a hydrothermal process followed with heat treatment. Ind Eng Chem Res 51:353–361
60. Jiang C, Xiao DA (2014) Nanosized zirconium dioxide particles as an efficient sorbent for lead removal in waters. Adv Mater Res 926–930:166–169
61. Bortun A, Bortun M, Pardini J, Khainakov SA, García JR (2010) Effect of competitive ions on the arsenic removal by mesoporous hydrous zirconium oxide from drinking water. Mater Res Bull 45:1628–1634
62. Luo X, Wang C, Wang L, Deng F, Luo S, Tu X, Au C (2013) NPs of graphene oxide hydrated zirconium oxide for simultaneous removal of As(III) and As(V) from water. Chem Eng J 220:98–106
63. Zhang Q, Jie T, Zhang Z, Nie G, Zhao H, Peng HQ, Jiao T (2015) Unique and outstanding cadmium sequestration by polystyrene-supported nanosized zirconium hydroxides: a case study. RSC Adv 5:55445–55452
64. Hua M, Jiang Y, Wu B, Pan B, Zhao X, Zhang Q (2013) Fabrication of a new hydrous Zr(IV) oxide-based nanocomposite for enhanced Pb(II) and Cd(II) removal from waters. ACS Appl Mater Interf 5:12135
65. Feng C, Aldrich C, Eksteen J, Arrigan D (2017) Removal of arsenic from alkaline process waters of gold cyanidation by use of γ-Fe_2O_3@ZrO_2 nanosorbents. Hydrometallurgy 174:71–77
66. Pan B, Li Z, Zhang Y, Xu J, Chen L, Dong H, Zhang W (2014) Acid and organic resistant nano-hydrated zirconium oxide (HZO)/polystyrene hybrid adsorbent for arsenic removal from water. Chem Eng J 248:290–296
67. Seisenbaeva GA, Geoffrey D, Kessler VG, Jean-Marie N (2015) General facile approach to transition-metal oxides with highly uniform mesoporosity and their application as adsorbents for heavy-metal-ion sequestration. Chemistry 20:10732–10736
68. Yalçınkaya Ö, Kalfa OM, Türker AR (2011) Chelating agent free-solid phase extraction (CAF-SPE) of Co(II), Cu(II) and Cd(II) by new nano hybrid material (ZrO2/B2O3). J Hazard Mater 195:332–339
69. Prabhakar R, Samadder SR (2018) Low cost and easy synthesis of aluminium oxide NP for arsenite removal from groundwater: A complete batch study. J Mol Liquids 250:192–201
70. Xie Y, Kocaefe D, Kocaefe Y, Cheng J, Liu W (2016) The effect of novel synthetic methods and parameters control on morphology of nano-alumina particles. Nanoscale Res Lett 11:259
71. Tabesh S, Davar F, Reza Loghman-Estarki M (2018) Preparation of -Al_2O_3 NP using modified sol-gel method and its use for the adsorption of lead and cadmium ions. J Alloys Comp 730:441–449
72. Stietiya MH, Wang JJ (2014) Zinc and cadmium adsorption to aluminum oxide NP affected by naturally occurring ligands. J Environ Qual 43:498
73. Patra AK, Dutta A, Bhaumik A (2012) Self-assembled mesoporous -Al_2O_3 spherical NP and their efficiency for the removal of arsenic from water. J Hazard Mater 201:170–177
74. Mahdavi S, Jalali M, Afkhami A (2015) Heavy metals removal from aqueous solutions by Al_2O_3 NP modified with natural and chemical modifiers. Clean Technol Environ 17:85–102
75. Poursani AS, Nilchi A, Hassani AH, Shariat M, Nouri J (2015) novel method for synthesis of nano–Al_2O_3: Study of adsorption behavior of chromium, nickel, cadmium and lead ions. Int J Environ Sci Technol 12:2003–2014

76. Liou JW, Chang H-H (2012) Bactericidal effects and mechanisms of visible lightresponsive titanium dioxide Photocatalysts on pathogenic Bacteria. Arch Immunol Ther Exp 60(4):267–275
77. Karunagaran V, Rajendran K, Sen S (2014) Antimicrobial activity of biosynthesized silver oxide NP. J Pure Appl Microbiol 4:3263–3268
78. Shah A, Haq S, Rehman W, Waseem M, Shoukat S, Rehman M (2019) Photocatalytic and antibacterial activities of paeonia emodi mediated silver oxide NP. Mater Res Express 6(4):45045
79. Manikandan V et al. (2017) Green synthesis of silver oxide NP and its antibacterial activity against dental pathogens. 3 Biotech 7(1):72
80. Nwanya AC et al. (2019) Industrial textile effluent treatment and antibacterial effectiveness of Zea mays L. Dry husk mediated bio-synthesized copper oxide NP. J Hazard Mater 375:281–289
81. Nabila MI, Kannabiran K (2018) Biosynthesis, characterization and antibacterial activity of copper oxide NP (CuO NPs) from actinomycetes, Biocatal. Agric Biotechnol 15:56–62
82. Moniri Javadhesari S, Alipour S, Mohammadnejad S, Akbarpour MR (2019) Antibacterial activity of ultra-small copper oxide (II) NP synthesized by mechanochemical processing against S. aureus and E. coli. Mater Sci Eng C 105:110011
83. Rajapaksha P, Cheeseman S, Hombsch S, Murdoch BJ, Gangadoo S, Blanch EW, Truong Y, Cozzolino D, McConville CF, Crawford RJ, Truong VK, Elbourne A, Chapman J (2019) Antibacterial properties of graphene oxide–copper oxide nanoparticle NPs. ACS Appl Bio Mater 2(12):5687–5696
84. Motshekga SC, Ray SS, Onyango MS, Momba MNB (2015) Preparation and antibacterial activity of chitosan-based NPs containing bentonite-supported silver and zinc oxide NP for water disinfection. Appl Clay Sci 114:330–339
85. Cruz GJF, Gomez MM, Solis JL, Rimaycuna J, Solis RL, Cruz JF, Rathnayake B, Keiski RL (2018) Composites of ZnO NP and biomass based activated carbon: adsorption, photocatalytic and antibacterial capacities. Water Sci Technol 2017(2):492–508
86. Esmailzadeh H, Sangpour P, Shahraz F, Hejazi J, Khaksar R (2016) Effect of nanocomposite packaging containing ZnO on growth of Bacillus subtilis and Enterobacter aerogenes. Mater Sci Eng C 58:1058–1063
87. El Saeed AM, El-Fattah MA, Azzam AM (2015) Synthesis of ZnO NP and studying its influence on the antimicrobial, anticorrosion and mechanical behavior of polyurethane composite for surface coating. Dye Pigment 121:282–289
88. Nwanya AC, Razanamahandry LC, Bashir AKH, Ikpo CO, Nwanya SC, Botha S, Ntwampe SKO, Ezema FI, Iwuoha EI, Maaza M (2019) Industrial textile effluent treatment and antibacterial effectiveness of Zea mays L. Dry husk mediated bio-synthesized copper oxide NP. J Hazard Mater 375:281–289
89. Naghizade Asl M, Mahmodi NM, Teymouri P, Shahmoradi B, Rezaee R, Maleki A (2016) Adsorption of organic dyes using copper oxide NP: isotherm and kinetic studies. Desalin Water Treat 57(52):25278–25287
90. Yogesh Kumar K, Archana S, Vinuth Raj TN, Prasana BP, Raghu MS, Muralidhara HB (2017) Superb adsorption capacity of hydrothermally synthesized copper oxide and nickel oxide nanoflakes towards anionic and cationic dyes. J Sci Adv Mater Devices 2(2):183–191
91. deMendonça VR, Mourão HAJL, Malagutti AR, Ribeiro C (2019) The role of the relative dye/photocatalyst concentration in TiO_2 assisted photodegradation process. Photochem Photobiol 90(1):66–72
92. Guesh K, Mayoral Á, Márquez-Álvarez C, Chebude Y, Díaz I (2016) Enhanced photocatalytic activity of TiO_2 supported on zeolites tested in real wastewaters from the textile industry of Ethiopia. Microporous Mesoporous Mater 225:88–97
93. Mortazavi K, Rajabi H, Ansari A, Ghaedi M, Dashtian K (2016) Preparation of silver nanoparticle loaded on activated carbon and its application for removal of malachite green from aqueous solution. Synth React Inorganic Met Nano-Metal Chem. https://doi.org/10.1080/15533174.2016.1228670

94. Jiang W, Wang XW, Wu Z, Yue X, Yuan S, Lu H, Liang B (2015) Silver oxide as superb and stable Photocatalyst under visible and nearinfrared light irradiation and its Photocatalytic mechanism. Ind Eng Chem Res 54(3):832–841
95. Shah A, Haq S, Rehman W, Waseem M, Shoukat S, Rehman M (2019) Photocatalytic and antibacterial activities of paeonia emodi mediated silver oxide NP. Mater Res Express 6(4):45045
96. Li R, Chen Z, Ren N, Wang Y, Wang Y, Yu F (2019) Biosynthesis of silver oxide NP and their photocatalytic and antimicrobial activity evaluation for wound healing applications in nursing care. J Photochem Photobiol B Biol 199:111593
97. Zafar MN, Dar Q, Nawaz F, Zafar MN, Iqbal M, Nazar MF (2019) Effective adsorptive removal of azo dyes over spherical ZnO NP. J Mater Res Technol 8(1):713–725
98. Monsef Khoshhesab Z, Souhani S (2018) Adsorptive removal of reactive dyes from aqueous solutions using zinc oxide NP. J Chin Chem Soc 65(12):1482–1490
99. Engates K, Shipley H (2011) Adsorption of Pb, cd, cu, Zn, and Ni to titanium dioxide NP: effect of particle size, solid concentration, and exhaustion. Environ Sci Pollut Res Int 18:386–395
100. Asztemborska M, Bembenek M, Jakubiak M, Stęborowski R, Bystrzejewska- Piotrowska G (2018) The effect of NP with sorption capacity on the bioaccumulation of divalent ions by aquatic plants. Int J Environ Res 12(2):245–253
101. Chen Z, Li Y, Guo M, Xu F, Wang P, Du Y, Na P (2016) One-pot synthesis of Mn-doped TiO_2 grown on graphene and the mechanism for removal of Cr(VI) and Cr(III). J Hazard Mater 310:188–198
102. Goswami A, Raul PK, Purkait MK (2012) Arsenic adsorption using copper (II) oxide NP. Chem Eng Res Des 90(9):1387–1396
103. Sreekala G, Fathima B, Beena B (2019) Adsorption of lead (Ii) ions by ecofriendly copper oxide NP orient. J Chem 35:1731–1736
104. Lucan RAE-D (2019) Polyaniline/ZNO nanocomposite: a novel adsorbent for the removal of Cr(VI) from aqueous solution. IntechOpen, Rijeka
105. Azizi S, Mahdavi Shahri M, Mohamad R (2017) Green synthesis of zinc oxide NP for enhanced adsorption of lead ions from aqueous solutions: equilibrium, kinetic and thermodynamic studies. Molecules 22(6):831
106. Yuvaraja G, Prasad C, Vijaya Y, Subbaiah MV (2018) Application of ZnO nanorods as an adsorbent material for the removal of as(III) from aqueous solution: kinetics, isotherms and thermodynamic studies. Int J Ind Chem 9(1):17–25
107. Sharma PR, Sharma SK, Antoine R, Hsiao BS (2019) Efficient removal of arsenic using zinc oxide nanocrystal-decorated regenerated microfibrillated cellulose scaffolds. ACS Sustain Chem Eng 7(6):6140–6615
108. Cui H, Li Q, Gao S, Shang JK (2012) Strong adsorption of arsenic species by amorphous zirconium oxide NP. J Ind Eng Chem 18:1418–1427
109. Inchaurrondo N, di Luca C, Mori F, Pintar A, Zˇerjav G, Valiente M, Palet C (2019) Synthesis and adsorption behavior of mesoporous alumina and Fe-doped alumina for the removal of dominant arsenic species in contaminated waters. J Environ Chem Eng 7:102901
110. Purwajanti S, Zhang H, Huang X, Song H, Yang Y, Zhang J, Niu Y, Meka AK, Noonan O, Yu C (2016) Synthesis of magnesium oxide hierarchical microspheres: a dual functional material for water remediation. ACS Appl Mater Interf 8:25306–25312
111. Wu P-Y, Jiang Y-P, Zhang Q-Y, Jia Y, Peng D-Y, Xu W (2016) Comparative study on arsenate removal mechanism of MgO and MgO/TiO_2 composites: FTIR and XPS analysis New. J Chem 40:2878–2885
112. Kumar S, Nair RR, Pillai PB, Gupta SN, Iyengar M, Sood A (2014) Graphene oxide-$MnFe_2O_4$ magnetic nanohybrids for efficient removal of lead and arsenic from water. ACS Appl Mater Interf 6:17426–17436
113. Chen J, Wang J, Zhang G, Wu Q, Wang D (2018) Facile fabrication of nanostructured cerium-manganese binary oxide for enhanced arsenite removal from water. Chem Eng J 334:1518–1526

114. Chen B, Zhu Z, Hong J, Wen Z, Ma J, Qiu Y, Chen J (2014) Nanocasted synthesis of ordered mesoporous cerium iron mixed oxide and its excellent performances for As(V) and Cr(VI) removal from aqueous solutions. Dalton Trans 43:10767–10777
115. Vazquez-Munoz R, Borrego B, Juarez-Moreno K, Garcia-Garcia M, Mota Morales JD, Bogdanchikova N, Huerta-Saquero A (2017) Toxicity of silver NP in biological systems: does the complexity of biological systems matter? Toxicol Lett 276:11–20
116. Valant J, Iavicoli I, Drobne D (2012) The importance of a validated standard methodology to define *in vitro* toxicity of nano-TiO_2. Protoplasma 249(3):493–502
117. Vandebriel RJ, De Jong WH (2012) A review of mammalian toxicity of ZnO NP. Nanotechnol Sci Appl 5:61–71
118. Kovvuru P, Mancilla PE, Shirode AB, Murray TM, Begley TJ, Reliene R (2015) Oral ingestion of silver NP induces genomic instability and DNA damage in multiple tissues. Nanotoxicology 9(2):162–171
119. Jarockyte G, Daugelaite E, Stasys M, Statkute U, Poderys V, Tseng TC, Hsu SH, Karabanovas V, Rotomskis R (2016) Accumulation and Toxicity of Superparamagnetic Iron Oxide NP in Cells and Experimental Animals. Int J Mol Sci 17(8):1193
120. Ottofuelling S, Von Der Kammer F, Hofmann T (2011) Commercial titanium dioxide np in both natural and synthetic water: comprehensive multidimensional testing and prediction of aggregation behavior. Environ Sci Technol. 45(23):10045–10052
121. Sun Q, Li Y, Tang T, Yuan Z, Yu CP (2013) Removal of silver NP by coagulation processes. J Hazard Mater 261:414–420
122. Popowich A, Zhang Q, Le XC (2015) Removal of NP by coagulation. J Environ Sci (China) 38:168–171
123. Thomas AH (2016) Regulating NMs: a case for hybrid governance. Bull Sci Technol Soc 36(4):219–228
124. Jarvis DS, Richmond N (2011) Regulation and governance of nanotechnology in China: regulatory challenges and effectiveness. EJLT 2
125. Montfort J-P, Indirli G, Georgieva D, Carrega C-M (2010) NMs under REACH: legal aspects: unless and until REACH is adapted to more specifically regulate NMs, is there scope for national measures to regulate these materials? EJRR 1(1):51–62
126. Madhwani KP (2013) Safe development of nanotechnology: a global challenge. Indian J Occup Environ Med 17(3):87–8

Application of Core–Shell Nanohybrid Structures in Water Treatment

Hirakendu Basu, Shweta Singh, Suresh Kumar Kailasa, and Rakesh Kumar Singhal

Abstract Core–shell nanoparticles have been the centre of attention of the researchers in various fields as the transition from bulk/micro particles to nanoparticles to core shell nanoparticles was seen to lead to enormous changes in the physical and chemical properties of the materials like increased surface to volume ratio, dominance of surface atoms over those in the core etc. Core–shell nanohybrid structures are nanocomposites which incorporates the advantages of both core–shell nanoparticles and other component of the hybrid material like polymer, ceramic, oxide structures. In recent times core–shell nanohybrid structures have gained wide attention in different energy and environment applications including sorption of pollutants from aquatic medium. Drinking water pollution is one of the major problems the world is facing today. Various technologies have been developed for removal of various contaminants from aquatic streams. Core–shell hybrid nanostructures can be tailored by chemical modification or by incorporation in the polymeric matrix. This not only makes them specific to some metal ions, radionuclides even nanoparticles but also enhances their sorption capacity. These materials can be synthesised by dispersion of building blocks in polymeric network, in situ polymerisation, sol–gel process, self-assembly of unit building blocks through layered structures or interpenetrating networks. In this book chapter, a series of core shell nanohybrid structures having application in water decontamination have been reviewed and discussedextensively. Synthesis strategies, sorptive properties of these core shell nanohybrid structures are summarised with emphasis on decontamination of conventional pollutants, radionuclides, dyes and organic pollutants.

Keywords Core–shell nanoparticle · Nanohybrid structure · Water treatment · Decontamination

H. Basu (✉) · S. Singh · R. Kumar Singhal
Analytical Chemistry Division, Bhabha Atomic Research Centre, Trombay, Mumbai 400085, India
e-mail: hirak@barc.gov.in

S. Kumar Kailasa
Department of Applied Chemistry, S. V. National Institute of Technology, Surat, Gujarat 395007, India

S. K. Swain (ed.), *Nanohybrid Materials for Water Purification*, Composites Science and Technology, https://doi.org/10.1007/978-981-19-2332-6_12

1 Introduction

Nanoparticle can be defined as any particle with size in the range of 1–100 nm. The properties of materials change drastically as one enters into nano world due to increase in surface area to volume ratio and onset of quantum confinement effects [1]. Due to their distinguished features, they find applications in various sectors. In healthcare area, nanoparticles are being synthesised to aid in the migration of chemotherapy drugs, straightway to the site of cancerous growth. Also, Au nanoparticles are commercially used as probe in order to detect the sequence of nucleic acids. Cosmetic sector makes use of TiO_2 nanoparticles in the sunscreens, which works by reflecting the visible light and absorbing the UV light, thus acting as a transparent barrier which shields the skin from harmful ultraviolet rays. Apart from this, nano TiO_2 also finds use in coatings which are applied to make self-cleaning surfaces [2]. Use of carbon nanotubes in baseball bats has made it lighter, thereby improving the performance. Transistors, which make all the modern computing possible have become smaller and smaller with the advent of nanotechnology [3]. Ultra-High Definition (UHD) displays are available that makes use of quantum dots to generate much vibrant colors and at the same time being more energy efficient [4]. Nanotechnology can also assist in meeting the requirements of clean drinking water by developing novel materials that could help in easy and affordable detection of harmful pollutants in water followed by their facile eradication [5]. In food industry, nanotechnology can bring improvement in food processing and packaging, food safety and in extending the shelf life of food products. Thus, it is clear that there is hardly any area which is not benefitted by the arrival of nanotechnology. Coming to the classification, Siegel classified the nanomaterials based on their dimensionality as 0D, 1D, 2D and 3D (Fig. 1).

(i) 0-dimensional nanomaterials: In 0-D nanomaterials, all the dimensions are in the nanorange. It comprises of nanospheres and nanoclusters.
(ii) 1-dimensional nanomaterials: In 1-D nanomaterials, one dimension is not in the nanorange. This category includes nanorods, nanowires, nanofibers etc.
(iii) 2-dimensional nanomaterials: In 2-D nanomaterials, two dimensions are not in the nanorange. Nanolayers and nanofilms come under this class of materials.
(iv) 3-dimensional nanomaterials: In 3-D nanomaterials, three dimensions are not in the nanorange. They are made up of multiple nanosize crystals arranged in different directions.

Core shell nanoparticles (CSNP) are concentric multilayer particles where dimension of each layer is in nanometre range [6]. Core–shell nanoparticles have been the centre of attention of the researchers in the field as the transition from bulk/micro particles to nanoparticles to core shell nanoparticles was seen to lead to enormous changes in the physical and chemical properties of the materials [7]. The properties of core shell nanohybrids are not only dependent on size but also on shape. Different physical and chemical features such as melting point, catalytic activity, coercivity, selectivity, optical properties are affected by shape [7–9]. In addition to

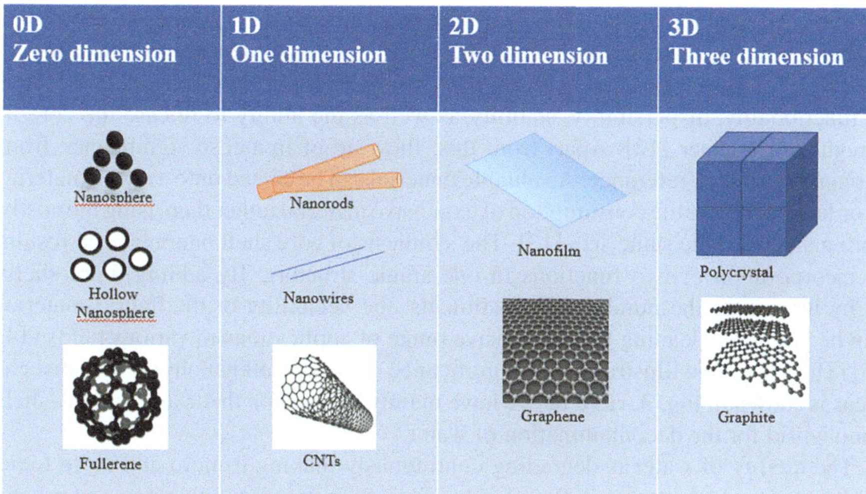

Fig. 1 Classification of nanomaterials on the basis of their dimensionality

these, the sensitivity towards Surface Enhanced Raman Scattering (SERS) is influenced by morphology of core–shell nanohybrids. PengzhenGuo et al. reported the synthesis of Au@Agnanohybrid and used it for the detection of pesticide thiram [10]. The detection limit was found to be different for nanocubes and nanocuboids. The schematic showing the different shapes of core–shell nanohybrid material is shown in Fig. 2 [11].

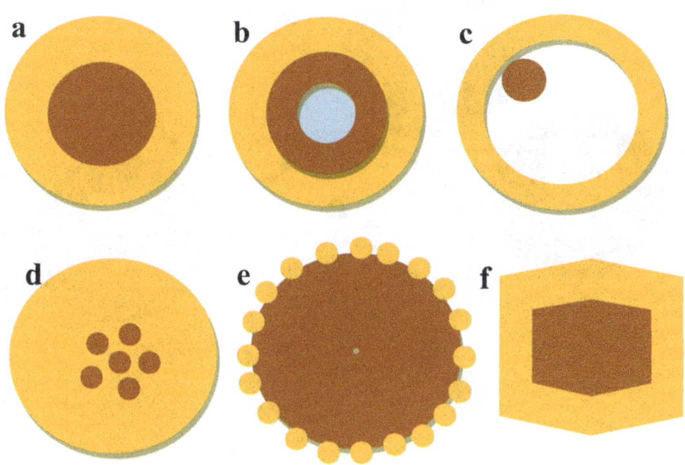

Fig. 2 Different core/shell nanoparticles: **a** spherical; **b** multicore spherical; **c** movable core in hollow shell; **d** multiple small cores coated by single shell; **e** Single core coated by multiple small particles as shell; **f** hexagonal (Reprinted with permission from Ref. [11])

The core shell nanohybrids are superior over conventional nanomaterials as their properties can be easily tailored either by varying the constituents or by changing the core to shell ratio. The shell coating provides multiple benefits such as an increase in functionality, dispersibility, stability as well as the ability to release the core in a regulated manner [12]. Apart from this, they are of immense significance from economic frame of reference. A valuable material can be coated onto a cheap material in order to decrease the consumption of expensive material rather than using the costly pure material of the same size [13]. The synthesis of core shell nanomaterial results in incorporation of two functions in one single structure. By adding more shells or by increasing the number of constituents, the versatility of the hybrid material can be increased, leading to an extensive range of applications in various fields [14, 15]. The schematic illustrating the significance of core–shell nanohybrids in diverse areas is shown in Fig. 3. Here in, we have mainly focused on the usage of core–shell nanohybrid for the decontamination of water.

The quality of water is degrading continuously making it more and more toxic for humans and environment. Rapid industrialisation, improper planning and urbanisation has resulted in the worldwide contamination of water resources [15]. Heavy metals, pharmaceuticals, pesticides, dyes, radionuclides are some of the important categories of pollutants which are deteriorating the water quality parameters. Even with the arrival and advancements of new technologies, the heavy metal concentration in potable water is still beyond the prescribed limits set be regulatory bodies in various parts of the globe [16, 17]. The exposure to the heavy metals results in increased risk towards cardiovascular disorders, renal injuries, immune system dysfunction,

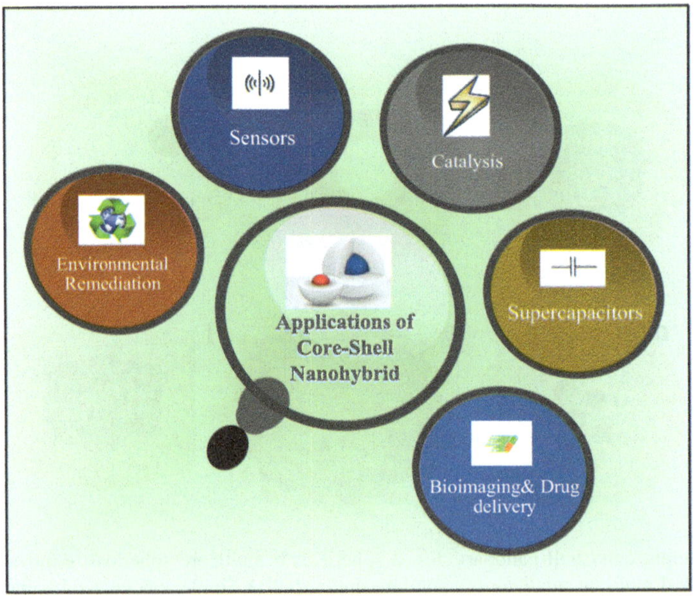

Fig. 3 Diverse applications of Core–Shell nanohybrids

gastrointestinal and neuronal problems, genomic instability and many more [18, 19]. Apart from heavy metals, the textile industry discharges huge quantities of dyes into water bodies resulting in an enhancement in the chemical and biochemical oxygen demand, inhibition in the growth of plant, bioaccumulation and an increased likelihood of mutagenicity and carcinogenicity. The indiscriminate usage of antibiotics on a large scale and their subsequent presence in water bodies has given birth to antibiotic resistance. Other pollutants like pesticides enter into water bodies through surface run off and are well known to adversely affect the health of aquatic species and humans. In view of the increasing load of pollutants, and the crisis for clean drinking water supply across globe, it is necessary to find suitable means to deal with the problem. Various methodologies have been adopted for the treatment of water such as membrane separation, advanced oxidation processes, photochemical degradation, sedimentation, flocculation, adsorption, ozonation etc. [20].

Among all these methods, adsorption is preferred widely because of its cost effectiveness, eco-friendly nature, easy handling and availability of large number of adsorbents. Various adsorbents such as graphene oxide, reduced graphene oxide, boron nitride, cellulose, chitosan, activated carbon, alumina, ferrites, iron oxide NPs etc. have been used by researchers for the removal of different kinds of pollutants [21–25]. Core–Shell Nanohybrids have proven to be a better adsorbent because of the high surface area and the ability to tailor the properties of core using suitable type of shell and its thickness [26, 27]. In view of the immense significance of core shell nanohybrids, here in we have discussed in detail their classification, synthetic strategies, general characterisation and finally their role in the removal of different kinds of pollutants.

2 Types of Core Shell Nanoparticles

Core shell nanoparticles (CSNP) may be defined as materials consisting of a core (inner material) and a shell (outer material). They can be categorised on the basis of composition of material as shown in Fig. 4 [7].

3 Inorganic/Inorganic CSNP

It is the most significant class among all the types as it finds applications in different fields ranging from information storage, biological labelling, optical bioimaging, catalysis, optoelectronics and many more. These CSNP can be further divided into following types.

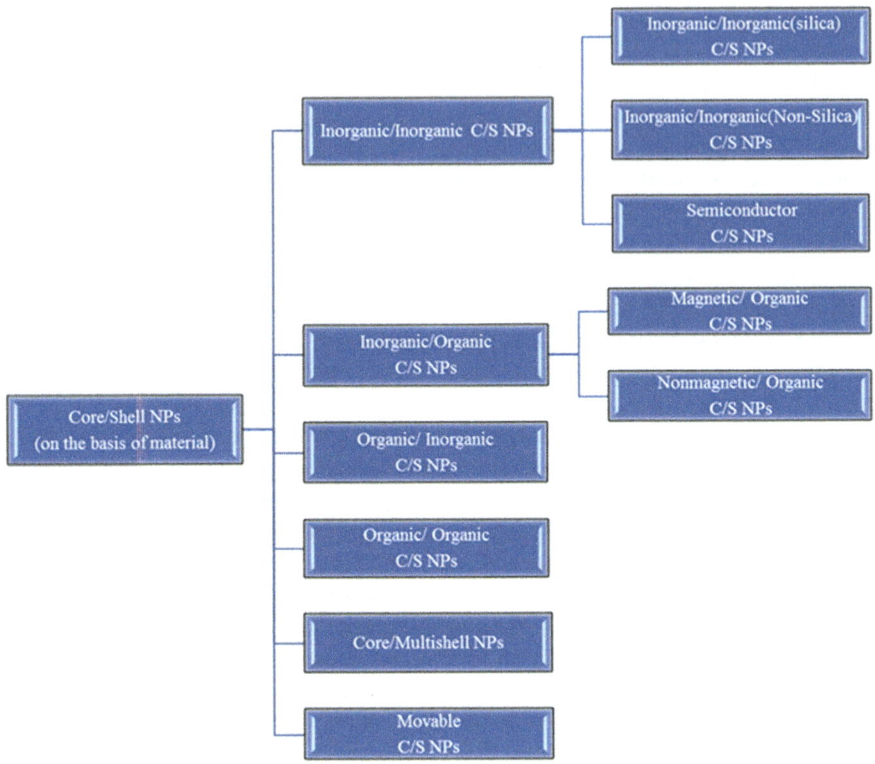

Fig. 4 Classification of Core–Shell Nanohybrids

3.1 Silica-Based Shell

The reasons for which silica is chosen by researchers for encapsulation of core nanomaterials are its extra ordinary colloidal stability especially in aquatic media, easy and controllable synthesis, modifiable porosity, chemical inertness and optical transparency [28–30]. Apart from this, due to the presence of silica shell, the core nanomaterial becomes biocompatible and the rich silica chemistry allows conjugation with functional groups easy which has resulted in widespread use of silica coated nanomaterials for various diagnostic and therapeutic purposes. Due to these distinguished properties, silica coatings have been made on different inorganic materials such as metals, metal oxides and metal salts. Silica coating is usually done by employing the classic Stober method involving hydrophobic silanes such as tetraethylorthosilicate (TEOS), tetramethylorthosilicate (TMOS) as silica precursor [31, 32]. Recently, water soluble silanes such as MPTES (3-(mercaptopropyl-triethoxysilane), MPTMS (3-(mercaptopropyl)-trimethoxysilane), MTMS (3-(methyltrimethoxysilane) have also been utilised [33].

3.2 Non Silica-Based Shell

Instead of Silica, various metals such as Au, Ni, Co, Pd, Pt, Cu, metal oxides, semiconductors can also be used as a shell material. Metal nanoshells consist of a dielectric core having a nano range metallic layer (usually of Au) over it. By changing the relative core and shell thickness, a broad variation can be seen in the color of nanoshells making it useful for biomedical applications [34]. Likewise, Fe_2O_3 coating on CaO and MgO nanoparticles can lead to an increase in the adsorption capacity of toxic materials compared to pure CaO and MgO [35].

3.3 Semiconductor CSNP

Here, either the core or the shell or both are made up of semiconducting material. They can be further classified as Nonsemiconductor/Semiconductor or/Semiconductor/Non-semiconductor Core Shell Nanoparticles. Among various types of nonsemiconductor/semiconductor materials, the ones consisting of a magnetic core having a semiconducting shell around them are very much versatile. The photocatalytic properties of magnetic core semiconductor shell is higher than pure semiconductor [36]. Researchers investigated the photocatalytic activity of TiO_2 and Fe_2O_3/TiO_2 core shell hybrid in malignant tumour therapy and they found out that survival of tumour cells is lesser for core/shell nanohybrid compared to pure semiconductor [37].

3.4 Semiconductor/Semiconductor CSNP

Both the core and shell are made up of semiconductor. The major advantage of these particles is that the outer coating of second semiconducting material over first leads to an increase in optical activity and stability towards photooxidation. Depending upon the band gap three subdivisions can be made: type-I, inverse type-1 and type-II [38]. In type I, the shell bandgap is larger than core and hence both hole (h) and electrons (e) are restricted within the core. In inverse type, shell bandgap is smaller than core; the h and e may be partially or wholly restricted to shell depending upon its thickness. In the last type, the edge of valence or conduction edge lies within the core bandgap. Hence, one carrier is chiefly restricted to the core and other to the shell.

4 Inorganic/Organic CSNP

These are made up of metal, metal oxide, metallic compound, or silica core having a polymeric shell or a shell build-up of any organic material of high density. Coating of organic material around metal decreases the susceptibility of metal towards surface oxidation and also makes it biocompatible for applications in biological field [39]. In many cases, the core particles are coated to enhance their stability in suspension media. On the basis of material properties, they can be further divided into following two groups.

4.1 Magnetic/Organic CSNP

Magnetic nanoparticles are coated with organic materials in order to reduce the agglomeration. Polysaccharides and hydrophilic polymers are commonly employed as coating materials. This class of materials find application in magnetic sealing, MRI, magnetic recording, electromagnetic shielding, magnetic shell separation and drug targeting [40–42].

4.2 Nonmagnetic/Organic CSNP

This class of materials can be further categorised on the basis of material constituting the nonmagnetic core- metal, metal or metalloid oxide, and metal salts. Among metals, silver and gold have been the centre of attraction because of their distinct optical properties. Au surrounded by polyaniline acts as biosensor for glucose [43]. In metal oxides, polystyrene, poly(methylmethacrylate), polyvinyl chloride are widely used for coating on silica core and the polymer coated silica nanoparticles find application in material additives, sensors, optical and electrical devices. For metal salts, coating is usually done with conducting polymers such as polyaniline, polythiophene, polypyrole etc. [44]. They usually find application in light emitting devices, chemical sensors and electronic devices.

5 Organic/Inorganic CSNP

The core is made up of polymeric material such as polystyrene, polyurethane, poly(ethylene oxide), poly(vinylpyrrolidone), poly(vinyl benzyl chloride), surfactants, dextrose, styrene-methyl methacrylate, poly(styrene-acrylic acid) etc. while the shell is composed of metals, metal chalcogenides, metal oxides or silica [45, 46].

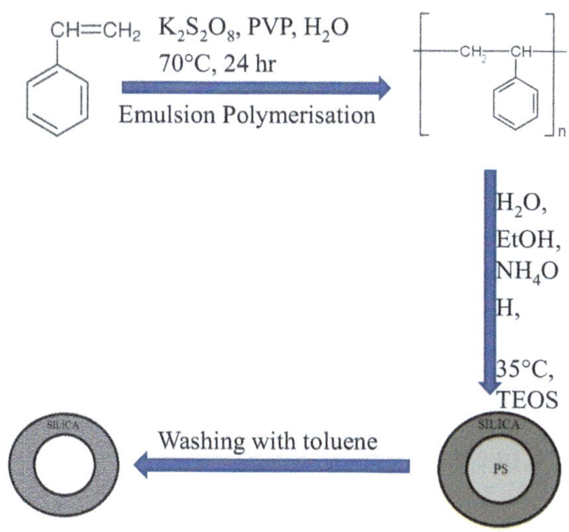

Fig. 5 Organic/inorganic core shell nanohybrid as template for the synthesis of hollow inorganic material

The coating of an inorganic material, mainly that of metal oxide over an organic material has several benefits such as increment in the strength of whole material, better colloidal and thermal stability and resistance towards oxidation and abrasion [47, 48]. Further, they could be used as a template for the synthesis of hollow inorganic material via dissolution of the core using an appropriate solvent or by calcination. The preparation of silica microspheres via formation of polystyrene@silica core shell nanohybrid is shown in Fig. 5 [48, 49].

6 Organic/Organic CSNP

Both the core and shell are composed of polymeric or some organic material. The profits gained by developing one polymer coating onto another is the modification in the physical properties of whole material, such as its toughness and glass transition temperature. Since, the glass transition temperature plays an important role in deciding whether a synthesised material is most suited for rigid or flexible application; tuning this property could help in getting desired properties. For instance, it has been reported that a high Tg core material improves the mechanical stability while a low Tg shell aids in the improvement of film formation capability. Apart from this, this class of materials also find application in controlled release of drugs. For instance, Dina M. Silva etal synthesised polymeric core shell nanohybrid and utilised it for the co-delivery of two drugs [50]. The curcumin drug is trapped in the core and ciprofloxacin was incorporated in the shell layer. The synthesis of nanohybrid material is shown in Fig. 6.

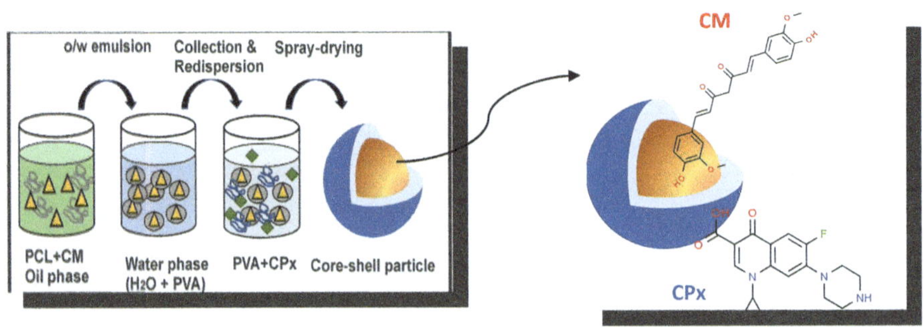

Fig. 6 Schematic representation of the processing steps of the core–shell particles. (Reprinted with permission from Ref. [50]).

7 Core/Multishell Nanoparticles

Bimetallics form an important part of this category of core shell NPs. Benito Rodríguez-González synthesised Au–Ag multishell nanoparticles [51]. The optical properties were found to change as second and third shells are placed. The color of pure gold was deep red, when silver shell is placed over it, the color changes to yellow, upon placement of second gold shell it changed to blue hue. Finally, when second silver shell is deposited over it, orange color was observed. When silica shell is placed onto CdSe/ZnS core–shell quantum dots, an enhancement in stability in biological buffers was observed apart from an increase in biocompatibility in case of fluorescence imaging [52].

8 Movable Core/Hollow Shell Nanoparticles

Template assist route is usually followed for the synthesis of movable core/shell NPs. In this methodology, first core with a double shell is prepared and then removal of double shell is done by using an appropriate technique such as calcination or dissolution. The synthetic scheme for the preparation of movable Au core inside a carbon shell following template assisted route is shown in Fig. 7 [53].

9 Synthesis Strategies for Core–Shell Nanohybrid Materials

For carrying out the synthesis of these nanomaterials two approaches are generally followed (a) Top-down approach (b) Bottom up approach. In the top down approach, the bulk material is broken down to produce nanosized particles. The techniques

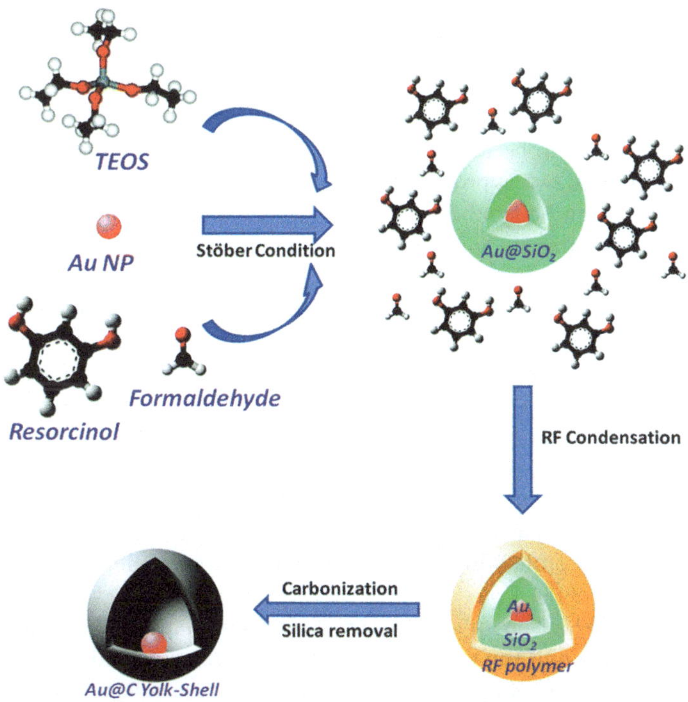

Fig. 7 One-step synthesis of Au–SiO_2–RF polymer CSS nanocomposite and its conversion to Au@C yolk–shell nanostructure. (Reprinted with permission from Ref. [53])

included in this approach are ball milling, electrospinning, lithographic technique, laser ablation, sputtering and arc discharge method. The main problem associated with this approach is the creation of imperfections on the surface structure and difficult to maintain uniformity in the core and shell size and dimensions [54]. Contrary to this, the bottom up approach involves the building up of a nanomaterial from bottom, that is either atom by atom or molecule by molecule (Fig. 8). Techniques like chemical vapor deposition, solvothermal methods, sol–gel method, reverse micelle

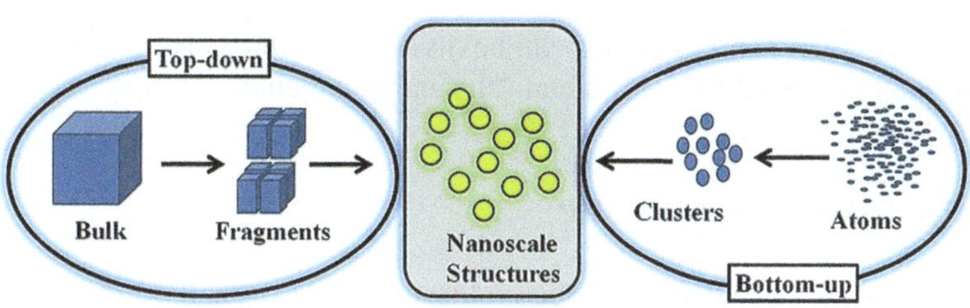

Fig. 8 Top-down and bottom-up approach for the synthesis of nanomaterials

methods, soft and hard templating methods, laser pyrolysis etc. come under the category of bottom up approach methodology [55, 56]. This approach is advantageous in the sense that it can give rise to smaller size nanoparticles as well as creation of less defects. As far as synthesis of core shell nanohybrid is concerned, bottom up approach is found to be more convenient as it gives better control in achieving uniform coating. The two approaches can also be used in combination, for example the core particles can be synthesised via top down approach and then coating can be carried out by bottom up approach so as to maintain more regular shell thickness [57].

A two-step process is followed for the synthesis of core/shell nanohybrids. The synthesis strategy can be categorised into two types depending on core particles availability: (i) In first method, synthesis of core particles is done separately followed by washing and drying. Subsequently, appropriate surface modification of the core is carried out in order to coat it with the shell material. (ii) In the second method, core particles are synthesised in situ and afterwards coating with shell is done. The main advantage offered by first method is that the core particles obtained are in the pure form and hence chances of impurities on its surface would be less. While, in the in situ approach, probability of having some impurities lying between core and shell increases [58, 59].

The two things which are of utmost concern in the synthesis of core/shell nanohybrids is to develop a uniform coating of shell around the core and to have a control over its thickness. The major problems encountered by researchers in achieving these goals is primarily because of the following reasons: (i) core particles get agglomerated in reaction media, (ii) separate formation of shell particles instead of onto the core, (iii) core surface is impartially covered, (iv) managing the rate of reaction. Various routes such as microemulsion, precipitation, polymerization, sol–gel condensation, layer by layer adsorption have been adopted for the synthesis of core shell nanohybrids [6, 60–62]. Considering the material properties only, synthetic methods utilised for the preparation of inorganic and organic materials irrespective of whether they are acting as core or shell are briefly described below:

9.1 Synthesis of Inorganic CSNP

Inorganic core–shell nano materials can be divided into three classes: (i) metal, (ii) oxides of metal or metalloid, (iii) metal salts and metal chalcogenides.

9.1.1 Synthesis of Metallic CSNP

Following strategies are adopted for the synthesis of cores or shells made up of metal.

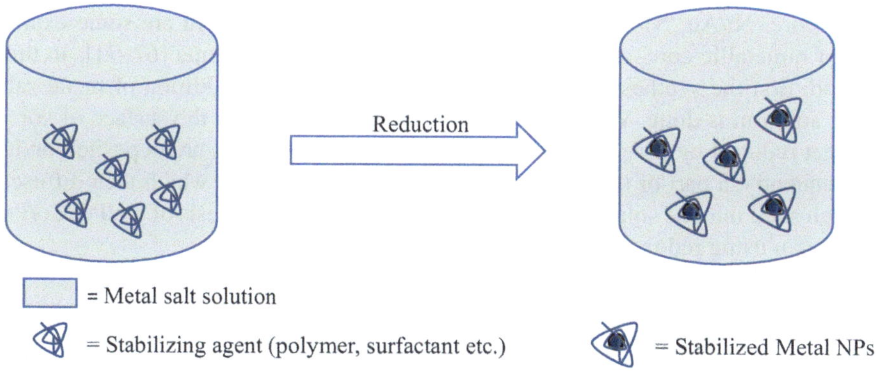

Fig. 9 Synthesis of metal NPs via reduction of metals salts

Reduction

It involves the reduction of a metal salt in the presence of a stabilizer. Various reducing agents have been used such as sodium borohydride, hydrazine, lithium aluminium hydride, tannic acid, sodium citrate; the former two being the most common [6, 63, 64]. Apart from the chemical reduction, greener reducing agents such as plant extract, bacteria and fungi have also been tried in order to reduce the use of harmful chemicals (Fig. 9). The main advantages associated with chemical reduction are its simplicity, cost-effectiveness and ability to control the size of particles by regulating various parameters such as nature of reducing agent, type of stabilizer, amount of reducing agent relative to salt precursor and the molar ratio of salt precursor and stabilizer [65, 66]. The choice of a reducing agent is a significant factor because size, shape as well as particle size distribution are strongly affected by it. For the reduction of metal salts, the reactivity of reducing agent needs to be adjusted appropriately. This is because if during the synthesis the rate of reaction is too fast, large number of nuclei will be formed rapidly resulting in generation of smaller size nanoparticles. Whereas if rate of reaction is very small, agglomeration is more likely to take place giving rise to bigger size nanoparticles.

Reduction Transmetallation

This technique is mainly used for the formation of bimetallic type of core–shell nanoparticles. Compared to common reduction methods, redox transmetallation offers numerous advantages for the synthesis of core/shell nanoparticles: (i) Additional reducing agent is not required (ii) Occurrence of spontaneous deposition of shell layer onto the surface of core (iii) Self Nucleation of shell is prevented. The major demerit associated with this method is that only those bimetallic core shell nanoparticles can be synthesised where reduction potential of shell is more than that of the core. Ni/Au, Ni/Ag, Ag/Au, Co/Pt, Co/Cu, Co/Pd, Co/Au are some examples of bimetallic core–shell structures prepared by this technique [67–71]. In this method, first the synthesis of core is carried out. Thereafter addition of metal salt to the solution is done. When these salts come in contact with the surface of core, they get reduced by core atoms present on the surface and thus are deposited on it. Consequently, a part of the metallic core is oxidised to the salt which then diffuses through shell into the solution. Woo-ram Lee reported the synthesis of Co/Pd, Co/Pt, and Co/Cu using redox tranmetallation route [67].

Thermal Decomposition

Metal nanoparticles especially that of core can also be synthesised by thermally decomposing the organometallic compounds at high temperature. The surfactants are generally employed for the stabilisation of nanoparticles. A very common example is thermal decomposition of $Co_2(CO)_8$. As metal is already in zero-valent state it is possible to dissociate M-CO bond in organic solvent through thermal activation in the presence of a stabilizer [72]. Hess et al. reported the synthesis of Co nanoparticles through decomposition of $Co_2(CO)_8$ at 110 °C in toluene [73]. Likewise, Fe NPs have been prepared by decomposition of $Fe(CO)_5$. Thereafter nanoparticles are precipitated and then stored in hexane so as to prevent the oxidation of surface. As the metal carbonyls are expensive and toxic, therefore other organometallic precursors such as [bis(salicylidene) cobalt(II)] oleylamine complex have also been tried for the synthesis of cobalt NPs [74]. Apart from organometallic compounds, metal nanoparticles are also made through thermal decomposition of metal complexes. MasoudSalavati-Niasari reported the synthesis of copper NPs by thermal decomposition of copper oxalate [75]. Similarly, synthesis of Ni and Au nanoparticles from thermal decomposition of their corresponding acetates have been reported [76, 77]. Once the metal is separated, it is washed with anhydrous alcohol for removing the stabilizer from the surface and then coating with shell is carried out. This technique is more helpful in the preparation of core shell nanohybrids of metal/metal oxide type. This is because after the synthesis of core, the only thing needed to be done is the removal of stabilizer from the surface of the core. After this surface atoms will get oxidised by atmospheric oxidation to metal oxide.

Wire Electrical Explosion (WEE)

It involves the application of a high voltage across a thin wire under inert conditions in a closed chamber having high pressure. This leads to the generation of an intense pulse with large current density. When such huge current passes through the wire, it becomes overheated causing it to evaporate rapidly. As a result, metal nanoparticles of different size are formed inside chamber. WEE is not commonly used in industry as it is highly expensive and secondly cannot be used for all metals. It can be employed for only those metals which have high electrical conductivity and can be easily

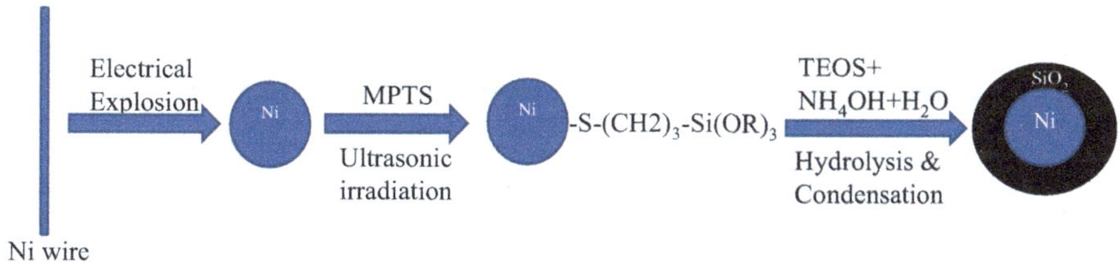

Fig. 10 Synthesis of Ni@SiO_2 via WEE

obtained in the form of thin wire. Nanoparticles of metals such as Cu, Al, Ni and core shell alloys such as Cu/Zn, Ti/Ni have been synthesised via WEE [78, 79]. The schematic diagram for the synthesis of Nickel nanoparticle using wire electrical explosion followed by coating with silica shell is shown in Fig. 10.

9.1.2 Synthesis of Oxide Nanoparticles

They form an important class of materials with interesting applications. The widely used methods for their synthesis are described below.

Sol–Gel Method

It comes under wet chemical techniques and is widely employed for the the synthesis of metal oxide nanoparticles. Hydrolysis followed by polycondensation are the two key steps leading to the generation of a solid phase network gradually. Metal alkoxides and metal salts are commonly used as precursors in solgel method. By varying various parameters, a wide range of materials such as fine powder, aerogel, xerogel, thin films and monoliths can be prepared [80, 81]. The overall process can be written as:

(1) Hydrolysis of precursors (usually alkoxides or chlorides)

$$M - OR + H_2O \rightarrow M - OH + R - OH$$

Here, M = metal, R = alkyl

(2) Polycondensation: This step involves the formation of metal oxide linkages with elimination of water or alcohol. The polymeric network gradually grows and ultimately reaches colloidal dimensions. The size of colloidal factors depends mainly on the pH of medium and ratio of precursor to water.

$$M - OX + M - OH \rightarrow H - OX + M - O - M$$

Here, X = alkyl or H.

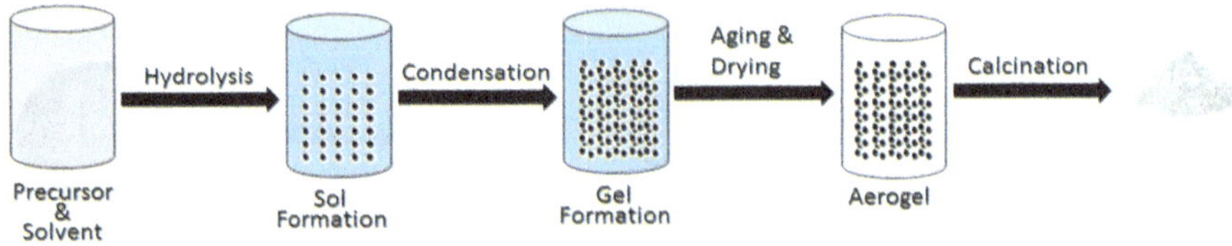

Fig. 11 Synthesis of metal oxide nanoparticles using sol–gel method

(3) Aging: During this process the polycondensation reaction continues to take place.
(4) Drying: There are different ways in which drying can be done: Supercritical, thermal, and freeze drying leading to formation of aerogel, xerogel and cryogel respectively.

The schematic for the process is shown in Fig. 11 [82, 83].

Coprecipitation

Precipitation reaction involves the reaction of two or more than two water soluble salts to generate at least one insoluble salt which precipitates out of the media. The solubility product is the key parameter for precipitation reaction. This is because initially the particles formed will be in liquid phase, and once their concentration exceeds solubility product only then particle formation will begin. This technique has been widely used for the synthesis of iron oxide nanoparticles and various ferrites. K. Petcharoen reported the synthesis of magnetite nanoparticles by the coprecipitation of $FeCl_2.4H_2O$ and $FeCl_3$ in the presence of ammonium hydroxide which is acting as precipitating agent [83]. Both hexanoic and oleic acid were used as coating agents for the stabilisation of formed nanoparticles. Finally, the formed coated nanoparticles were filtered and subsequently washed. First washing is carried out with water in order to remove Cl^- ions and then with ethanol so as to get rid of excess coating agent. Similarly, Yeong Il Kim prepared $CoFe_2O_4$ nanoparticles by the coprecipitation of $CoCl_2.6H_2O$ and $FeCl_3.6H_2O$ using sodium hydroxide as the precipitating agent [84]. They also found that particle size can be tuned by properly regulating the temperature during coprecipitation. Apart from ferrites, other metal oxide nanoparticles such as ZnO have been prepared following this route. The extension of this route for the synthesis of core/shell nanohybrids is simple. For instance, after the synthesis of Fe_3O_4 core by co-precipitation, urea and titanium sulfate are added in the same solution. Decomposition of urea generate ammonia which undergoes reaction with $Ti(SO_4)_2$ to form TiO_2, thus ultimately forming Fe_3O_4/TiO_2 core shell structure [85].

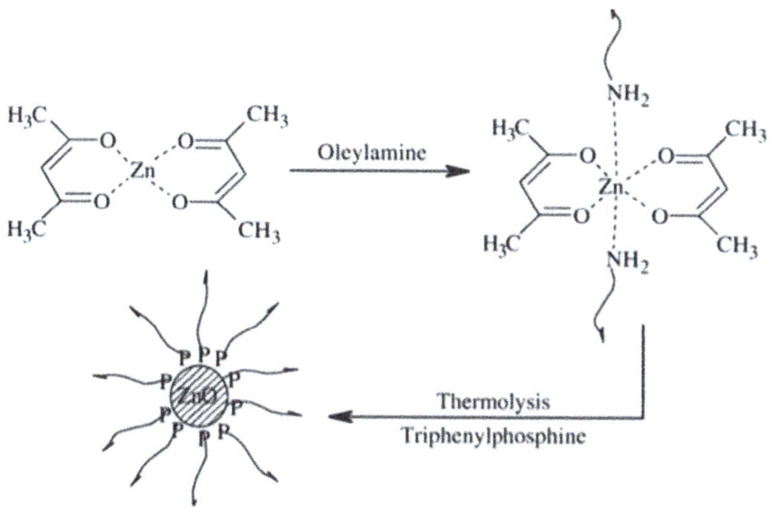

Fig. 12 Synthesis of ZnO NPs by the thermal decomposition of Zn-Oleylamine (Reprinted with permission from Ref. [88])

Thermal Decomposition

Metal complexes can be decomposed by heat treatment to generate metal oxide, a process also known as thermolysis. This methodology is mainly adopted for the synthesis of core particles and is quite simple. Various transition metal oxides have been synthesised following this route. Teresa Palacios-Hernández synthesised Co_3O_4 and CuO nanoparticles by performing calcination of tartarate complexes of Co and Cu complexes respectively for 5 h at a temperature of 500 °C [86]. Similarly, Sun et al. prepared Fe_3O_4 nanoparticles by thermal decomposition of $Fe(acac)_3$. Oleic acid and oleyl amine were added in the reaction mixture as stabilizer [87]. They also found that particle size can be tuned by changing the reaction conditions or by following seed mediated growth. Likewise, ZnO nanoparticles are prepared by thermally decomposing [bis(acetylacetonato)zinc(II)]–oleylamine complex. Oleyl amine and triphenylphosphine were used for restricting the particle size. The synthetic scheme is shown in Fig. 12 [88].

9.1.3 Synthesis of Metal Salt and Metal Chalcogenide Nanoparticles

Metal chalcogenides, especially sulfides of transition elements are employed as semiconducting materials in electronic industry and metal salts of rare earth metals are utilized for bioimaging. The synthesis of this class of materials is usually carried out by the use of precipitation reaction. Various semiconducting core shell nanohybrids which have been prepared via precipitation are CdS/ZnS, CdS/PbS, CdS/CdSe, ZnS, CdSe etc. [89–92].

9.2 Synthesis of Organic Nanoparticles

Most of the organic nanoparticles come under the class of polymers. Therefore, the technique most commonly used for the synthesis of organic core or shell is the addition or condensation polymerisation. There are various techniques by which polymerisation can be done- bulk, solution, suspension and emulsion polymerization. Generally, the synthesis of organic core is carried out by emulsion polymerisation and then it is used as a template for the preparation of core shell nanostructure [93, 94]. The development of polymeric shell on either the organic or inorganic material is usually done insitu. For the enhancement of shell coating the surface of core is commonly modified with a polyelectrolye in case of an organic core or via a surfactant for the inorganic core. After surface modification is achieved, polymerisation proceeds on the surface of the core by the technique of solution or bulk polymerisation [95]. A variety of nanohybrids have been synthesised where polymeric nanoparticle is playing the role of either core or shell or both. For instance, Chunlei Wang et al. prepared $Fe_3O_4@CaCO_3@PMMA$ core shell nanohybrid by insitu emulsion polymerisation on the surface of oleic acid altered $Fe_3O_4@CaCO_3$ nanoparticles. The synthetic scheme is shown in Fig. 13 [96]. The advantage gained by having a polymeric coating of hydrophilic nature around an inorganic core is an enhancement in the biocompatibility of the core.

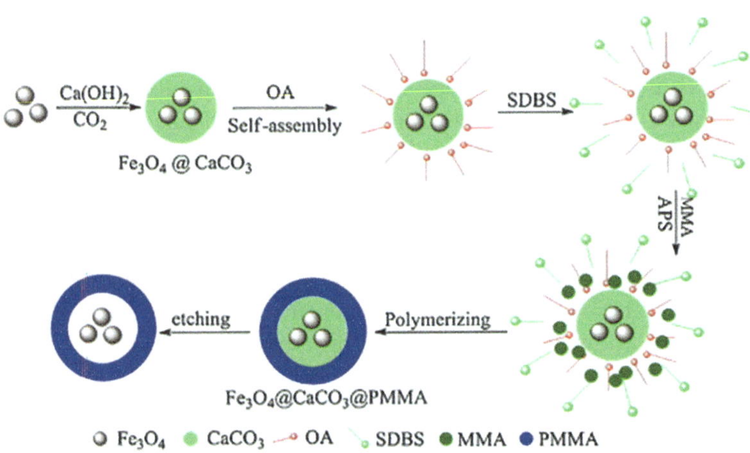

Fig. 13 $Fe_3O_4@CaCO_3@PMMA$ core shell nanohybrid prepared by insitu emulsion polymerisation on the surface of oleic acid altered $Fe_3O_4@CaCO_3$ nanoparticles [OA: Oleic Acid, SDBS: Sodium dodecylbenzene sulfonate, MMA: Methyl methacrylate, PMMA: Poly Methyl methacrylate] (Reprinted with permission from Ref. [97])

9.3 General Characterisation of Core Shell Nanohybrid

After the successful synthesis of core shell nanohybrid, the next task is to analyse it with respect to size, shape, particle size distribution, morphology and structural composition. Dynamic light scattering (DLS) which is also known by the name of photon correlation spectroscopy or quasi-elastic light scattering is used for the measurement of size of particles. The technique works by measuring the Brownian motion and correlate it with the size of particle. The bigger the size of particle, the slower will be its Brownian motion. The diameter that is obtained by DLS is actually hydrodynamic diameter because it comes from the way in which a particle diffuses in a fluid. Also, the diameter acquired refers to the diameter of a sphere having identical translational diffusion coefficient to that of the particle. Ionic strength of the medium, alternation in surface structure and degree of non-sphericity in particle have a significant influence on the result of DLS measurements. Further, temperature should be kept stable, otherwise convection currents will result in having a non-random movement in fluid leading to erroneous results. For the determination of shape as well as surface morphology, scanning electron microscopy (SEM) is employed. It is usually done in conjunction with energy dispersive spectrometry (EDS) in order to have a better picture of structural composition. Whereas if one wants to know information about inner structure, transmission electron microscopy (TEM) proves to be very useful. Atomic force microscopy (AFM) is beneficial for checking the uniformity of shell coating onto the core. In order to study the surface functionalisation, Fourier transform infrared (FTIR) spectroscopy in different modes-transmittance, reflectance and attenuated total reflectance (ATR) as well as Raman spectroscopy is widely used. For the identification of crystal structure, X-Ray diffractomers are utilized. If the core shell nanohybrid is to be used for water decontamination or for any other application relying on surface, determination of surface area becomes very much crucial. Brunauer–Emmett–Teller (BET) isotherm is usually used by researchers for finding the surface area of a nanosorbent. Zeta potential is an important parameter which determines the stability of nanohybrids and its variation with pH aids in the mechanistic understanding of sorption. The measurements are done with the help of a zetasizer. For characterising the composition, numerous techniques such as FTIR, X-Ray fluorescence, inductively coupled plasma atomic emission spectrometry, inductively coupled plasma mass spectrometry, flame and graphite furnace atomic absorption spectrometry are available.

9.4 Application of Core–Shell Nanohybrids for Water Treatment

Water contains numerous kinds of pollutants ranging from organic dyes, heavy elements, radionuclides, nanoparticles, pharmaceuticals, pesticides etc. Here, we have discussed the role of nanohybrids in the treatment of water.

9.5 *Removal of Conventional Pollutants*

Heavy metals are naturally present in earth but anthropogenic activities have caused tremendous increase in their concentration thus posing a major risk on the health of plants, animals and humans. Typical conventional pollutants like As, F, Cr, Cu, Hg, Ni etc. can enter the human body through drinking water causing serious health effects [97, 98]. Several regions of the world are already facing higher concentrations of arsenic and fluoride in groundwater. Various core shell nanohybrids have been reported for the removal of conventional pollutants. Abdel Halim A. Saad et al. fabricated ZnO/Chitosan core shell nanocomposite and used it for the removal of Cu^{2+}, Cd^{2+} and Pb^{2+} [99]. The reusability of the material was found to be high. Meng Zhang prepared Ni/Mg(OH)2 and used it for the adsorption of Zn^{2+}, Cu^{2+} and Cd^{2+} ions. Magnesium hydroxide is an ideal sorbent for heavy metal ions and has been already used for the removal of nickel, lead, cadmium, zinc, copper and cobalt [11, 100, 101]. But the problem associated with it is its poor recyclability. In view of this, the researchers synthesised $Ni/Mg(OH)_2$ core shell nanostructure as it can be easily removed from the media by application of external magnetic field [102]. Apart from this the shell acts as a protective layer and aids in preventing the oxidation of Ni to NiO, thereby enabling it to retain its magnetism. Further, removal efficiency of all the metals was found to be close to 100% and remained 95% even after 5 cycles. The recycling strategy is shown as a schematic in Fig. 14.

Another example of a nanohybrid where the shell leads to enhancement in the adsorptive capacity is a core double shell structure prepared by KairuZhe where in the core is made up of magnetic Fe_3O_4 nanoparticles having an inner polydopamine shell and an outer shell made up of zeoliticidazolate frameworks-8 (MP@ZIF-8). The adsorption capacity of MP@ZIF-8 was found to be 136.56 mg g^{-1} for Cr(VI) which is greater as compared to MP (92.27 mg g^{-1}). Further from XPS studies it

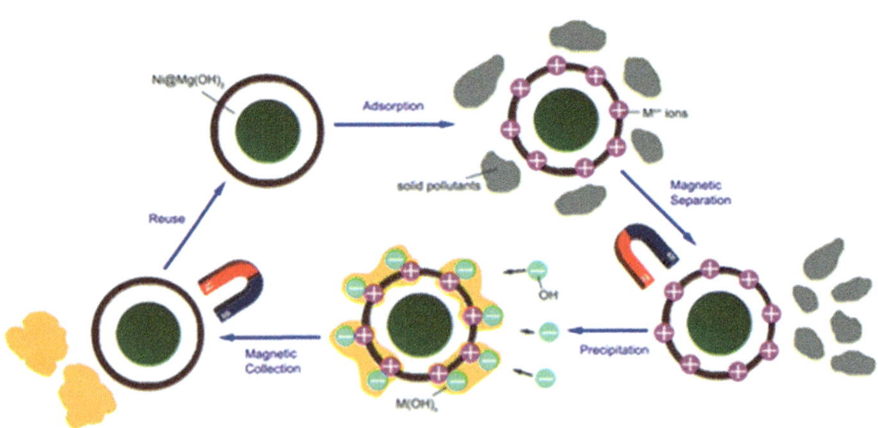

Fig. 14 The recycling strategy of $Ni@Mg(OH)_2$ water treatment agent (Reprinted with permission from Ref. [103])

was found that Cr(VI) was partially reduced to Cr(III) during the adsorption process. The authors attributed this to the presence of nitrogen atom groups on polydopamine and ZIF8 which are acting as reducing agents [103]. In another study, core shell silica coated Fe_3O_4 functionalised with diglycolic acid (FGA-1) was found to be a better adsorbent for the removal of Pb(II) and Cr(VI) ions [104]. The magnetic core has the advantage of ease of isolation by the application of external magnetic field. Further, the presence of different functional groups around the core protects it against agglomeration and also enhances its selectivity. Different sort of chemical modification of core can be done in order to make it selective for the capture of desired metal ions. As the prepared nanosorbent (FGA-1) bears terminal acidic moieties and since it is known that Pb(II) and Cr(VI) ions exhibit excellent binding towards acidic functionalities, they are chosen for carrying out further uptake measurements. From the zeta potential studies, it was found that the surface of nanosorbent is negatively charged at neutral pH, which was attributed to the ionisation of carboxylic acid. As a result, adsorption of Pb(II) is preferred over Cr(VI) ions. However, at lower pH the uptake of Cr(VI) was found to be higher, which the authors thought might be due to the protonation of sorbent. The adsorption capacity for both the ions was found to be high under optimised conditions with reasonably good recyclability upto 5 cycles.

Numerous core–shell nanohybrids have been reported in the literature highlighting their immense significance in the adsorption of a wide range of ions (Zn(II), Cd(II), Hg(II), Pb(II), Cu((II), Cr(VI), Cr(III), Ni(II), Co(II), Co(III), As(VI), As(III) etc.), a comprehensive list of which is outlined in Table 1 [99–141]. Though actinides are too heavy elements, they are dealt in a separate subsection which is exclusively dedicated to role of core shell nanohybrids in radionuclide removal. The table also enlists the sorption capacities of different sorbents for the sake of comparison. But it should be noted that the maximum adsorption capacity reported necessarily need not be at room temperature and the working pH is different for different nanohybrids.

9.6 Radionuclide Removal

Water resources can get contaminated by radioactive materials. It may occur due to nuclear weapon testing, nuclear accidents, leakage from nuclear power and reprocessing plants. The devastating earthquake and tsunami that struck Japan has caused drastic damage to the Fukushima Nuclear Power Plant resulting in the release of tons of water contaminated with radionuclides into the ocean in March, 2011. Other than this, various steps involved in oil extraction also poses threat to the radioactive contamination of water. As hospitals also use numerous isotopes for diagnostic and therapeutic purposes; improper disposal of hospital waste also contributes to the generation of radioactive waste into water. The challenge to meet the energy demand is rising throughout the globe and nuclear power industry is more likely to grow in the coming years, with the risk of generating more and more radioactive waste into water bodies. In view of this, researchers have synthesised various core–shell nanohybrids for the removal of radionuclides [144].

Table 1 Core–shell nanohybrids for removal of conventional pollutants

Core–shell nanohybrids	Pollutant removed	Sorption capacity (mg g^{-1})	References
ZnO/Chitosan	Pb (II)	476.1	[100]
(Fe_3O_4@[ECH-co-TETA]n-g-CSSNa)		293.8	[106]
$MnFe_2O_4$@Mn–Co oxide		481.2	[107]
CdS/ZnS			[108]
Fe_3O_4@PDA@L-Cys		46.95	[109]
Carboxymethyl β-cyclodextrin bonded Fe_3O_4@SiO_2–NH_2		170	[110]
Fe_3O_4@ZIF-8		719.42	[111]
Ni/NiO		308 m	[112]
FeO@SiO_2-DNA		Not reported	[113]
Fe_3O_4-mesoporous magnesium silicate (FMMS)		223.2	[114]
Fe_3O_4@C		Not reported	[115]
PHCS-15@ZIF-8		462.9	[116]
$CaCO_3$/HPC		677.6	[117]
Fe_3O_4@DAPF CSFMNRs		83.3	[118]
Fe_3O_4@TA–Fe^{3+}		1115.3	[119]
$CoFe_2O_4$@SiO_2–NH_2		181.6	[120]
Fe_3O_4@MnO_2		666.67	[121]
MgO@mesoporous silica spheres		3297	[122]
MnO_2@Fe_3O_4/PmPD		438.6	[123]
Fe_3O_4@SBA-15		243.9	[124]
ZnO/Chitosan	Cd (II)	135.1	[101]
Ni/Mg(OH)2			[104]
(Fe_3O_4@[ECH-co-TETA]n-g-CSSNa)		256.69	[106]
$MnFe_2O_4$@Mn–Co oxide		345.5	[107]
Fe_3O_4@SiO_2@MO		59.7	[125]
Ni/NiO		72	[112]
Fe_3O_4@C		Not reported	[115]
$CoFe_2O_4$@SiO_2–NH_2		199.9	[120]
Fe_3O_4@SiO_2@polypyrrole		120	[126]
MnO_2@Fe_3O_4/PmPD		121.5	[123]
Porphyrin-functionalized Fe_3O_4@SiO_2	Hg (II)		[127]
CdS/ZnS		Not reported	[108]
$CoFe_2O_4$@SiO_2–SH,		641.0	[128]
$CoFe_2O_4$@SiO_2-EDTA		103.3 mg	[129]

(continued)

Table 1 (continued)

Core–shell nanohybrids	Pollutant removed	Sorption capacity (mg g^{-1})	References
Fe_3O_4@SiO_2@PTL		701.51	[130]
FeO@SiO_2-DNA		Not reported	[113]
Fe_3O_4@TA–Fe^{3+}		279.4	[119]
Fe_3O_4@silica-NH_2		126.7	[131]
ZnO/Chitosan	Cu (II)	117.6	[100]
Ni/$Mg(OH)_2$		Not reported	[104]
(Fe_3O_4@[ECH-co-TETA]n-g-CSSNa)		277.93	[106]
$MnFe_2O_4$@Mn–Co oxide		386.2	[107]
PMMA/PEI		14	[132]
$MnFe_2O_4$@TiO_2 decorated on rgo		225.99	[133]
Fe_3O_4@SiO_2@MO		71.6	[125]
Fe_3O_4@ZIF-8		301.33	[111]
Fe_3O_4-mesoporous magnesium silicate (FMMS)		53.5	[114]
Fe_3O_4@C		Not reported	[115]
$CoFe_2O_4$@SiO_2–NH_2		177.8	[120]
Fe_3O_4@Carbon@ZIF-8		234.74	[134]
Ni/$Mg(OH)_2$	Zn(II)		[104]
(Fe_3O_4@[ECH-co-TETA]n-g-CSSNa)		225.07	[106]
Fe_3O_4@mSiO_2@CS@PANI	Cr (VI)	249.6	[135]
SiO_2–TiO_2		13.91	[136]
CoO/MoO_2/Mo_2C		294.1	[137]
Fe_3O_4@C		Not reported	[115]
Fe_3O_4@mesoPDA		574.71	[138]
Fe_3O_4@PmPDs		246.09	[139]
Fe_3O_4@SiO_2-DTPA-DTC	Co (II)	98.47	[140]
Fe_3O_4@CD MNPs		38.03	[141]
$CaCO_3$/HPC	Co (III)	308.5	[117]
Fe_3O_4@SiO_2@polypyrrole	Ni (III)	98	[126]
Fe_3O_4@ZIF-8	As(III)	100	[142]
FeO@SiO_2-DNA		Not reported	[113]
CMNP@PmPDs	As (V)	95.2	[143]

For instance, Yan Liu et al. prepared $FeS@Fe_3O_4$ core shell nanohybrid and utilised it for the adsorption of uranyl ions [148]. The synthesis was carried out via ultrasonic assisted strategy. The nanohybrid proved to be a better adsorbent as compared to FeS and Fe_3O_4 alone. This is supported by their adsorption capacities. For $FeS@Fe_3O_4$, the adsorption capacity came out to be 229.03 mg g^{-1} which is considerably higher than FeS (211.0 mg g^{-1}) and Fe_3O_4 (154.0 mg g^{-1}). This happens because bare FeS can get easily aggregated, leading to reduction in its sorption capacity. Once it gets encapsulated by Fe_3O_4 layer, it becomes more stable and dispersible; thereby enhancing its adsorption capacity. The results from surface area measurement are also in agreement with this. The surface area of hybrid material is 82.47 m^2g^{-1} which is significantly larger than that of FeS. Another example highlighting the immense importance of core shell nanohybrid material is the synthesis of silica coated Fe_3O_4 functionalised with amidoxime. The magnetic core was selected for facilitating the separation of nano-sorbent [150]. But since, the magnetic nanoparticles are susceptible towards oxidation and agglomeration, surface modifications are carried out in order to overcome these limitations. In view of this, the researchers coated the surface of magnetite nanoparticles with silica because of its high stability and ability to functionalise it with different kinds of groups due to the presence of hydroxyl groups on the surface. Finally the core shell material, $Fe_3O_4@SiO_2$ was functionalised with amidoxime to prepare $Fe_3O_4@SiO_2$-AO and various studies pertaining to adsorption of uranyl ions were conducted. From the speciation diagram and pH studies, it was proposed that oxime groups and amino nitrogen are responsible for binding with uranyl ion. A consolidated list of various core–shell materials developed for radionuclides removal from water have been listed in Table 2 [143–155].

9.7 *Removal of Dyes*

Water pollution because of dyeing industry is a matter of grave concern since a huge amount of it is released into water bodies every year. As the solubility of dyes in water is high, it becomes really difficult to eradicate them by conventional means. The presence of dyes in water leads to a reduction in the penetration of light into water bodies, thereby causing a decrease in the photosynthesis rate and hence lowering the dissolved oxygen levels. All this adversely affects the life of entire aquatic organisms. The dyes are also reported to have mutagenic and carcinogenic effects and once they enter the food chain can readily undergo biomagnification and thus species at higher trophic level are more likely to get affected. Thus, it becomes very necessary to find measures to deal with dye pollution. Among various ways, adsorption of dyes by nanomaterials can prove to be a promising approach. In particular, core shell naonybrids have emerged as recent adsorbents for tackling dye pollution. Various core–shell nanohybrids have been used for the adsorption of diverse kinds of dyes ranging from acidic to basic. Zikang Xiong et al. synthesised $Fe_3O_4@Carbon@ZIF$-8 core shell nanomaterial and used it for the adsorption of Congo red dye [133].

Table 2 Core–Shell nanohybrids for removal of radionuclides

Core- shell nanohybrid	Radionuclide removed	Sorption capacity ($mg\ g^{-1}$ unless specified)	References
Fe@FeO	U (VI)	353	[145]
MNPs@SCSOEt		216.45	[146]
Fe_3O_4@SiO_2@Ni-L		129.26	[147]
Fe_3O_4@C–KO		38.7	[148]
Fe_3O_4@PDA@PAO MNPs		162.5	[149]
FeS@Fe_3O_4		229.03	[150]
Fe_3O_4@C@ASA		46.2	[151]
Fe_3O_4@SiO_2-AO		0.441	[152]
Alginate-Ca/attapulgite clay		199.345	[153]
Fe_3O_4@PDA		56.39	[154]
Nano-PY-MCM-41@Fe_3O_4 (NPMF)		246.322	[155]
PDA@MgAl-LDHs		142.86	[156]
C@$Na_2Ti_3O_7 \cdot 9H_2O$		8.151	[157]
Fe_3O_4@TiO_2		313.6	[158]
h-Fe_3O_4@mC		0.566	[159]
Fe_3O_4@ZIF-8		539.7	[160]
Fe_3O_4@TiO_2		118.8	[161]
C@$Na_2Ti_3O_7 \cdot 9H_2O$	Eu (III)	4.846 $mmol\ g^{-1}$	[157]
PDA@MgAl-LDHs		76.02	[156]
h-Fe_3O_4@mC		1.013 mmol g – 1	[159]
Fe_3O_4@HA MNPs		6.95×10^{-5} mol g–1	[162]
Fe_3O_4@ZIF-8		255.6	[160]
h-Fe_3O_4@mC	Sr (II)	0.733 mmol g – 1	[159]
zeolite@Alg–Ca		83.31	[163]
MW@AMS		80.01	[164]
Fe_3O_4@WO_3@GO		61.42	[165]
h-F_e3O_4@mC	Co (II)	0.860 mmol g – 1	[159]
MW@AMS		87.72	[164]
C@$Na_2Ti_3O_7 \cdot 9H_2O$	Cs (I)	5.757 $mmol\ g^{-1}$	[157]

The adsorption capacity of the prepared nanosorbent was found to be very high (806.45 mg g^{-1}). For the easy removal of adsorbents, magnetic cores are commonly preferred and are protected using a shell. The researchers chose amorphous carbon as the inner shell as it is easier to synthesise and more environmentally benign as compared to silica or other polymeric shells. Apart from this, it also plays the role of linker as well as stabilizer between magnetic core and outer shell layer. Metal Organic frameworks [Zeolitic Imidazolate Framework-8 (ZIF-8)] were selected as the outer shell material because of its high surface area, availability of numerous unsaturation sites, tuneable pore size and outstanding chemical stability. Further ZIF-8 was particularly chosen because of its high selectivity and superior water stability. To gain better insight into mechanistic aspects, FTIR and XPS studies are carried out after and before adsorption. It was concluded that electrostatic interaction, pi-pi bonding and hydrogen bonding are the driving forces behind the adsorption of congo red dye. Finally selectivity of the adsorbent was examined in the presence of other anionic and cationic dyes and it was revealed that Fe_3O_4@Carbon@ZIF-8 can be successfully used for the selective adsorption of Congo Red in complex wastewater system. Similarly, Fe_3O_4@lignosulfonate/phenolic core–shell microspheres has been used for the adsorption of methylene blue [164]. The nanohybrid material can be separated easily on account of the presence of magnetic core and functional groups of lignosulfonate (shell) are responsible for high adsorption capacity [165]. A list of different core shell nanohybrid materials those have been used for the adsorption of different dyes with their sorption capacities is summarized in Table 3 [165–177].

9.8 Removal of Organic Pollutants

Organic pollutants comprise of pharmaceuticals, pesticides, insecticides, organic solvents and are harmful for human beings because of the concentration in which they are found in aquatic medium. Pharmaceutical pollution also knows as drug pollution occurs when drugs or their metabolites reach the water bodies. This leads to various adverse consequences. To mention a few, it has been revealed from numerous studies that estrogen and chemicals which behave similar to it result in a feminizing effect on fishes thus causing an increase in the population of intersex and female fishes. Apart from this, various commonly used antidepressants were found to be accumulated in the tissues of brain in fishes. Further, as antibiotics are largely used for the treatment of variety of infections, their presence in sewage treatment systems can lead to the inhibition of sewage bacteria, thereby deteriorating the decomposition of organic matter. Some studies also indicate that presence of antibiotics in water has given rise to antibiotic resistance. Besides this, pesticides also poses threats to environment due to their presence in drinking water in significant concentration. Organic solvents are widely used in almost all industries and are too a major contributor in water pollution. A variety of methods are available for the removal of organic pollutants, the particular method being chosen depends upon the nature of pollutant [178–180]. Core–Shell nanohybrids also provide a means of getting rid of these organic pollutants from water bodies. Most of the core shell nanohybrids either employ adsorption or photocatalytic degradation for removing the pollutant a list of which is given in Table 4 [178–188] and. As already discussed, many times, most of the core–shell nanosorbents synthesised use magnetite as a core for its facile recovery. The shell material and its further functionalisation is carried out on the basis of nature of pollutant to be removed.

Table 3 Core–shell nanohybrids for removal of dyes

Core- shell nanohybrid	Dye removed	Sorption capacity (mg g^{-1})	References
Co-MOF@AC	Methylene blue	128.72	[166]
Ni-MOF@AC	Methylene blue	127.08	[167]
Fe_3O_4@Carbon@ZIF-8	Congo red	806.45	[134]
$CoO/MoO_2/Mo_2C$	Methyl orange	555.6	[137]
$CoO/MoO_2/Mo_2C$	Methylene blue	384.6	[137]
$CoO/MoO_2/Mo_2C$	Congo red	219.8	[137]
Fe_3O_4/NiO	Congo red	210.78	[168]
Fe_3O_4@PDA	Methylene blue	10	[169]
Fe_3O_4@Tb/AMP ICP	Alizarin Red	357.14	[170]
Fe_3O_4@Tb/AMP ICP	Congo Red	909.09	[170]
Fe_3O_4@DAPF	Amaranth	142.05	[171]
Fe_3O_4@DAPF	Orange II	121.07	[172]
Fe_3O_4@DAPF	Acid red 18	99.60	[171]
Fe_3O_4@C	Methyl orange	2.76	[172]
Fe_3O_4/C	Methylene blue	44.38	[172]
Fe_3O_4/C	Cresol red	11.22	[172]
Fe_3O_4@C	Methylene blue	117	[172]
Fe@G	Basic Yellow 28	52.36	[173]
Fe@G	Basic Red 46	46.73	[173]
Fe_3O_4@SiO_2/PEI	Methyl orange	231.0	[174]
Fe_3O_4@SiO_2/PEI	Congo red	134.6	[174]
Fe_3O_4@MnO_2	Methylene blue	Not reported	[175]
Fe@G-N	AcidRed 88	63.7	[176]
Fe@G-N	DirectOrange 26	42.7	[176]
Fe_3O_4@lignosulfonate/phenolic microsphere	Methylene blue	283.6	[166]
Al_2O_3@ZnO	Congo red	714	[177]
Fe_3O_4@MIL-100(Fe)	Methylene blue	73.8	[178]
Fe_3O_4@SiO_2@Zn – TDPAT	Methylene blue	58.67	[179]
Fe_3O_4@SiO_2@Zn – TDPAT	Congo red	17.73	[179]

9.9 Conclusion

In this book chapter, various core shell nanohybrid structureshaving application in water decontamination have been discussed. Core shell nanohybrid structuresare obtained through self-assembling techniques, sol–gel method, using polymers for

Table 4 Core–shell nanohybrids for removal of organic pollutants

Core- shell nanohybrid	Dye removed	Sorption capacity (mg g^{-1})	References
NFO@SiO_2@APTS	Ibuprofen	59	[180]
NFO@SiO_2@APTS	Acetaminophen	58	[180]
NFO@SiO_2@APTS	Streptomycin	49	[180]
MMIPs	Dichlorophen	50.45	[181]
TiO_2@C	Tetracycline	240	[182]
TiO_2@C	Ofloxacin	232	[182]
TiO_2@C	Norfloxacin	190	[182]
HTCC	Diclofenac	240.4	[183]
Fe_3O_4@MON-NH_2	Tetrabromobisphenol A	135.9	[184]
Fe_3O_4@polyaniline	Bisphenol A	23.1	[185]
Fe_3O_4@polyaniline	α-naphthol	28.7	[185]
Fe_3O_4@polyaniline	β-naphthol	9.1	[185]
Fe_3O_4@SiO_2–MWCNTs	pentachlorophenol	96.4	[186]
Fe_3O_4@CD MNPs	1-naphthol	235.06	[187]
Fe_3O_4@COFs	Diphenylamine	246.25	[188]
Fe_3O_4@COFs	Benzidine	95.20	[188]
Fe_3O_4@COFs	1-naphthylamine	85.85	[188]
Fe_3O_4@COFs	4-phenylphenol	107.20	[188]
Fe_3O_4@COFs	O-tolidine	123.55	[188]
Fe_3O_4@COFs	Triclosan	3.5	[189]
Fe_3O_4@COFs	Triclocarban	1.5	[189]
Fe_3O_4@PANI-GO	Bisphenol A	14.43	[190]
Fe_3O_4@PANI-GO	α-naphthol	13.19	[190]
Fe_3O_4@PANI-GO	t-octyl-phenol	24.15	[190]

dispersion of building blocks. While major percentage of these nanohybrid structuresare reported for conventional pollutants removal (mainly inorganic, organic pollutants);nanohybrid structures are comparatively less for removal of radionuclides, nanoparticles. Many core shell nanohybrid structures discussed here are potential alternate to conventional water treatment processes due to their advance properties, environment friendliness and cost effectiveness. Most nanostructures, however, are in nascent stage for industrial and large-scale applications due to technical challenges in scaling up, reusability and sometimes owing to lack of stability under ambient environment.

References

1. Roduner E (2006) Size matters: why nanomaterials are different. Chem Soc Rev 35(7):583–592
2. Banerjee S, Dionysiou DD, Pillai SC (2015) Self-cleaning applications of TiO2 by photo-induced hydrophilicity and photocatalysis. Appl Catal B 176:396–428
3. Kim J, Lee HC, Kim KH, Hwang MS, Park JS, Lee JM ... Park HG (2017) Photon-triggered nanowire transistors. Nat Nanotechnol 12(10):963–968
4. Liu Z, Lin CH, Hyun BR, Sher CW, Lv Z, Luo B ... He JH (2020) Micro-light-emitting diodes with quantum dots in display technology. Light: Sci Appl 9(1):1–23
5. Jassby D, Cath TY, Buisson H (2018) The role of nanotechnology in industrial water treatment. Nat Nanotechnol 13(8):670–672
6. Gawande MB, Goswami A, Asefa T, Guo H, Biradar AV, Peng DL ... Varma RS (2015) Core–shell nanoparticles: synthesis and applications in catalysis and electrocatalysis. Chem Soc Rev 44(21):7540–7590
7. Ghosh Chaudhuri R, Paria S (2012) Core/shell nanoparticles: classes, properties, synthesis mechanisms, characterization, and applications. Chem Rev 112(4):2373–2433
8. Kwizera EA, Chaffin E, Shen X, Chen J, Zou Q, Wu Z ... Huang X (2016) Size-and shape-controlled synthesis and properties of magnetic–plasmonic core–shell nanoparticles. J Phys Chem C 120(19):10530–10546
9. Radi A, Pradhan D, Sohn Y, Leung KT (2010) Nanoscale shape and size control of cubic, cuboctahedral, and octahedral Cu– Cu2O core– shell nanoparticles on Si (100) by one-step, templateless, capping-agent-free electrodeposition. ACS Nano 4(3):1553–1560
10. Guo P, Sikdar D, Huang X, Si KJ, Xiong W, Gong S, ... Cheng W (2015) Plasmonic core–shell nanoparticles for SERS detection of the pesticide thiram: size-and shape-dependent Raman enhancement. Nanoscale 7(7):2862–2868
11. Basu H, Saha S, Kailasa SK, Singhal RK (2020) Present status of hybrid materials for potable water decontamination: a review. Environ Sci: Water Res Technol 6(12):3214–3248
12. Singh R, Bhateria R (2021) Core–shell nanostructures: A simplest two-component system with enhanced properties and multiple applications. Environ Geochem Health 43(7):2459–2482
13. Leng Z, Li L, Zhang D, Li G (2018) Tunable green/red dual-mode luminescence via energy management in core-multishell nanoparticles. Mater Des 152:119–128
14. Lei L, Dai X, Cheng Y, Wang Y, Xiao Z, Xu S (2019) Dual-mode color tuning based on upconversion core/triple-shell nanostructure. J Mater Chem C 7(11):3342–3350
15. Dhanalakshimi T, Devi MS (2020) Water pollution. A primer on earth pollution: pollution types and disposal 97
16. WHO (2006) Guidelines for drinking-water quality [Electronic Resource]: Incorporating First Addendum (vol. I) Recommendations, pp 375–377, 448–1456
17. WHO (1993) Guidelines for drinking water quality recommendations. Geneva
18. Yakamercan E, Ari A, Aygün A (2021) Land application of municipal sewage sludge: human health risk assessment of heavy metals. J Clean Prod 319:128568
19. Engwa GA, Ferdinand PU, Nwalo FN, Unachukwu MN (2019) Mechanism and health effects of heavy metal toxicity in humans. Poisoning in the modern world-new tricks for an old dog 10
20. Crini G, Lichtfouse E (2019) Advantages and disadvantages of techniques used for wastewater treatment. Environ Chem Lett 17(1):145–155
21. Rathi BS, Kumar PS (2021) Application of adsorption process for effective removal of emerging contaminants from water and wastewater. Environ Pollut 280:116995
22. Suárez-Iglesias O, Collado S, Oulego P, Díaz M (2017) Graphene-family nanomaterials in wastewater treatment plants. Chem Eng J 313:121–135
23. Peng B, Yao Z, Wang X, Crombeen M, Sweeney DG, Tam KC (2020) Cellulose-based materials in wastewater treatment of petroleum industry. Green Energy Environ 5(1):37–49

24. Yu S, Wang X, Pang H, Zhang R, Song W, Fu D ... Wang X (2018) Boron nitride-based materials for the removal of pollutants from aqueous solutions: a review. Chem Eng J 333, 343–360
25. Wong S, Ngadi N, Inuwa IM, Hassan O (2018) Recent advances in applications of activated carbon from biowaste for wastewater treatment: a short review. J Clean Prod 175:361–375
26. Wan X, Zhan Y, Long Z, Zeng G, He Y (2017) Core@ double-shell structured magnetic halloysite nanotube nano-hybrid as efficient recyclable adsorbent for methylene blue removal. Chem Eng J 330:491–504
27. Alijani H, Shariatinia Z (2018) Synthesis of high growth rate SWCNTs and their magnetite cobalt sulfidenanohybrid as super-adsorbent for mercury removal. Chem Eng Res Des 129:132–149
28. Meaney SP, Follink B, Tabor RF (2018) Synthesis, characterization, and applications of polymer-silica core–shell microparticle capsules. ACS Appl Mater Interf 10(49):43068–43079
29. Guo X, Liu X, Xu B, Dou T (2009) Synthesis and characterization of carbon sphere-silica core–shell structure and hollow silica spheres. Colloids Surf, A 345(1–3):141–146
30. Tang F, Liu L, Alva G, Jia Y, Fang G (2017) Synthesis and properties of microencapsulated octadecane with silica shell as shape–stabilized thermal energy storage materials. Sol Energy Mater Sol Cells 160:1–6
31. Rao KS, El-Hami K, Kodaki T, Matsushige K, Makino K (2005) A novel method for synthesis of silica nanoparticles. J Colloid Interf Sci 289(1):125–131
32. Khanal A, Inoue Y, Yada M, Nakashima K (2007) Synthesis of silica hollow nanoparticles templated by polymeric micelle with core– shell– corona structure. J Am Chem Soc 129(6):1534–1535
33. Deng TS, Zhang QF, Zhang JY, Shen X, Zhu KT, Wu JL (2009) One-step synthesis of highly monodisperse hybrid silica spheres in aqueous solution. J Colloid Interf Sci 329(2):292–299
34. Basu H, Saha S, Mahadevan IA, Pimple MV, Singhal RK (2019) Humic acid coated cellulose derived from rice husk: a novel biosorbent for the removal of Ni and Cr. J Water Process Eng 32:100892
35. Decker S, Klabunde KJ (1996) Enhancing effect of Fe2O3 on the ability of nanocrystalline calcium oxide to adsorb SO2. J Am Chem Soc 118(49):12465–12466
36. Basu H, Saha S, Pimple MV, Singhal RK (2019) Novel hybrid material humic acid impregnated magnetic chitosan nano particles for decontamination of uranium from aquatic environment. J Environ Chem Eng 7(3):103110
37. Basu H, Pimple MV, Saha S, Patel A, Dansena C, Singhal RK (2020) TiO2 microsphere impregnated alginate: a novel hybrid sorbent for uranium removal from aquatic bodies. New J Chem 44(10):3950–3960
38. Nayak MK, Singh J, Singh B, Soni S, Pandey VS, Tyagi S (2017) Introduction to semiconductor nanomaterial and its optical and electronics properties. In Metal semiconductor core-shell nanostructures for energy and environmental applications, pp 1–33. Elsevier
39. Chiozzi V, Rossi F (2020) Inorganic–organic core/shell nanoparticles: progress and applications. Nanoscale Adv 2(11):5090–5105
40. Zhou Z, Liu R (2017) Fe_3O_4@ polydopamine and derived Fe_3O_4@ carbon core–shell nanoparticles: comparison in adsorption for cationic and anionic dyes. Colloids Surf, A 522:260–265
41. Gutierrez AM, Bhandari R, Weng J, Stromberg A, Dziubla TD, Hilt JZ (2019) Novel magnetic core-shell nanoparticles for the removal of polychlorinated biphenyls from contaminated water sources. Mater Chem Phys 223:68–74
42. Karsten S, Nan A, Turcu R, Liebscher J (2012) A new access to polypyrrole-based functionalized magnetic core-shell nanoparticles. J Polym Sci, Part A: Polym Chem 50(19):3986–3995
43. Xian Y, Hu Y, Liu F, Xian Y, Wang H, Jin L (2006) Glucose biosensor based on Au nanoparticles–conductive polyaniline nanocomposite. Biosens Bioelectron 21(10):1996–2000
44. Yan W, Feng X, Chen X, Li X, Zhu JJ (2008) A selective dopamine biosensor based on AgCl@ polyaniline core–shell nanocomposites. Bioelectrochemistry 72(1):21–27

45. Nakate UT, Ahmad R, Patil P, Bhat KS, Wang Y, Mahmoudi T, … Hahn YB (2019) High response and low concentration hydrogen gas sensing properties using hollow ZnO particles transformed from polystyrene@ ZnO core-shell structures. Int J Hydrogen Energy 44(29):15677–15688
46. Zhao Y, Chen Z, Zhu X, Möller M (2016) A facile one-step approach toward polymer@ SiO2 core–shell nanoparticles via a surfactant-free miniemulsion polymerization technique. Macromolecules 49(5):1552–1562
47. Basu H, Singhal RK, Pimple MV (2016) Highly efficient removal of TiO2 nanoparticles from aquatic bodies by silica microsphere impregnated Ca-alginate. New J Chem 40:3177–3186
48. Basu H, Singhal RK, Pimple MV, Reddy AVR (2015) Synthesis and characterization of silica microsphere and their application in removal of Uranium and Thorium from water. Int J Environ Sci Technol 12(6):1899–1906
49. Basu H, Singhal RK, Pimple MV, Saha S (2018) Graphene-prussian blue nanocomposite impregnated in alginate for efficient removal of cesium from aquatic environment. J Environ Chem Eng 6(4):4399-4407
50. Silva DM, Liu R, Gonçalves AF, da Costa A, Gomes AC, Machado R ... Sencadas V (2021) Design of polymeric core-shell carriers for combination therapies. J Colloid Interf Sci 587:499–509
51. Rodríguez-González B, Burrows A, Watanabe M, Kiely CJ, Marzán LML (2005) Multishell bimetallic AuAg nanoparticles: synthesis, structure and optical properties. J Mater Chem 15(17):1755–1759
52. Gerion D, Pinaud F, Williams SC, Parak WJ, Zanchet D, Weiss S, Alivisatos AP (2001) Synthesis and properties of biocompatible water-soluble silica-coated CdSe/ZnS semiconductor quantum dots. J Phys Chem B 105(37):8861–8871
53. Liu R, Qu F, Guo Y, Yao N, Priestley RD (2014) Au@ carbon yolk–shell nanostructures via one-step core–shell–shell template. Chem Commun 50(4):478–480
54. Arole VM, Munde SV (2014) Fabrication of nanomaterials by top-down and bottom-up approaches-an overview. J Mater Sci 1:89–93
55. Indiarto R, Indriana LPA, Andoyo R, Subroto E, Nurhadi B (2021) Bottom–up nanoparticle synthesis: a review of techniques, polyphenol-based core materials, and their properties. Eur Food Res Technol 1–24
56. Ealia SAM, Saravanakumar MP (2017, November). A review on the classification, characterisation, synthesis of nanoparticles and their application. In IOP conference series: materials science and engineering, Vol 263, No 3, p 032019. IOP Publishing
57. Mehta VN, Desai ML, Basu H, Singhal RK, Kailasa SK (2021) Recent developments on fluorescent hybrid nanomaterials for metal ions sensing and bioimaging applications: a review. J Mol Liq 333:115950
58. Basu H, Singhal RK, Pimple MV, Saha S (2018) Graphene oxide encapsulated in alginate beads for enhanced sorption of uranium from different aquatic environments. J Environ Chem Eng 6(2):1625–1633
59. Basu H, Singhal RK, Pimple MV, Kumar A, Reddy AVR (2014) Association and migration of uranium and thorium with silica colloidal particles in saturated subsurface zone. J Radio Anal Nucl Chem 303:2283–2290
60. Malik MA, O'Brien P, Revaprasadu N (2002) A simple route to the synthesis of core/shell nanoparticles of chalcogenides. Chem Mater 14(5):2004–2010
61. Ban Z, Barnakov YA, Li F, Golub VO, O'Connor CJ (2005) The synthesis of core–shell iron@ gold nanoparticles and their characterization. J Mater Chem 15(43):4660–4662
62. Khatami M, Alijani HQ, Nejad MS, Varma RS (2018) Core@ shell nanoparticles: greener synthesis using natural plant products. Appl Sci 8(3):411
63. Khan A, Rashid A, Younas R, Chong R (2016) A chemical reduction approach to the synthesis of copper nanoparticles. Int Nano Lett 6(1):21–26
64. Guzmán MG, Dille J, Godet S (2009) Synthesis of silver nanoparticles by chemical reduction method and their antibacterial activity. Int J Chem Biomol Eng 2(3):104–111

65. Jana NR, Gearheart L, Murphy CJ (2001) Evidence for seed-mediated nucleation in the chemical reduction of gold salts to gold nanoparticles. Chem Mater 13(7):2313–2322
66. Kailasa SK, Koduru JR, Desai ML, Park TJ, Singhal RK, Basu H (2018) Recent progress on surface chemistry of plasmonic metal nanoparticles for colorimetric assay of drugs in pharmaceutical and biological sample., Trends Anal Chem 105:106–120
67. Basu H, Singhal RK, Pimple MV, Saha S (2017) Association and migration behavior of trace metals with humus colloidal particles in aquatic subsurface medium. J Radio Anal Nucl Chem 311(1): 503–551
68. Lee WR, Kim MG, Choi JR, Park JI, Ko SJ, Oh SJ, Cheon J (2005) Redox– transmetalation process as a generalized synthetic strategy for core– shell magnetic nanoparticles. J Am Chem Soc 127(46):16090–16097
69. Chen D, Liu S, Li J, Zhao N, Shi C, Du X, Sheng J (2009) Nanometre Ni and core/shell Ni/Au nanoparticles with controllable dimensions synthesized in reverse microemulsion. J Alloy Compd 475(1–2):494–500
70. Park JI, Cheon J (2001) Synthesis of "solid solution" and "core-shell" type cobalt– platinum magnetic nanoparticles via transmetalation reactions. J Am Chem Soc 123(24):5743–5746
71. Ji Y, Yang S, Guo S, Song X, Ding B, Yang Z (2010) Bimetallic Ag/Au nanoparticles: a low temperature ripening strategy in aqueous solution. Colloids Surf, A 372(1–3):204–209
72. Lee CC, Chen DH (2006) Large-scale synthesis of Ni–Ag core–shell nanoparticles with magnetic, optical and anti-oxidation properties. Nanotechnology 17(13):3094
73. Singhal RK, Gangadhar B, Basu H, Venkatesh M, Naidu GRK, Reddy AVR (2012) Remediation of malathion contaminated soil using zero valent iron nano-particles. Am J Anal Chem 3:76–82
74. Hess PH, Parker PH Jr (1966) Polymers for stabilization of colloidal cobalt particles. J Appl Polym Sci 10(12):1915–1927
75. Salavati-Niasari M, Davar F, Mazaheri M, Shaterian M (2008) Preparation of cobalt nanoparticles from [bis (salicylidene) cobalt (II)]–oleylamine complex by thermal decomposition. J Magn Magn Mater 320(3–4):575–578
76. Salavati-Niasari M, Davar F, Mir N (2008) Synthesis and characterization of metallic copper nanoparticles via thermal decomposition. Polyhedron 27(17):3514–3518
77. Wang L, Wang L, Luo J, Fan Q, Suzuki M, Suzuki IS ...Zhong CJ (2005) Monodispersed core– shell Fe_3O_4@ Au nanoparticles. J Phys Chem B 109(46):21593–21601
78. De Jesus JC, González I, Quevedo A, Puerta T (2005) Thermal decomposition of nickel acetate tetrahydrate: an integrated study by TGA, QMS and XPS techniques. J Mol Catal A: Chem 228(1–2):283–291
79. Fu Y, Shearwood C (2004) Characterization of nanocrystallineTiNi powder. Scripta Mater 50(3):319–323
80. Wang Q, Yang H, Shi J, Zou G (2001) Preparation and characterization of nanocrystalline powders of Cu–Zn alloy by wire electrical explosion method. Mater Sci Eng, A 307(1–2):190–194
81. Basu H, Singh S, Venkatesh M, Pimple MV, Singhal RK (2021) Graphene oxide-MnO2-goethite microsphere impregnated alginate: a novel hybrid nanosorbent for As (III) and As (V) removal from groundwater. J Water Process Eng 42:102129
82. Basu H, Singhal RK, Pimple MV, Manisha V, Bassan MKT, Reddy AVR, Mukherjee T (2011) Development of naturally occurring siliceous material for the preferential removal of thorium from U-Th from aquatic environment. J Radio Anal Nucl Chem 289:231–237
83. Parashar M, Shukla VK, Singh R (2020) Metal oxides nanoparticles via sol–gel method: a review on synthesis, characterization and applications. J Mater Sci: Mater Electron 31(5):3729–3749
84. Petcharoen K, Sirivat A (2012) Synthesis and characterization of magnetite nanoparticles via the chemical co-precipitation method. Mater Sci Eng, B 177(5):421–427
85. Kim YI, Kim D, Lee CS (2003) Synthesis and characterization of CoFe2O4 magnetic nanoparticles prepared by temperature-controlled coprecipitation method. Physica B 337(1–4):42–51

86. He Q, Zhang Z, Xiong J, Xiong Y, Xiao H (2008) A novel biomaterial—Fe_3O_4: TiO_2 core-shell nano particle with magnetic performance and high visible light photocatalytic activity. Opt Mater 31(2):380–384
87. Palacios-Hernández T, Hirata-Flores GA, Contreras-López OE, Mendoza-Sánchez ME, Valeriano-Arreola I, González-Vergara E, Méndez-Rojas MA (2012) Synthesis of Cu and Co metal oxide nanoparticles from thermal decomposition of tartrate complexes. Inorganica Chimica Acta 392:277–282
88. Sun S, Zeng H (2002) Size-controlled synthesis of magnetite nanoparticles. J Am Chem Soc 124(28):8204–8205
89. Salavati-Niasari M, Davar F, Mazaheri M (2008) Preparation of ZnO nanoparticles from [bis (acetylacetonato) zinc (II)]–oleylamine complex by thermal decomposition. Mater Lett 62(12–13):1890–1892
90. Tian Y, Newton T, Kotov NA, Guldi DM, Fendler JH (1996) Coupled composite CdS–CdSe and core– shell types of (CdS) CdSe and (CdSe) CdS nanoparticles. J Phys Chem 100(21):8927–8939
91. Korgel BA, Monbouquette HG (2000) Controlled synthesis of mixed core and layered (Zn, Cd) S and (Hg, Cd) S nanocrystals within phosphatidylcholine vesicles. Langmuir 16(8):3588–3594
92. Zhou HS, Honma I, Komiyama H, Haus JW (1993) Coated semiconductor nanoparticles; the cadmium sulfide/lead sulfide system's synthesis and properties. J Phys Chem 97(4):895–901
93. Kortan AR, Hull R, Opila RL, Bawendi MG, Steigerwald ML, Carroll PJ, Brus LE (1990) Nucleation and growth of CdSe on ZnS quantum crystallite seeds, and vice versa, in inverse micelle media. J Am Chem Soc 112(4):1327–1332
94. Xie XL, Li RKY, Liu QX, Mai YW (2004) Structure-property relationships of in-situ PMMA modified nano-sized antimony trioxide filled poly (vinyl chloride) nanocomposites. Polymer 45(8):2793–2802
95. Ballauff M, Lu Y (2007) "Smart" nanoparticles: preparation, characterization and applications. Polymer 48(7):1815–1823
96. Singh S, Basu H, Bassan MKT, Singhal RK (2022) Thiol functionalised silica microsphere loaded polymeric hydrogel: development of a novel hybrid sorbent for removal of lead and cadmium. Chemosphere 286:131659
97. Wang C, Yan J, Cui X, Cong D, Wang H (2010) Preparation and characterization of magnetic hollow PMMA nanospheres via in situ emulsion polymerization. Colloids Surf, A 363(1–3):71–77
98. Mohod CV, Dhote J (2013) Review of heavy metals in drinking water and their effect on human health. Int J Innov Res Sci, Eng Technol 2(7):2992–2996
99. Fu Z, Xi S (2020) The effects of heavy metals on human metabolism. Toxicol Mech Methods 30(3):167–176
100. Saad AHA, Azzam AM, El-Wakeel ST, Mostafa BB, Abd El-latif MB (2018) Removal of toxic metal ions from wastewater using ZnO@ Chitosan core-shell nanocomposite. Environ Nanotechnol Monitor Manag 9:67–75
101. Mehta VN, Desai ML, Basu H, Singhal RK, Kailasa SK (2021) Recent developments on fluorescent hybrid nanomaterials for metal ions sensing and bioimaging applications: a review. J Mol Liq 333:115950
102. Basu H, Singhal RK, Pimple MV, Reddy AVR (2015) Arsenic removal from ground water by goethite impregnated calcium alginate beads. Water, Air, Soil Pollut 226:22
103. Zhang M, Song W, Chen Q, Miao B, He W (2015) One-pot synthesis of magnetic Ni@ Mg (OH) 2 core–shell nanocomposites as a recyclable removal agent for heavy metals. ACS Appl Mater Interf 7(3):1533–1540
104. Zhu K, Chen C, Xu H, Gao Y, Tan X, Alsaedi A, Hayat T (2017) Cr (VI) reduction and immobilization by core-double-shell structured magnetic polydopamine@ zeoliticidazolate frameworks-8 microspheres. ACS Sustain Chem Eng 5(8):6795–6802
105. Nawaz T, Zulfiqar S, Sarwar MI, Iqbal M (2020) Synthesis of diglycolic acid functionalized core-shell silica coated Fe 3 O 4 nanomaterials for magnetic extraction of Pb (II) and Cr (VI) ions. Sci Rep 10(1):1–13

106. Liu L, Liu S, Zhao L, Su G, Liu X, Peng H, ... Tang A (2020) Fabrication of novel magnetic core-shell chelating adsorbent for rapid and highly efficient adsorption of heavy metal ions from aqueous solution. J Mol Liq 313:113593
107. Ma Z, Zhao D, Chang Y, Xing S, Wu Y, Gao Y (2013) Synthesis of $MnFe_2O_4$@ Mn–Co oxide core–shell nanoparticles and their excellent performance for heavy metal removal. Dalton Trans 42(39):14261–14267
108. Amiri O, Hosseinpour-Mashkani SM, Rad MM, Abdvali F (2014) Sonochemical synthesis and characterization of CdS/ZnS core–shell nanoparticles and application in removal of heavy metals from aqueous solution. Superlattices Microstruct 66:67–75
109. Yang W, Wang Y, Wang Q, Wu J, Duan G, Xu W, Jian S (2021) Magnetically separable and recyclable Fe_3O_4@ PDA covalent grafted by l-cysteine core-shell nanoparticles toward efficient removal of Pb^{2+}. Vacuum 189:110229
110. Jahanbakhsh Z, Hosseinzadeh H, Masoumi B (2021) Synthesis of carboxymethyl β-cyclodextrin bonded Fe_3O_4@ SiO_2–NH_2 core-shell magnetic nanocomposite adsorbent for effective removal of Pb (II) from wastewater. J Sol-Gel Sci Technol 1–13
111. Jiang X, Su S, Rao J, Li S, Lei T, Bai H, ... Yang X (2021) Magnetic metal-organic framework (Fe_3O_4@ ZIF-8) core-shell composite for the efficient removal of Pb (II) and Cu (II) from water. J Environ Chem Eng 9(5):105959
112. Bhardwaj S, Sarkar T (2020) Core–shell type magnetic Ni/NiO nanoparticles as recyclable adsorbent for Pb (II) and Cd (II) ions: one-pot synthesis, adsorption performance, and mechanism. J Taiwan Inst Chem Eng 113:223–230
113. Patiño-Ruiz D, Rehmann L, Mehrvar M, Quiñones-Bolaños E, Herrera A (2020) Synthesis of FeO@ SiO 2–DNA core–shell engineered nanostructures for rapid adsorption of heavy metals in aqueous solutions. RSC Adv 10(64):39284–39294
114. Zhao Z, Zhang X, Zhou H, Liu G, Kong M, Wang G (2017) Microwave-assisted synthesis of magnetic Fe_3O_4-mesoporous magnesium silicate core-shell composites for the removal of heavy metal ions. Microporous Mesoporous Mater 242:50–58
115. Ji S, Miao C, Liu H, Feng L, Yang X, Guo H (2018) A hydrothermal synthesis of Fe 3 O 4@ C hybrid nanoparticle and magnetic adsorptive performance to remove heavy metal ions in aqueous solution. Nanoscale Res Lett 13(1):1–10
116. Chen J, Liu K, Jiang M, Han J, Liu M, Wang C, Li C (2019) Controllable preparation of porous hollow carbon sphere@ ZIF-8: Novel core-shell nanomaterial for Pb^{2+} adsorption. Colloids Surf, A 568:461–469
117. Fathy M, Zayed MA, Moustafa YM (2019) Synthesis and applications of CaCO3/HPC core–shell composite subject to heavy metals adsorption processes. Heliyon 5(8):e02215
118. Venkateswarlu S, Yoon M (2015) Core–shell ferromagnetic nanorod based on amine polymer composite (Fe_3O_4@ DAPF) for fast removal of Pb (II) from aqueous solutions. ACS Appl Mater Interf 7(45):25362–25372
119. Luo H, Zhang S, Li X, Liu X, Xu Q, Liu J, Wang Z (2017) Tannic acid modified Fe_3O_4 core–shell nanoparticles for adsorption of Pb^{2+} and Hg^{2+}. J Taiwan Inst Chem Eng 72:163–170
120. Ren C, Ding X, Fu H, Li W, Wu H, Yang H (2017) Core–shell superparamagnetic monodisperse nanospheres based on amino-functionalized $CoFe_2O_4$@ SiO_2 for removal of heavy metals from aqueous solutions. RSC Adv 7(12):6911–6921
121. Ashrafi A, Rahbar-Kelishami A, Shayesteh H (2017) Highly efficient simultaneous ultrasonic assisted adsorption of Pb (II) by Fe_3O_4@ MnO_2 core-shell magnetic nanoparticles: synthesis and characterization, kinetic, equilibrium, and thermodynamic studies. J Mol Struct 1147:40–47
122. Gui CX, Li QJ, Lv LL, Qu J, Wang QQ, Hao SM, Yu ZZ (2015) Core–shell structured MgO@ mesoporous silica spheres for enhanced adsorption of methylene blue and lead ions. RSC Adv 5(26):20440–20445
123. Xiong T, Yuan X, Cao X, Wang H, Jiang L, Wu Z, Liu Y (2020) Mechanistic insights into heavy metals affinity in magnetic MnO_2@ Fe_3O_4/poly (m-phenylenediamine) core– shell adsorbent. Ecotoxicol Environ Saf 192:110326

124. Wang S, Wang K, Dai C, Shi H, Li J (2015) Adsorption of Pb^{2+} on amino-functionalized core–shell magnetic mesoporous SBA-15 silica composite. Chem Eng J 262:897–903
125. Wan S, Yu C, Li Y, Lu Z, Wang Y, Wang Y, He F (2021) Highly selective and ultrafast removal of cadmium and copper from water by magnetic core-shell microsphere. Chem Eng J 405:126576
126. Abolhasani J, Khanmiri RH, Ghorbani-Kalhor E, Hassanpour A, Asgharinezhad AA, Shekari N, Fathi A (2015) An Fe_3O_4@ SiO_2@ polypyrrole magnetic nanocomposite for the extraction and preconcentration of Cd (II) and Ni (II). Anal Methods 7(1):313–320
127. Sun L, Li Y, Sun M, Wang H, Xu S, Zhang C, Yang Q (2011) Porphyrin-functionalized Fe_3O_4@ SiO_2 core/shell magnetic colorimetric material for detection, adsorption and removal of Hg^{2+} in aqueous solution. New J Chem 35(11):2697–2704
128. Zhu H, Shen Y, Wang Q, Chen K, Wang X, Zhang G ... Bai R (2017) Highly promoted removal of Hg (II) with magnetic $CoFe_2O_4$@ SiO_2 core–shell nanoparticles modified by thiol groups. RSC Adv 7(62):39204–39215
129. Xia K, Guo Y, Shao Q, Zan Q, Bai R (2019) Removal of mercury (II) by EDTA-functionalized magnetic $CoFe_2O_4$@ SiO_2 nanomaterial with core-shell structure. Nanomaterials 9(11):1532
130. Removal of Hg(II) by poly(1-vinylimidazole)-grafted Fe_3O_4@SiO_2 magnetic nanoparticles Water Res 69C:252–260 (2014) https://doi.org/10.1016/j.watres.2014.11.030
131. Afraz A, Hajian A, Niknam Z, Mosayebi E, Yusefi A, Sillanpää M (2017) Amin-functionalized magnetic-silica core-shell nanoparticles for removal of Hg^{2+} from aqueous solution. J Dispers Sci Technol 38(5):750–756
132. Wu A, Jia J, Luan S (2011) Amphiphilic PMMA/PEI core–shell nanoparticles as polymeric adsorbents to remove heavy metal pollutants. Colloids Surf, A 384(1–3):180–185
133. Chang L, Pu Y, Jing P, Cui Y, Zhang G, Xu S ... Qiao C (2021). Magnetic core-shell MnFe2O4@ TiO2 nanoparticles decorated on reduced graphene oxide as a novel adsorbent for the removal of ciprofloxacin and Cu (II) from water. Appl Surf Sci 541:148400
134. Xiong Z, Zheng H, Hu Y, Hu X, Ding W, Ma J, Li Y (2021) Selective adsorption of Congo red and Cu (II) from complex wastewater by core-shell structured magnetic carbon@ zeoliticimidazolate frameworks-8 nanocomposites. Separ Purif Technol 119053
135. Zuo B, Deng Q, Shao H, Cao B, Fan Y, Li W, Huang M (2021) Fe_3O_4@ Mesoporous-SiO_2@ Chitosan@ polyaniline core-shell nanoparticles as recyclable adsorbents and reductants for hexavalent chromium. ACS Appl Nano Mater 4(2):1831–1840
136. Ekka B, Dhar G, Sahu S, Mishra M, Dash P, Patel RK (2021) Removal of Cr (VI) by silica-titania core-shell nanocomposites: in vivo toxicity assessment of the adsorbent by Drosophila melanogaster. Ceram Int 47(13):19079–19089
137. Zhang Y, Xu S, Zhou H, Qi H, Wang H (2021) Adsorption of organic pollutants and heavy metal by Co-doped core-shell MoO2/Mo2C adsorbent. J Solid State Chem 293:121801
138. Yang Q, Wang H, Li F, Dang Z, Zhang L (2021) Rapid and efficient removal of Cr (vi) by a core–shell magnetic mesoporous polydopamine nanocomposite: roles of the mesoporous structure and redox-active functional groups. J Mater Chem A 9(22):13306–13319
139. Wang T, Zhang L, Li C, Yang W, Song T, Tang C ... Luo J (2015) Synthesis of core–shell magnetic Fe_3O_4@ poly (m-phenylenediamine) particles for chromium reduction and adsorption. Environ Sci Technol 49(9):5654–5662
140. Facile synthesis of layered core-shell structure Fe_3O_4 magnetic composites and its application for the Co^{2+} removal. J Mol Liq 325:114517 (2020). https://doi.org/10.1016/j.molliq.2020.114517
141. Zhang X, Wang Y, Yang S (2014) Simultaneous removal of Co (II) and 1-naphthol by core–shell structured Fe_3O_4@ cyclodextrin magnetic nanoparticles. Carbohyd Polym 114:521–529
142. Huo JB, Xu L, Yang JCE, Cui HJ, Yuan B, Fu ML (2018) Magnetic responsive Fe_3O_4-ZIF-8 core-shell composites for efficient removal of As (III) from water. Colloids Surf, A 539:59–68
143. Wu J, Zhu H, Liu G, Tan L, Hu X, Chen C ... Tan X (2017) Fabrication of core–shell CMNP@ PmPD nanocomposite for efficient As (V) adsorption and reduction. ACS Sustain Chem Eng 5(5):4399–4407

144. Singhal RK, Basu H, Manisha V, Reddy AVR, Mukherjee T (2011) Removal of low level americium-241 from potable water originated from different geochemical environments by calcium alginate. Desalination 280:313–318
145. Wang Z, Wang J, Zhu L, He Y, Duan T (2020) Scalable Fe@ FeO core-shell nanoparticle-embedded porous wood for high-efficiency uranium (VI) adsorption. Appl Surf Sci 508:144709
146. Dai Z, Zhang H, Sui Y, Ding D, Hu N, Li L, Wang Y (2018) Synthesis and characterization of a novel core–shell magnetic nanocomposite via surface-initiated RAFT polymerization for highly efficient and selective adsorption of uranium (VI). J Radio Anal Nucl Chem 316(1):369–382
147. Tan L, Zhang X, Liu Q, Wang J, Sun Y, Jing X ... Liu L (2015) Preparation of magnetic core–shell iron oxide@ silica@ nickel-ethylene glycol microspheres for highly efficient sorption of uranium (VI). Dalton Trans 44(15):6909–6917
148. Liu Q, Li W, Zhao W, Tan L, Jing X, Liu J ... Wang J (2016) Synthesis of ketoxime-functionalized Fe 3 O 4@ C core–shell magnetic microspheres for enhanced uranium (vi) removal. RSC Adv 6(26):22179–22186
149. Yang Y, Wang J, Wu F, Ye G, Yi R, Lu Y, Chen J (2016) Surface-initiated SET-LRP mediated by mussel-inspired polydopamine chemistry for controlled building of novel core–shell magnetic nanoparticles for highly-efficient uranium enrichment. Polym Chem 7(13):2427–2435
150. Liu Y, Yang P, Li Q, Liu Y, Yin J (2019) Preparation of FeS@ Fe_3O_4 core–shell magnetic nanoparticles and their application in uranyl ions removal from aqueous solution. J Radio Anal Nucl Chem 321(2):499–510
151. Li P, Wang J, Wang X, He B, Pan D, Liang J ... Fan Q (2018) Arsenazo-functionalized magnetic carbon composite for uranium (VI) removal from aqueous solution. J Mol Liq 269:441–449
152. Zhao Y, Li J, Zhao L, Zhang S, Huang Y, Wu X, Wang X (2014) Synthesis of amidoxime-functionalized Fe_3O_4@ SiO_2 core–shell magnetic microspheres for highly efficient sorption of U (VI). Chem Eng J 235:275–283
153. Hu S, Lin X, Zhang Y, Huang R, Qu Y, Luo X, Zhou J (2017) Preparation and application of alginate-Ca/attapulgite clay core/shell particle for the removal of uranium from aqueous solution. J Radio Anal Nucl Chem 314(1):307–319
154. Feng J, Cai Y, Wang X, Wang X, Zhu M, Fang M ... Tan X (2021) Designed Core–Shell Fe 3 O 4@ polydopamine for effectively removing uranium (VI) from aqueous solution. Bull Environ Contamin Toxicol 106:165–174
155. Li L, Lu W, Ding D, Dai Z, Cao C, Liu L, Chen T (2019) Adsorption properties of pyrene-functionalized nano-Fe_3O_4 mesoporous materials for uranium. J Solid State Chem 270:666–673
156. Zhu K, Lu S, Gao Y, Zhang R, Tan X, Chen C (2017) Fabrication of hierarchical core-shell polydopamine@ MgAl-LDHs composites for the efficient enrichment of radionuclides. Appl Surf Sci 396:1726–1735
157. Tan X, Fang M, Tan L, Liu H, Ye X, Hayate T, Wang X (2018) Core–shell hierarchical C@Na2Ti3O7·9H2O nanostructures for the efficient removal of radionuclides. Environ Sci: Nano 5:1140–1149
158. Zhao M, Cui Z, Pan D, Fan F, Tang J, Hu Y ... Wu W (2021) An efficient uranium adsorption magnetic platform based on amidoxime-functionalized flower-like Fe_3O_4@ TiO_2 Core–Shell microspheres. ACS Appl Mater Interf 13(15):17931–17939
159. Xu S, Yingguo Z, Yingguo Z, Zheng F, Zheng F, Zhan Y (2016) Hollow Fe_3O_4@mesoporous carbon core–shell microspheres for efficient sorption of radionuclides, Springer. J Mater Sci 51(5). https://doi.org/10.1007/s10853-015-9567-y
160. Wu Y, Li B, Wang X, Yu S, Pang H, Liu Y ... Wang X (2019) Magnetic metal-organic frameworks (Fe_3O_4@ ZIF-8) composites for U (VI) and Eu (III) elimination: simultaneously achieve favorable stability and functionality. Chem Eng J 378:122105
161. Tan L, Zhang X, Liu Q, Jing X, Liu J, Song D ... Wang J (2015) Synthesis of Fe_3O_4@ TiO2 core–shell magnetic composites for highly efficient sorption of uranium (VI). Colloids Surf A: Physicochem Eng Aspects 469:279–286

162. Yang S, Yang S, Zong P, Zong P, Ren X, Ren X, Wang X, Wang X (2012) Rapid and highly efficient preconcentration of Eu(III) by Core–Shell structured Fe_3O_4@Humic acid magnetic nanoparticles, ACS Appl Mater Interf 4(12):6891–6900
163. Zhang Y, Lin X, Hu S, Zhang X, Luo X (2016) Core–shell zeolite@ Alg–Ca particles for removal of strontium from aqueous solutions. RSC Adv 6(78):73959–73973
164. Fakhri H, Mahjoub AR, Aghayan H (2019) Effective adsorption of Co^{2+} and Sr^{2+} ions by 10-tungsten-2-molybdophosphoric acid supported amine modified magnetic SBA-15. J Radio Anal Nucl Chem 321(2):449–461
165. Mou W, Du S, Yu Q, Li X, Wei H, Yang Y (2018) Efficient capture of radioactive strontium from water using magnetic WO3 assembled on grapheme oxide nanocomposite. Chem Select 3(24):6992–6997
166. Wang G, Liu Q, Chang M, Jang J, Sui W, Si C, Ni Y (2019) Novel Fe_3O_4@ lignosulfonate/phenolic core-shell microspheres for highly efficient removal of cationic dyes from aqueous solution. Ind Crops Prod 127:110–118
167. Didehban A, Zabihi M, Faghihi M, Akbarbandari F, Akhtarivand H (2021) Design and fabrication of core-shell magnetic and non-magnetic supported carbonaceous metal organic framework nanocomposites for adsorption of dye. J Phys Chem Solids 152:109930
168. Review on nickel-based adsorption materials for Congo red. J Hazardous Mater 403:123559 (August 2020). https://doi.org/10.1016/j.jhazmat.2020.123559
169. Yang S, Zong P, Ren S, Wang X (2014) Highly regenerable mussel-inspired Fe_3O_4@polydopamine-Ag core-shell microspheres as catalyst and adsorbent for methylene blue removal. ACS Appl Mater Interf 6(11):8845–8852, Jun 11. https://doi.org/10.1021/am501632f
170. Huang W, Xu J, Lu D, Deng J, Shi G, Zhou T (2018) Rational design of magnetic infinite coordination polymer core-shell nanoparticles as recyclable adsorbents for selective removal of anionic dyes from colored wastewater. Appl Surf Sci 2018–12–01. https://doi.org/10.1016/j.apsusc.2018.08.122
171. Huo Y, Wu H, Wang Z, Wang F, Liu Y, Feng Y, Zhao Y (2018) Preparation of core/shell nanocomposite adsorbents based on amine polymer-modified magnetic materials for the efficient adsorption of anionic dyes, Author links open overlay panel. Colloids Surf A: Physicochem Eng Aspects 549, 20 July 2018:174–183
172. Zhang Z, Kong J (2011) Novel magnetic Fe_3O_4@C nanoparticles as adsorbents for removal of organic dyes from aqueous solution, Author links open overlay panel. J Hazardous Mater 193:325–329, 15 October
173. Konickia W, Hełminiakb A, Arabczykb W, Mijowska E (2018) Adsorption of cationic dyes onto Fe@graphite core–shell magnetic nanocomposite: equilibrium, kinetics and thermodynamics, Author links open overlay panel. Chem Eng Res Design 129:259–270, Jan 2018
174. Huang B, Liu Y, Li B, Wang H, Zeng G (2019) Adsorption mechanism of polyethyleneimine modified magnetic core–shell Fe_3O_4@SiO_2 nanoparticles for anionic dye removal. RSC Adv 9:32462–32471
175. Wang W, Jiao T, Zhang Q, Luo X, Hu J, Chen Y, Peng Q, Yan X, Li B (2015) Hydrothermal synthesis of hierarchical core–shell manganese oxide nanocomposites as efficient dye adsorbents for wastewater treatment. RSC Adv 5:56279–56285
176. Konickia W, Hełminiak A, Arabczyk W, Mijowska E (2017) Removal of anionic dyes using magnetic Fe@graphite core-shell nanocomposite as an adsorbent from aqueous solutions, Author links open overlay panel. J Colloid Interf Sci 497:155–164, 1 July
177. Zheng Y, Liub J, Cheng B, You W, Ho W, Tang H (2019) Hierarchical porous Al2O3@ZnO core-shell microfibres with excellent adsorption affinity for Congo red moleculeAuthor links open overlay panel. Appl Surf Sci 473:251–260, 15 April
178. Shaoa Y, Zhou L, Bao C, Ma J, Liu M, Wang F (2016) Magnetic responsive metal–organic frameworks nanosphere with core–shell structure for highly efficient removal of methylene blue, This paper is dedicated to the memory of Professor Yanfeng Li.Author links open overlay panel. Chem Eng J 283:1127–1136, 1 Jan

179. Wo R, Li Q-L, Zhu C, Zhang Y, Qiao G-f, Lei K-y, Du P, Jiang W (2019) Preparation and characterization of functionalized metal–organic frameworks with core/shell magnetic particles (Fe_3O_4@SiO_2@MOFs) for removal of congo red and methylene blue from water solution. J Chem Eng Data 64(6):2455–2463
180. Kollarahithlu SC, Balakrishnan RM (2021) Adsorption of pharmaceuticals pollutants, Ibuprofen, Acetaminophen, and Streptomycin from the aqueous phase using amine functionalized superparamagnetic silica nanocomposite. Author links open overlay panel. J Clean Prod 294:12615, 20 April
181. Guo J, Yu M, Wei X, Huang L (2018) Preparation of core–shell magnetic molecularly imprinted polymer with uniform thin polymer layer for adsorption of dichlorophen. J Chem Eng Data 63(8):3068–3073
182. Wang Z, Tang H, Li W, Li J, Xu R, Zhang K, He G, Shearing PR, Brett DJL (2019) Core–shell TiO_2@C ultralong nanotubes with enhanced adsorption of antibiotics. J Mater Chem A 7:19081–19086
183. Soares SF, Fernandes T, Sacramento M, Trindade T, Daniel-da-Silva AL (2019) Magnetic quaternary chitosan hybrid nanoparticles for the efficient uptake of diclofenac from water-Author links open overlay panel. Carbohydr Polym 203:35–44, 1 Jan
184. Du ZD, Cui YY, Yang CX, Yan X-P (2019) Core–Shell magnetic amino-functionalized microporous organic network nanospheres for the removal of tetrabromobisphenol A from aqueous solution. ACS Appl Nano Mater 2(3):1232–1241
185. Zhoua Q, Wang Y, Xiao J, Fan H (2016) Adsorption and removal of bisphenol A, α-naphthol and β-naphthol from aqueous solution by Fe_3O_4@polyaniline core–shell nanomaterialsAuthor links open overlay panel. Synthetic Metals 212:113–122, Feb
186. Zhou L, Pan S, Chen X, Zhao Y, Zou B, Jin M (2014) Kinetics and thermodynamics studies of pentachlorophenol adsorption on covalently functionalized Fe_3O_4@SiO_2–MWCNTs core–shell magnetic microspheres. Chem Eng J 257:10–19, 1 Dec
187. Zhang X, Wang Y, Yang S (2014) Simultaneous removal of Co(II) and 1-naphthol by core–shell structured Fe_3O_4@cyclodextrin magnetic nanoparticles. Carbohydrate Polymers 114:521–529, 19 Dec
188. Gao M, Deng L, Kang X, Fua Q, Zhang K, Wang M, Xia Z, Gao D (2020) Core-shell structured magnetic covalent organic frameworks for magnetic solid-phase extraction of diphenylamine and its analogs, J Chromatogr A 1629:461476, 11 Oct 2020
189. Li Y, Zhang H, Chen Y, Huang L, Lin Z, Cai Z (2019) Core–Shell structured magnetic covalent organic framework nanocomposites for Triclosan and Triclocarban adsorption. ACS Appl Mater Interf 11(25):22492–22500
190. Zhou Q, Wang Y, Xiao J, Fan H (2017) Fabrication and characterisation of magnetic graphene oxide incorporated Fe_3O_4@polyaniline for the removal of bisphenol A, t-octyl-phenol, and α-naphthol from water. Sci Rep 7. Article number: 11316

Future Challenges and Perspectives in Water Purification by Hybrid Materials

Soulaima Chkirida, Nadia Zari, Abou el kacem Qaiss, and Rachid Bouhfid

Abstract One of the main emerged trends in the water purification sector is hybrid materials adoption. Due to their peculiarity of combining several components into one formulation, hybrid materials are effective in removing a panoply of pollutants from contaminated water. Thereby, they provide a conceivable alternate to conventional water purification. Nevertheless, considerable challenges are remaining in the industrial process scale-up, including, stability, lifecycle, sustainability, and cost-effectiveness. The main objective of this book chapter is to allow scrutiny and gain an appropriate understanding of future perspectives and challenges faced by hybrid materials, for the simple reason that a proper understanding of the challenges will add to the understanding of measures to be taken.

1 Introduction

Despite the great number of existed materials families, it was found that it cannot fulfill all the technological and scientific required functions, which underlines the need for a new class of materials with improved properties. Hybrid materials are one of the most successful examples. In a broader sense and as their name suggests, hybrid materials are built by combining two or more materials in a single polymeric matrix to give rise to super properties compared with their individual counterparts [1].

There are various other definitions of hybrid materials from different viewpoints currently in use in the literature that are all perfectly satisfactory; going from describing hybrid materials formation as a consequence of electron orbitals

S. Chkirida · N. Zari · A. Qaiss · R. Bouhfid (✉)
Moroccan Foundation for Advanced Science, Innovation and Research (MAScIR), Composites and Nanocomposites Center, Rabat Design Center, Rue Mohamed El Jazouli, Madinat El Irfane, 10100 Rabat, Morocco
e-mail: r.bouhfid@mascir.com

S. Chkirida
Faculty of Science, Laboratoire de Chimie organique et Hétérocyclique, Mohammed V University Mohammed in Rabat, Rabat, Morocco

S. K. Swain (ed.), *Nanohybrid Materials for Water Purification*, Composites Science and Technology, https://doi.org/10.1007/978-981-19-2332-6_13

arrangement of two or more materials [2], to their description based on different chemical bonds development between two or more materials [3]. There is no defined classification exist for hybrid materials since they could be classified according to countless different criteria. Hybrid materials could be a mixture of inorganic materials, organic materials, or their combination where all components are called on to render their contribution.

For organic–inorganic hybrid materials, two types of organic components could be used; small organic molecules or organic macromolecules as polymers/biopolymers. These latter along with inorganic particles such as metals are the most used components for organic–inorganic hybrid materials production. Each material has its own emphasis and role. For example, polymers/biopolymers matrices are often responsible for the hybrid material structural properties, components cohesion, and shape flexibility, while inorganic reinforcements are responsible for rigidity, thermal and mechanical stability, electric and magnetic properties [4].

Generally speaking, two types of hybrid organic–inorganic materials may be distinguished based on their strength interactions and synthesis approaches as seen in Fig. 1;

(1) hybrid materials with chemical bonds between their components synthesized throughthe chemical process giving rise to covalent cross-linking polymers.
(2) hybrid materials with weak bonds synthesized through the inorganic component's incorporation into the polymeric matrix giving rise to blends and interpenetrating networks [4].

To obtain stable and homogeneous hybrid materials over long periods, it is recommended and preferable to properly implement chemical interactions as crosslinking, and covalent bonds between the different hybrid materials components. Otherwise, potential shortcomings may exist as phase separation, loss of material integrity, or leaching out of the material's constituents. To tackle this issue, in situ preparation

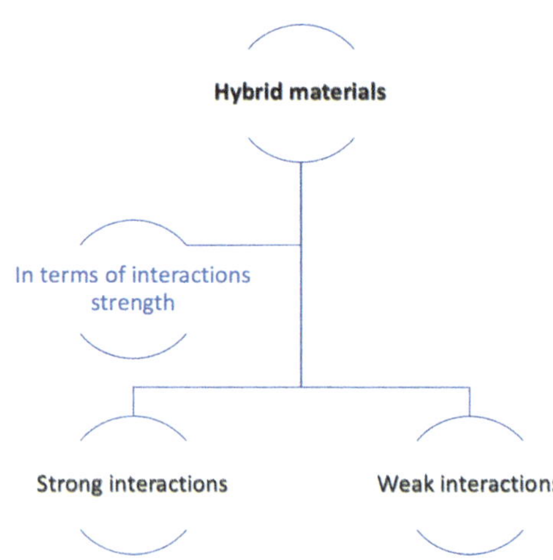

Fig. 1 Classification of organic–inorganic hybrid materials

in presence of both organic and inorganic components is the most used method to enable greater compatibility among the hybrid material's components.

Since the hybrid materials, applications are controlled by their features including physical, chemical, thermal, structural, and morphological properties. It is noteworthy that these properties can be tuned in a controlled way by adequately changing their structure, components, and their preparation approaches. It lies therein all the interest of elaborating hybrid materials as materials that are fit for their purposes.

There exist different routes for hybrid materials elaboration as briefly described below [1];

- In situ approach based on the chemical transformation of the used precursors leading to
- *The sol–gel* method is based on wet chemical methods to give rise to highly homogeneous hybrid materials by encapsulating the organic components within the derived inorganic components.
- *Building block* approach based on maintaining the molecular integrity of starting components.
- *The hydrothermal* method is based on crystallizing techniques at high temperatures and pression to engender hybrid materials with better nucleation control.

2 Hybrid Materials for Water Decontamination

Hybrid materials would have served many purposes, but mainly in treating contaminated water to eliminate its pollutants. According to the literature, hybrid materials have several outstanding properties that make them excellent tools for water decontamination, including high surface area, remarkable stability, high porosity, tunable properties, and a greater affinity towards a panoply of water pollutants. On that account, hybrid materials provide an unusually efficient option for meeting the growing needs for sustainable water treatment [5].

Broadly speaking, the most emergent water pollutants could be divided into three major categories: inorganic pollutants (heavy metals, nitrates, nitrites, phosphate…), organic pollutants (phenols, phenols, COD, BOD…), and microbiological pollutants (*E. Coli, S.Aureus…*) [6]. This section provides a summary of the recent hybrid material's achievement in water depollution.

2.1 Elimination of Inorganic Pollutants

Efficient sequestering action on heavy metals using hybrid materials was reported by several studies. As an illustration, Ni (II) and Pb (II) were effectively removed using a hybrid Xanthan gum- Glutathione/Zeolite with a high uptake of the order of 85% and 93% respectively [7]. Another hybrid material based on bentonite and guar gum was elaborated for lead removal from real wastewaters obtained from electroplating and battery manufacturing. The study has revealed an uptake removal of 83.5% using ion exchange and electrostatic interactions between the hybrid and the heavy metal [8]. Hybrid materials based on copolymers were also investigated for heavy metals removal. Hybrid copolymer chitosan-g- PMMA silica gel was elaborated via an emulsifier free emulsion polymerization to remove toxic Cr(VI) from waters with a removal efficiency of 98% at an optimum pH of 4 [9].

Besides heavy metals, nitrogen and phosphate compounds are also among the most frequently encountered inorganic contaminants. For removing both phosphate and nitrate, a hybrid composite based on Fe_3O_4/ZrO_2/chitosan was elaborated at mild conditions leading to a maximum adsorption amount of the order of 26.5 mg P/g and 89.3 mg/g, respectively [10]. Another study has reported that a fast and efficient removal of nitrate (74% in 5 min) was achieved using a hybrid material based on chitosan incorporated with Al_2O_3 and Ag-doped TiO_2 nanoparticles.

2.2 *Elimination of Organic Pollutants*

Since organic pollutants are countless, the researchers tend to assess the effectiveness of their hybrid materials using the three parameters' indicators of organic pollution namely; TOC (Total organic carbon), COD (chemical oxygen demand), and BOD (biochemical oxygen demand). Elaborating hybrid materials capable of removing these three parameters at once would be ideal. The best illustration is magnetic Gluten/Pectin/Fe_3O_4 hybrid material that has demonstrated its ability to remove simultaneously 80% of BOD, 60% of COD, and 50% of TOC [11].

Another hybrid material based on zinc oxide nanoparticles and chitosan was prepared to remove organic dissolved matter from milky wastewater. The study carried out has yielded great results; COD removal around 97% [12].

Phenol compounds are among the most toxic and recalcitrant organic contaminants. Many studies have already demonstrated the hybrid material's capability of removing recalcitrant phenols from water. The hybrid material based on trimethyl chitosan-loaded cerium oxide CeO_2 particles is a case on point. It shows an effective removal of three phenolic compounds of the order of 78%, 90%, and 60% for 2-chlorophenol, 4-chlorophenol, and phenol, respectively.

Loads of studies have underlined the removal of dyes using hybrid materials. Photocatalytic hybrid material based on TiO_2, clay, and alginate biopolymer was elaborated in order not only to adsorb the dye chromophore groups but to destruct them using UV irradiations. In other words, the prepared hybrid material has shown an ability to effectively removing color with 98% uptake along with dissolved organic matter COD with 93% uptake [13]. Similarly, a hybrid material based on alginate and iron modified TiO_2 was elaborated with the ability to remove methylene blue under both UV and Visible irradiations depending on the treatment conditions [14].

2.3 Elimination of Microbiological Pollutants

Water microbiological contamination is defined as the presence and prevalence of pathogenic microorganisms. Fortunately, enough, hybrid materials have made it possible to cope with microbiological pollution and benefit from their tuned antibacterial and antifungal activities. As many studies have proven, hybrid materials based on chitosan have been shown due to their natural prominent antibacterial activity against a broad spectrum of bacteria and fungi [15]. Likewise, hybrid materials based on silver nanoparticles have also shown potent antibacterial activity due to the strong reaction of silver nanoparticles with the bacteria proteins and DNA besides generating free radicals which would surely lead to bacterial membrane damage [16].

A chitosan hybrid material based on cellulose, titania, and silver nanoparticles was elaborated via a one-pot synthesis method and shown a maintaining great antibacterial activity up to 12 h [17]. Two different fungal strains; *Rhizoctonia solani* and *Alternaria alternate* were used to assess the antibacterial and antifungal activity of a novel hybrid material based on cellulose, Nickel metallic nanoparticles, and polyaniline. Considerable inhibitions of the fungal growth were recorded of the order of 42% and 50% for R. solani and A. Alternate, respectively [18].

Combining antibacterial activity with magnetic properties would guarantee a strong recoverable hybrid material for water treatment. The hybrid material based on Poly (aniline-co-pyrrole), alginic acid, and magnetic nanoparticles is a case on point. After being synthesized via an in situ co-precipitation, the magnetic hybrid material was tested against five bacteria; *Bacillus subtilis*, *Staphylococcus aureus*, *Candida albicans*, *Pseudomonas aeruginosa*, and*Escherichia coli* where it has shown good antifungal and antibacterial activities [19].

The following table summarizes the various applications of hybrid materials in water decontamination (Table 1).

3 Challenging Aspects of Hybrid Materials in Water Purification

The employment of hybrid materials in water purification has pursued a quite long development path. In fact, an abundant number of research studies at bench-scale testing have been conducted on the issue. Nevertheless, it may seem a paradoxical finding that only a very modest number of systems based on hybrid materials have been practically used in water purification.

The major constraint lays in the gap between the laboratory-scale and industrial-scale in terms of hybrid materials efficiencies. Actually, in academic research, there is a tendency to idealize, overvalue, and overestimate the performances of the hybrid materials as well as making a lot of good promises concerning their industrial scaling-up that would fall by the wayside if no real consideration is taken of their potential

Table 1 The summary of the applications of hybrid materials in water decontamination

Hybridmaterials	Pollutant	Removal	References
Inorganicpollutants			
Chitosangrafted- PMMA/Silica	Cr(VI)	98%	[9]
Xanthangum-Glutathione/ Zeolite	Ni (II) Pb(II)	85% 93%	[7]
Guargum/bentonite	Pb(II)	83.5%	[8]
Fe_3O_4/ZrO_2/chitosan	Nitrate Phosphate	89.3 mg/g 26.5 mgP/g	[10]
Organicpollutants			
Chitosan-zinc oxide	COD	97%	[12]
Alginate/Bentonite impregnated TiO_2 beads	Methylene Blue COD	98% 93%	[13]
Chitosan-Guar gum blend silver nanoparticle	TOC	82–84%	[20]
Trimethyl chitosan-loaded cerium oxide particles	2-chlorophenol 4-chlorophenol phenol	78% 90% 60%	[21]
Magnetic Gluten/Pectin/Fe_3O_4 hydrogel	COD BOD TOC	60% 80% 50%	[11]
Microbiologicalpollutants			
Hybridmaterials	Bacteria	Inactivation/inhibition	References
Nanometer-thick titania/chitosan/Ag-NP film	Escherichia coli Staphylococcus aureus	≅ 100% ≅ 100%	[17]
Poly(aniline-copyrrole)@ Fe_3O_4@alginic acid	Candida albicans Staphylococcus aureus Escherichia coli Pseudomonas aeruginosa Bacillus subtilis	Good antibacterial and antifungal activities	[19]
Hybrid nickel doped polyaniline/cellulose	Alternariaalternate Rhizoctoniasolani	50% 42%	[18]
Chitosan-manganese dioxide	Escherichia coli Staphylococcus aureus	50%	[22]

limitations, as technical practicability and feasibility. Therefore, there is an overriding need to prevent the research and industrial community from being misled attributable to the shift to industrial large scale. And this by identifying the most frequent challenges and roadblocks that may be found.

Herein, we shall emphasize the main future challenges regarding hybrid material's future applications in the water purification sector.

3.1 Scaling Up

Currently, nearly all the research studies based on hybrid materials for water treatment are conducted on a batch scale. Scaling up is referring to moving from the laboratory to industrial applications which is a major part of research and development (R&D). Scaling up is not only a matter of quantitative concept by increasing in the batch size, but also a matter of feasibility of the process (scalable or not), its productiveness, and its effectiveness that need to be fully checked and explored. These batch experiments play a pivotal role in allowing a prior statement and initial understanding of the depollution process using hybrid materials but don't reflect fully the real yields at larger scale. The slow-going shift from lab scale to large practice is due to the intricacy of the large system proceedings.

In fact, the design and preparation of hybrid materials for certain pollutants removal from simulated contaminated water at lab scale requires limited quantities of materials and chemicals besides reduced energy consumptions. That is not the case for larger scale, where a lot of variables may interfere and demand additional processing, besides involving considerable process modifications. On top of that, other sciences may get in line as engineering, economics, and materials science to ensure the maintaining of the effectiveness and meet the growing demands and expectations of industrial applications.

For better profitability of hybrid materials performances and smoother sustainable implantation in the water depollution sector, several actions can be undertaken:

- Examination of synthesis methods for a better outcome.
- Structuring of scalable models for engineering suitability studies and economic analysis.
- Effective quality control and relevant analytical methodology during all the process steps for quality guarantee.
- Evaluation of purity profiles of the starting materials and the final materials.
- Production of large sufficient and homogeneous quantities of hybrid materials well characterized for the critical prototyping phase testing, thus industrial assessment.
- Good documentation of technology transfer methodologies.

3.2 Stability

Good sorption or catalytic efficiencies of hybrid materials are not enough to meet the expected demand at larger scale. Good structural stability and high fatigue resistance of hybrid materials even in extreme water conditions are the key factor that is driving the progress and the perception of water purification quality.

Since hybrid materials are made up of two or more parts, their stability reflects first and foremost a maintained assembly of all their components and their active parts over a long life without any sort of leaching in the reaction medium. The second priority is ensuring identical reliability and effectiveness over long periods.

Recognizing and understanding the factors that may lead to the non-stability of hybrid materials are pivotal to prevent their occurrence and bring stability to the materials even after multiple reuses.

According to most of the literature, the performance and stability evaluation of hybrid materials is usually done under simulated conditions closely resembling the real ones but inshort-term experimental studies. In fact, the sufficient ins and outs of long-term effect on hybrid materials decontamination mechanism, and their stability are missing and need to be fully explored. All this is intended to serve as the basis for future larger-scale implantation.

Checking out the stability of the hybrid materials is usually done by these means:

- Exploring the leaching behavior of hybrid materials into the water medium using analytical tools as atomic absorption spectroscopy (SAA) and inductive coupled plasma emission (ICP) [13].
- Examining the hybrid materials changes before and after treating polluted water through X-ray photoelectron spectroscopy (XPS) to assess the binding energies, chemical composition, and atomic percentages of hybrid materials components [23, 24].
- Determining the saturation magnetization in the case of magnetic hybrid materials [24, 25].

The good stability of hybrid materials opens a double-wide window of attractive opportunities for their practical applications at an industrial scale. Reusability and regeneration for longtime cycling are the perfect examples.

3.3 Reusability and Regeneration

The reusability and regeneration abilities of hybrid materials are the very sine qua non not just for their economic feasibility but even for their straightforward application on larger scale. Indeed, these two criteria are a major source of value addition to hybrid materials that it is advisable to fully take into account when developing these materials as this assists in making the most of the significant time and money necessary to manufacture them.

Materials reusability or recycling depicts their ability to be reused several times in a row until their saturation and overload with water pollutants, translating into a decline in their effectiveness and depollution performances. One solution comes to mind immediately: restoring the material's primary depollution features and this by their regeneration. This latter helps benefit from the hybrid materials for subsequent reuses which serve as the basis for potential cost-reduction opportunities.

Aside from the economic benefits of materials regeneration, recovering the adsorbate pollutants is indispensable especially in the case of toxic heavy metals that need to be desorbed and concentrated for further industrial purposes instead of their leaching in the environment. For materials reusability assessment at lab scale, hybrid materials are subject to repeating usage for multiple cycles under identical conditions

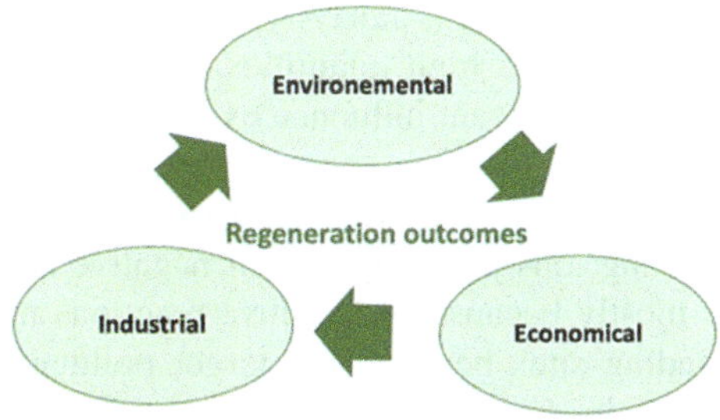

Fig. 2 The outcomes of hybrid materials regeneration

to enable a clear insight of hybrid materials shelf life and maximum performances. Whereas, regeneration is ensured by several techniques to revive the spent exhausted hybrid materials to diminish significantly virgin materials usage in favor of regenerated materials (Fig. 2). Depending on the hybrid material type and formulation, the regeneration could be using ultrasounds [26], thermal energy [27], electrochemical [28], and chemical treatments [15]…etc.

Testing the hybrid material's reusability and its ability to be regenerated is an essential prerequisite at the lab scale. However, the usefulness of the data collected from these experiments is limited and doesn't reflect necessarily what may take place at larger scale. By way of explanation, most, if not all, of the lab-scale experiments are done in batches. The transition from static mode (lab-scale) to dynamic continuous processes (larger scale) is usually associated with many risk factors that come into play: the effect of flowing rate, and the severe operating conditions of larger scale that may be conducive to degradation of the hybrid materials. Thus, shortening their shelf life to a much-reduced duration compared to what was estimated when in static mode.

In the light of the above considerations and for better practicability, hybrid materials should be able to stand up to the toughest use and conditions. Thereby, more relevant consideration should therefore be taken into account while transiting from a static, closed, and controllable system to a dynamic, open, and uncontrollable large system.

3.4 Interferences

Contaminated water has at its disposal a broad panoply of co-existing pollutants none of which are considered harmless to human health, viz. pathogenic organisms, organic and inorganic contaminants, metallic elements, oils, radioactive elements, etc. [6]. Knowing this, hybrid material's efficacy is better appraised if they have a demonstrated ability to remove various contaminants parallelly.

Most studies carried out on this vein using hybrid materials focused on the removal of water pollutants taken separately in solutions. Nevertheless, few studies shed light on the importance of assessing the effect of the co-existing ions on the pollutants removal [29]. It is fairly typical for inorganic co-existing ions to predominate in wastewater in high quantities. According to current scientific studies, these ions have a significant influence over pollutants removal mechanisms: they may help to speed up, slow down, or even stop the decontamination process.

Predictably and according to the literature, the simultaneous presence of co-existing ions has a significant negative influence on pollutants removal [13]. This is mostly because competitive reactions may take place over the hybrid materials binding sites between the target pollutants and co-existing ions, which leads to a certain decline in terms of hybrid materials affinity and their decontamination performances.

As mentioned above, polluted waters like wastewaters (sewage) are a complex cocktail of pollutants. Ergo, to permit a more accurate evaluation of the prepared hybrid materials and considering possible interferences within polluted water, simulating wastewater at lab scale is an indispensable step not to mispresent the contaminated water conditions, but basically to fully state them to infer the hybrid material's behavior. Then pass to the second step that is to test the hybrid materials on real wastewater samples for a thorough investigation of their real ability to decontaminate water. To this extent, all possible interferences are being already taken into consideration at the lab scale as well as understanding the functioning of these materials; this makes it possible to make conclusions based on sound science and valid data.

3.5 Cost-Effectiveness

In comparing hybrid materials, account shall be taken of their effectiveness in removing pollutants, and their costs. In fact, the cost-effectiveness analysis is a crucial component of environmental initiatives to maximize environmental benefits at the lowest possible cost. Thus, ensuring an economically viable alternative for water treatment at larger scale.

Of all the literature reviewed, the cost estimation of hybrid materials for water treatment is seldom or never clearly defined. Generally, estimating the cost of certain elaborated material for a certain application may depend on a range of factors, including precursors accessibility and their pre-treatment, elaboration methods and processing, the materials recyclability, energy consumption, etc.

In the case of hybrid materials based on bio-resources like biopolymers, their starting materials are commonly available, natural, and inexpensive, making their larger-scale practicability possible [30, 31]. However, more focus must be put upon optimal natural resources management and exploitation to benefit from the environment, yet preserving its resources.

For the case where hybrid materials are based on nanoparticles and nanomaterials, providing large quantities of these nanomaterials at reasonable costs is among the main concerns of nanotechnology application for water decontamination. Hence, finding suppliers that fill the needs while offering lower prices or at least in reasonable limits would be beneficial for these materials implementation on the existing conventional wastewater sector as effective, and attractive alternatives.

3.6 Sustainability and Toxicity

Biodegradable, non-toxic, biocompatible, eco-friendly, ecological, etc. are the attractive most-used terms describing hybrid materials for water treatment in the literature [13, 15, 32]. Tobridge the gap between these good promises and the reality, measuring the sustainability and the "non-toxicity" of these materials is a must before their larger-scale applications. However, there is a lack of studies regarding this issue.

Only a few reports addressed the biodegradability or the sustainability metrics of already prepared hybrid materials by analyzing quantitatively and qualitatively their potential environmental influences [32], or by evaluating their biodegradability [33]. In the case of adsorption, the spent hybrid materials are considered secondary waste as long as they are not regenerated. Indeed, adsorbents full of toxic pollutants as heavy metals even on a small scale are deemed to pose many threats to human health and the environment, and this risk would be magnified by switching to the larger scale.

Further research and risk/benefit analysis are required to corroborate the non-toxicity assumptions of hybrid materials. Using polymers or natural resources in the processing of hybrid materials does not endow them with a guaranteed sustainability label. There is a significant concern over the usage of nanoparticles within the biopolymeric matrix. These nanoparticles exert their effect making the hybrid materials high-performing yet less benign as expected. That is where the concern stems from, and ironically no studies are reasserting specifically about the potential toxicity of hybrid materials incorporating nanoparticles for water decontamination. All the focus is usually geared toward the effectiveness of removing pollutants without paying enough attention to the harmful effects of nanoparticles. Another issue that has to be addressed is the possible leaching of these persistent insoluble nanoparticles into water and hence seeping into the soil [5, 34].

4 Conclusion and Future Perspectives

The rising number of studies devoted to water decontamination employing hybrid materials is the best proof of their suitability and high effectiveness which is reflected in encouraging results over the other conventional materials. Despite this, hybrid materials testing so far have been at a small lab-scale, requiring further consideration and research for the assessment of their potential applications at larger-scale on the one hand, and determining the optimal procedures besides risk management measures on the other hand. The remaining key challenges for evolving hybrid materials from lab scale to industrial wastewater sector are diversified, i.e. the mass scale synthesis should be preferably continuous and scalable for a high yield sustainable production, improving the selectivity of these materials towards target pollutants, and evaluating their effectiveness over long periods using representative samples of real wastewater.

Last but not least, the underlying objective behind conducting these researches is elaborating new materials for water decontamination and restoring its purity for further reuse towards a sustainable future with the least possible environmental damage. Otherwise, we will be introducing new sorts of pollutants to the environment.

References

1. Rejab MRBM, Hamdan MHBM, Quanjin M, Siregar JP, Bachtiar D, Muchlis Y (2020). Hist Dev Hybrid Mater. https://doi.org/10.1016/b978-0-12-803581-8.10546-6
2. Yamada A, Sasabe H, Osada Y, Shiroda Y (1989) Concepts of hybrid materials, hybrid materials–Concept and case studies. ASM international Ohio
3. Singh A, Verma N, Kumar K (2019) Hybrid composites: a revolutionary trend in biomedical engineering. In: Materials for biomedical engineering: bioactive materials, properties, and applications, Elsevier Inc., pp 33–46. https://doi.org/10.1016/B978-0-12-818431-8.00002-7
4. Kickelbick G (2014) Hybrid materials—Past, present and future. Hybrid Mater 1:39–51. https://doi.org/10.2478/hyma-2014-0001
5. Chkirida S, Zari N, Qaiss AEK, Bouhfid R (2019) Nanocomposite materials based on TiO_2/Clay for wastewater treatment, pp 363–380. https://doi.org/10.1007/978-3-030-02381-2_16
6. Ntombenhle G, Thabang P, Meyyappan M, Moothi K (2019) Simultaneous removal of pollutants from water using nanoparticles: a shift from single pollutant control to multiple pollutant control. Sci Total Environ 656:808–833. https://doi.org/10.1016/j.scitotenv.2018.11.257
7. Ahmad R, Mirza A (2018) Adsorptive removal of heavy metals and anionic dye from aqueous solution using novel Xanthan gum-Glutathione/ Zeolite bionanocomposite. Groundw Sustain Dev 7:305–312. https://doi.org/10.1016/j.gsd.2018.07.002
8. Ahmad R, Mirza A (2018) Synthesis of Guar gum/bentonite a novel bionanocomposite: Isotherms, kinetics and thermodynamic studies for the removal of Pb(II) and crystal violet dye. J Mol Liq 249:805–814. https://doi.org/10.1016/j.molliq.2017.11.082
9. Sethy TR, Sahoo PK (2019) Highly toxic Cr(VI) adsorption by (chitosan-g-PMMA)/silica bionanocomposite prepared via emulsifier-free emulsion polymerisation. Int J Biol Macromol 122(Vi): 1184–1190. https://doi.org/10.1016/j.ijbiomac.2018.09.069
10. Jiang H, Chen P, Luo S, Tu X, Cao Q, Shu M (2013) Synthesis of novel nanocomposite Fe_3O_4/ZrO_2/chitosan and its application for removal of nitrate and phosphate. Appl Surf Sci 284:942–949. https://doi.org/10.1016/j.apsusc.2013.04.013
11. Pirsa S, Asadzadeh F, Karimi Sani I (2020) Synthesis of magnetic Gluten/Pectin/Fe_3O_4 nano-hydrogel and its use to reduce environmental pollutants from lake urmia sediments. J Inorganic Organometall Polymers Mater 30(8):3188–3198. https://doi.org/10.1007/s10904-020-01484-y
12. Thirugnanasambandham K, Sivakumar V (2016) Modeling and optimization of treatment of milk industry wastewater using chitosan—Zinc oxide nanocomposite. Desalin Water Treat 57:18630–18638. https://doi.org/10.1080/19443994.2015.1102089
13. Chkirida S et al (2021) Highly synergic adsorption/photocatalytic efficiency of Alginate/Bentonite impregnated TiO_2 beads for wastewater treatment. J Photochem Photobiol, A 412:113215. https://doi.org/10.1016/j.jphotochem.2021.113215
14. Chkirida S, Zari N, Achour R, el kacemQaiss A, Bouhfid R (2020) Efficient hybrid bionanocomposites based on iron-modified TiO_2 for dye degradation via an adsorption-photocatalysis synergy under UV-Visible irradiations. Environ Sci Pollut Res. https://doi.org/10.1007/s11356-020-11664-5

15. Chkirida S, Zari N, Bouhfid R, el kacemQaiss A (2021) Insight into the bionanocomposite applications on wastewater decontamination: review. J Water Process Eng 43(August): 102198. https://doi.org/10.1016/j.jwpe.2021.102198
16. Lok CN et al (2007) Silver nanoparticles: partial oxidation and antibacterial activities. J Biol Inorg Chem 12:527–534. https://doi.org/10.1007/s00775-007-0208-z
17. Xiao W, Xu J, Liu X, Hu Q, Huang J (2013) Antibacterial hybrid materials fabricated by nanocoating of microfibril bundles of cellulose substance with titania/chitosan/silver-nanoparticle composite films. J Mater Chem B 1(28):3477–3485. https://doi.org/10.1039/c3tb20303d
18. Ahmad N, Sultana S, Kumar G, Zuhaib M, Sabir S, Khan MZ (2019) Polyaniline based hybrid bionanocomposites with enhanced visible light photocatalytic activity and antifungal activity. J Environ Chem Eng 7(1). https://doi.org/10.1016/j.jece.2018.11.048
19. Zare EN, Lakouraj MM, Mohseni M, Motahari A (2015) Multilayered electromagnetic bionanocomposite based on alginic acid: characterization and biological activities. Carbohyd Polym 130:372–380. https://doi.org/10.1016/j.carbpol.2015.05.020
20. Vanaamudan A, Sadhu M, Pamidimukkala P (2018) Chitosan-Guar gum blend silver nanoparticle bionanocomposite with potential for catalytic degradation of dyes and catalytic reduction of nitrophenol. J Mol Liq 271:202–208. https://doi.org/10.1016/j.molliq.2018.08.136
21. Mohammad F, Ar T, Al-lohedan HA (2017) Enhanced biological activity and biosorption performance of trimethyl chitosan-loaded cerium oxide particles. J Ind Eng Chem 45:33–43. https://doi.org/10.1016/j.jiec.2016.08.029
22. Anwar Y (2018) Antibacterial and lead ions adsorption characteristics of chitosan-manganese dioxide bionanocomposite. Int J Biol Macromol 111:1140–1145. https://doi.org/10.1016/j.ijbiomac.2018.01.096
23. Bilal M, Ihsanullah I, Younas M, Ul Hassan Shah M (2021) Recent advances in applications of low-cost adsorbents for the removal of heavy metals from water: a critical review. Separ Purific Technol 278(August): 119510. https://doi.org/10.1016/j.seppur.2021.119510
24. Pang Y, Yu J, Tang L, Zeng G, Zhu C, Wei X (2018) Magnetic nanohybrid materials for water-pollutant removal. Elsevier Inc., https://doi.org/10.1016/B978-0-12-814154-0.00001-3
25. Fan C, Li K, Wang Y, Qian X, Jia J (2016) The stability of magnetic chitosan beads in the adsorption of Cu^{2+}. RSC Adv 6:2678–2686. https://doi.org/10.1039/c5ra20943a
26. Salvador F, Martin-Sanchez N, Sanchez-Montero MJ, Montero J, Izquierdo C (2013) Regeneration of activated carbons contaminated by phenol using supercritical water. J Supercrit Fluids 74:1–7. https://doi.org/10.1016/j.supflu.2012.11.025
27. Sun Z, Li C, Wu D (2010) Removal of methylene blue from aqueous solution by adsorption onto zeolite synthesized from coal fly ash and its thermal regeneration. J Chem Technol Biotechnol 85:845–850. https://doi.org/10.1002/jctb.2377
28. Khas H, Delhi N (2013) Steam regeneration of adsorbents: an experimental and technical review. Chem Sci Trans 2:1078–1088. https://doi.org/10.7598/cst2013.545
29. Wu Z, Deng W, Zhou W, Luo J (2019) Novel magnetic polysaccharide/graphene oxide @Fe_3O_4 gel beads for adsorbing heavy metal ions. Carbohyd Polym 216:119–128. https://doi.org/10.1016/j.carbpol.2019.04.020
30. Shchipunov Y (2012) Bionanocomposites: green sustainable materials for the near future. Pure Appl Chem 84(12):2579–2607. https://doi.org/10.1351/PAC-CON-12-05-04
31. Darder M, Aranda P, Ferrer ML, Gutiérrez MC, Del Monte F, Ruiz-Hitzky E (2011) Progress in bionanocomposite and bioinspired foams. Adv Mater 23(44):5262–5267. https://doi.org/10.1002/adma.201101617
32. Karthikeyan P, Vigneshwaran S, Meenakshi S (2020) Al^{3+} incorporated chitosan-gelatin hybrid microspheres and their use for toxic ions removal: Assessment of its sustainability metrics. Environ Chem Ecotoxicol 2:97–106. https://doi.org/10.1016/j.enceco.2020.07.004

33. Pavlidou S, Papaspyrides CD (2008) A review on polymer-layered silicate nanocomposites. Progr Polymer Sci (Oxford) 33(12):1119–1198. https://doi.org/10.1016/j.progpolymsci.2008.07.008
34. Bundschuh M, et al. (2018) Nanoparticles in the environment: where do we come from, where do we go to? Environ Sci Europe 30(1). https://doi.org/10.1186/s12302-018-0132-6

The manufacturer's authorised representative in the EU is Springer Nature Customer Service Centre GmbH, Europaplatz 3, 69115 Heidelberg, Germany. If you have any concerns regarding our products, please contact ProductSafety@springernature.com

Printed and bound by CPI Group (UK) Ltd, Croydon, CR0 4YY
30/09/2026
02225250-0008